SOLID STATE PHYSICS

VOLUME 48

Founding Editors

FREDERICK SEITZ
DAVID TURNBULL

SOLID STATE PHYSICS

Advances in
Research and Applications

Editors

HENRY EHRENREICH

FRANS SPAEPEN

Division of Applied Sciences
Harvard University, Cambridge, Massachusetts

VOLUME 48

ACADEMIC PRESS, INC.

Harcourt Brace & Company, Publishers

Boston   San Diego   New York
London   Sydney   Tokyo   Toronto

COPYRIGHT © 1994 BY ACADEMIC PRESS, INC.

ACADEMIC PRESS, INC.
525 B STREET, SUITE 1900, SAN DIEGO, CA 92101-4403

UNITED KINGDOM EDITION PUBLISHED BY
ACADEMIC PRESS LIMITED
24–28 OVAL ROAD, LONDON NW1 7DX

LIBRARY OF CONGRESS CATALOG CARD NUMBER: 55–12200

ISBN 0-12-607748-7 (hardcover)
ISBN 0-12-606048-7 (paperback)

PRINTED IN THE UNITED STATES OF AMERICA
94 95 96 97 98 99 BB 9 8 7 6 5 4 3 2 1

Contents

Solid State Properties of Fullerenes and Fullerene-Based Materials

J. H. WEAVER AND D. M. POIRIER

Preparation of Fullerenes and Fullerene Based Materials

CHARLES M. LIEBER AND CHIA-CHUN CHEN

Structure and Dynamics of Crystalline C_{60}

J. D. AXE, S. C. MOSS, AND D. A. NEUMANN

Electrons and Phonons in C_{60}-Based Materials

WARREN E. PICKETT

Phsyical Properties of Metal-Doped Fullerene Superconductors

CHARLES M. LIEBER AND ZHE ZHANG

Contributors

Numbers in parenthesis indicate the pages on which the authors' contributions begin.

J. D. AXE (149), *Physics Department, Brookhaven National Laboratory, Upton, New York 11973*

CHIA-CHUN CHEN (109), *Division of Applied Sciences and Department of Chemistry, Harvard University, Cambridge, Massachusetts 02138*

CHARLES M. LIEBER (109, 349), *Division of Applied Sciences and Department of Chemistry, Harvard University, Cambridge, Massachusetts 02138*

S. C. MOSS (149), *Physics Department, University of Houston, Houston, Texas 77204-5506*

D. A. NEUMANN (149), *Materials Science and Engineering Laboratory, National Institute of Standards and Technology, Gaithersburg, Maryland 20899*

WARREN E. PICKETT (225), *Complex Systems Theory Branch, Naval Research Laboratory, Washington D.C. 20375-5345*

D. M. POIRIER (1), *Department of Materials Science and Chemical Engineering, University of Minnesota, Minneapolis, Minnesota 55455*

J. H. WEAVER (1), *Department of Materials Science and Chemical Engineering, University of Minnesota, Minneapolis, Minnesota 55455*

ZHE ZHANG (349), *Division of Applied Sciences and Department of Chemistry, Harvard University, Cambridge, Massachusetts 02138*

Preface

New fields in solid state physics often emerge as a result of the discovery of new materials. The past decade provides several striking examples: quasicrystals (Shechtman and collaborators, 1984), oxide superconductors (Bednorz and Müller, 1986), and, most recently, polyhedral forms of carbon that became known as fullerenes. The first two fields are discussed in Volume 42 of this series; the present volume is a status report on the last one.

Crucial to the establishment of a such a new field is a simple method for producing reproducibly, large amounts of sample material. That is why fullerene research has blossomed since 1990, when Krätschmer, Huffman, and collaborators discovered a method for producing and isolating large amounts of fullerenes in both solution and crystalline form. The idea that carbon atoms may form polyhedral shells had arisen much earlier.

The angle between two sp^3 hybrid orbitals of a carbon atom in diamond is 109.5°; therefore, their bonds easily form planar pentagons that have internal angles of 108°. The three planar sp^2p orbitals in graphite make 120° angles, and their bonds naturally form hexagons. Intermediate hybridization allows the hexagons and pentagons to be assembled into polyhedral surfaces, on which each carbon has three nearest neighbors. Application of Euler's rule to such a polyhedron shows that the number of pentagons is always equal to twelve, and that the number of hexagons is ten less than half the number of atoms. The best known fullerene molecule, C_{60}, for example, consists of twelve (nonadjacent) pentagons and twenty hexagons arranged with icosahedral symmetry, as on a soccer ball. Similar topological considerations were used by the architect R. Buckminster Fuller, eponym of these materials, in the design of geodesic domes.

Speculation about the existence of polyhedral carbon molecules can be found as far back as 1966. The first demonstration of their existence was made by Kroto, Smalley, and collaborators in 1985, who observed a strong C_{60} signal in the mass spectrum of carbon clusters formed by laser ablation into a pulsed jet of helium. This method, modified to lengthen the cluster assembly time for a higher yield, is one of several described in the article by Lieber and Chen, along with methods for cluster separation (mainly by chromatography), characterization and crystallization into powders, thin films, and single crystals. Among the fullerene-related materials, special attention is paid to preparation of metal-doped C_{60} crystals, which can be superconducting.

C_{60} clusters crystallize in a close-packed structure. At high temperature the individual clusters are orientationally disordered, and the crystal structure is face-centered cubic. At about 260 K, a first order orientational phase transition occurs that makes the structure simple cubic. At the lowest temperatures (<90 K), an orientational glass has been postulated. The article by Axe, Moss, and Neumann describes how x-ray and neutron scattering from powders and single crystals can be used to determine the orientational order parameters in these phases, and how these in turn are used in statistical mechanics to derive orientational potentials. Neutron scattering and other studies of rotational diffusion, librational dynamics, and (intermolecular) phonons show that at no temperature the C_{60} molecule can rotate freely.

The most extensively studied of the doped superconducting fullerenes have the composition M_3C_{60}, where M is an alkali metal, such as K or Rb. The alkali atoms reside in the octahedral and tetrahedral interstices of the face-centered cubic C_{60} lattice. Superconducting transition temperatures as high as 31 K have been observed. The article by Lieber and Zhang reviews the phenomenology and mechanisms of superconductivity in these fullerenes. They appear to be extreme type II superconductors with upper critical fields as high as 50 T and correspondingly short coherence lengths (25 Å). The proposed theories are based on electron pairing mediated either by the "conventional" BCS-type phonon mechanism or by electron correlation effects. Although more work is needed to test the theories fully, the preponderance of evidence at this time points towards a BCS-type mechanism. The authors' own measurements of the isotope shift of the transition temperature in $K_3{}^{13}C_{60}$ are particularly convincing in this respect.

Because they can occur in molecular or solid form and can have tubular, capsular, and onion-like structures, fullerenes have a variety of interesting properties. These properties have already been elucidated, both experimentally and theoretically, to a remarkable extent. The overviews by Weaver and Poirier and by Pickett provide detailed experimentally- and theoretically-oriented summaries, respectively. In the first of these articles we find discussions of fullerenes in the gas phase, in solution, in the solid state, and of doped fullerenes. The discussion is amplified by the inclusion of a wealth of experimental data and literature references. The second of these articles emphasizes electronic structures, molecular and lattice vibrations, charge density contours, Fermi surfaces of metallic phases, and many-electron effects, with explicit reference to experiments. The two articles are complementary and have sufficient overlap to facilitate intercomparison.

The volume thus furnishes the reader with a complete current survey of the field. Practical applications, alas, are only barely visible on the horizon. But, just as in the case of the semiconductor injection laser whose utility was not recognized until a decade or so after its discovery, or high temperature superconductivity, few doubt that they will come about. For the present we must content ourselves with a characterization presented in the British House of Lords, which is delightfully described by John Weaver (see page 107). In response to a question concerning fullerene applicability, one of its members asked rhetorically, "Could it be said that it does nothing in particular and it does it extremely well?" As this volume demonstrates, the scientific base exists. Invention will surely lead to a robust technology.

Henry Ehrenreich
Frans Spaepen

SOLID STATE PHYSICS, VOLUME 48

Solid State Properties of Fullerenes and Fullerene-Based Materials

J. H. WEAVER
D. M. POIRIER

Department of Materials Science and Chemical Engineering
University of Minnesota, Minneapolis, Minnesota

I. Introduction

1. CHAPTER OVERVIEW AND GOALS

A new family of carbon molecules dubbed the "fullerenes" has captured the attention of a diverse group, including scientists, medical

1

professionals, and laymen.[1-6] By focusing on the fullerenes in their various forms, this chapter and this book seek to convey the excitement evident in the fullerene community and to touch upon some of their fascinating properties.

The fullerenes are all-carbon molecules with a closed–cage structure and nearly spherical appearance. These cages are derived from 12 pentagons and an appropriate number of hexagons. The best known is C_{60}, the molecule with 60 equivalent carbon atoms arranged on a nearly spherical shell, approximately 7 Å in diameter. Of the growing number of identified fullerenes, C_{60} is unique in that it has been given a name, buckminsterfullerene, and even a nickname, buckyball.[7] C_{60} possesses the highest possible finite symmetry, point group I_h, and its arrangement of hexagons and pentagons is just that found on a soccer ball, as depicted in Fig. 1(a).[8,9] There are other stable fullerenes, including C_{70}, C_{76}, C_{78}, and C_{84}. C_{70} resembles a rugby ball, as in Fig. 1(b).[10,11] Higher fullerenes such as C_{78} and C_{84} can form with multiple isomeric structures having different pentagon–hexagon placements.[12] C_{76} also exhibits chirality.[13]

The existence of the fullerenes as closed-cage molecules was first proposed by Kroto et al.[8] Interest in the fullerenes expanded dramatically after 1990 when Krätschmer et al.[14] developed techniques to produce and separate fullerenes in sufficient quantity that their various properties could be examined. The fullerenes have been attractive to the nonscientific community as well, perhaps because of their high symmetry and the prospects for future applications. Scientists and nonscientists alike

---

[1]See, for example, papers in (a) Special Issue of *Accounts of Chemical Research*, Vol. **25**, March (1992); (b) Special Issue of *Journal of Physics and Chemistry of Solids*, Vol. **53**, November (1992); and (c) *Carbon*, Special Issue on Fullerenes, Vol. **30**, No. 8 (1992).

[2]R. F. Curl and R. E. Smalley, *Scientific American*, October (1991), p. 54

[3]H. W. Kroto, A. W. Allaf, and S. P. Balm, *Chem. Rev.* **91**, 1213 (1991).

[4]K. Prassides and H. Kroto, *Physics World,* April (1992).

[5]A. F. Hebard, *Physics Today* **45**, 26 (1992).

[6]A. F. Hebard, *Ann. Rev. Mater. Sci.* **23**, 159 (1993).

[7]R. Baum, *C&E News,* December 23 (1985), p. 20.

[8]H. W. Kroto, J. R. Heath, S. C. O'Brien, R. F. Curl, and R. E. Smalley, *Nature* **318**, 162 (1985).

[9]S. Saito and A. Oshiyama, *Phys. Rev. Lett.* **66**, 2637 (1991).

[10]J. R. Heath, S. C. O'Brien, Q. Zhang, Y. Liu, R. F. Curl, H. W. Kroto, F. K. Tittel, and R. E. Smalley, *J. Am. Chem. Soc.* **107**, 7779 (1985).

[11]D. R. McKenzie, C. A. Davis, D. J. H. Cockayne, D. A. Muller, and A. M. Vassallo, *Nature* **355**, 622 (1992).

[12]F. Diederich and R. L. Whetten, *Accts. Chem. Res.* **25**, 119 (1992).

[13]R. Ettl, I. Cho, F. Diederich, R. L. Whetten, *Nature* **353**, 149 (1991).

[14]W. Krätschmer, L. D. Lamb, K. Fostiropoulos, and D. R. Huffman, *Nature* **347**, 354 (1990).

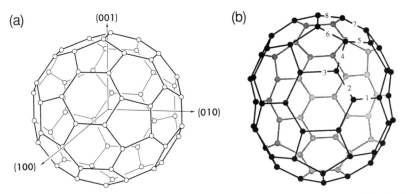

FIG. 1. Representation of (a) C_{60} and (b) C_{70}. For C_{60}, the atoms are identically coordinated and closure is achieved with 12 pentagons and 20 hexagons. The orientation of the molecule on a Cartesian frame anticipates the Fm$\bar{3}$ symmetry of a crystalline array. For C_{70}, the addition of 10 carbon atoms elongates the molecule by ~1 Å. The numbering identifies distinguishable bond lengths. From Saito and Oshiyama[9] and McKenzie et al.,[11] respectively.

have found it remarkable that a new form of carbon could be discovered late in the twentieth century. There is a general feeling that new carbon-based (fullerene-based) materials will be important in the science and technology of the twenty-first century.

Today, just three years after the Krätschmer–Huffman (K–H) breakthrough in fullerene synthesis, the materials communities are aggressively exploring the properties of pure fullerenes, fullerides (salts such as K-C_{60}), fullerene derivatives (fullerenes with chemical attachments), endofullerenes (fullerenes with atoms captured within the closed carbon shells), nanotubes (closed cylinders resembling chicken wire with capped ends), and onions (nearly spherical concentric multilayers). There has been enormous progress, and time after time the fullerenes in general and C_{60} in particular have revealed a fascinating array of novel properties.

To understand the interest from the scientific community, one need only note that the fullerenes are molecular crystals that are stable at room temperature and are bound largely by van der Waals forces; that the molecules in crystalline C_{60} rotate at several gigahertz at high temperature but exhibit a variety of ordering transitions when cooled; that some of the alkali metal fullerides are insulators, that others are metals, and that the metals are also superconductors with transition temperatures reaching 33 K; and that these materials show promise as the building blocks of a host of new chemical species. The fullerenes have drawn together an extended family of materials researchers from the physics, chemistry, materials, and engineering communities. Those active in fullerene research have found this interdisciplinary character to be particularly rewarding.

The purpose of this chapter is to provide an overview. Other chapters will delve into the synthesis of the fullerenes and their derivatives, the structural properties of fullerene-based solids, superconductivity, and theoretical treatments that provide a basis for interpreting the unique properties of fullerenes. Here, we will introduce the reader to the fullerenes by comparing gas phase and solid state properties, examining thin film and bulk crystal growth, and doping with alkali metals and other atoms. We will focus on the electronic structure of these correlated-electron systems, relating the electronic properties and vibrational properties to superconductivity. Finally, we will consider the latest additions to the fullerene family, namely the endofullerenes, tubes, and the buckyonions.

2. HISTORICAL PERSPECTIVES

The shape of C_{60}, a truncated icosahedron, has long intrigued mankind. Drawings of the object by da Vinci date from the fifteenth century and others may predate him. Figure 1 shows a modern rendition of C_{60} where every vertex is identical and a closed shell is achieved by combining 20 hexagons with 12 pentagons. The strain associated with bending a planar array of hexagons is then evenly distributed over a nuclear cage of diameter 7 Å. Our discussion of these molecules condensed into a crystal of Fm$\bar{3}$ symmetry is anticipated by the crystallographic axes drawn and their relationship to the molecular symmetry.[9] The deviation from planar character is achieved by forming bonds with hybridization intermediate between sp^3 and sp^2p, as discussed by Haddon[15] for C_{60} and other higher fullerenes. The soccer-ball symmetry is easily recognized. On a larger scale, curved geodesic structures derived from pentagons and hexagons have been associated with the work of R. Buckminster Fuller, an architect and inventor. It was in the architect's honor that this new family of molecules was named.

While artists, architects, and athletes have long been familiar with (or at least acquainted with) the truncated icosahedron, the first association of the soccer-ball shape with a molecule was made by Kroto et al.[8] in

---

[15]R. C. Haddon, *Accts. Chem. Res.* **25**, 127 (1992).

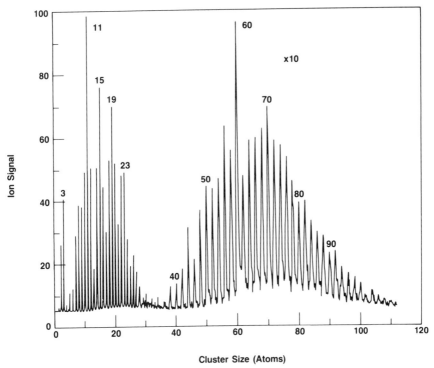

Cluster Size (Atoms)

FIG. 2. Distribution of carbon species as a function of cluster number, *n*, showing the presence of clusters of both even and odd numbers for small *n* and then only fullerenes with even number *n* above about 38. C_{60} and C_{70} are identifiable in this spectra but are not dominant. Under different conditions of growth, the exceptional stabilities of C_{60} and C_{70} are apparent. From Rohlfing *et al.*[16]

1985. In their studies of all-carbon species, they formed clusters by laser ablation from a graphite target into an exchange gas of helium. Using mass spectrometry, Kroto *et al.* found a rich array of even-numbered carbon clusters for sizes greater than about 32 atoms, as had Rohlfing *et al.*[16] when they earlier reported the mass distribution shown in Fig. 2. Kroto *et al.* and Rohlfing *et al.* demonstrated that odd-numbered carbon structures were not stable above ~30 atoms. By optimizing the growth conditions, Kroto *et al.* found that the relative abundance of C_{60} could be greatly enhanced, and in their landmark paper they reasoned that this must be due to the high stability of the molecule. In that paper, they demonstrated the dominance of C_{60} under certain clustering

[16]E. A. Rohlfing, D. M. Cox, and A. Kaldor, *J. Chem. Phys.* **31**, 3322 (1984).

conditions, proposed the soccer-ball shape, and christened this molecule Buckminsterfullerene. They showed subsequently that it was possible to capture atoms within the cage[10] and that the cage could be shrunk by laser-induced C_2 ejection until it was too small to accommodate the endohedral atom and burst.[17]

The carbon cluster investigation of Kroto *et al.* and also those of Krätschmer *et al.* were motivated by an interest in unidentified carbon species believed to be responsible for certain interstellar absorption lines.[18] Carbon was assumed to be important because of its natural abundance, but the structure of any carbon species present in the interstellar dust was (and still is) elusive.[19] Kroto *et al.* were trying to form what they expected to be carbon chains that might account for this interstellar absorption when they detected the large-number carbon species and deduced their closed-shell structure. Unfortunately, the laser-assisted cluster beam apparatus that the Smalley group had pioneered produced relatively small quantities of the fullerenes, and, while important gas phase studies were undertaken at Rice and elsewhere,[10,17,20,21] the materials research communities had insufficient material to be particularly enamored with C_{60}.

It was in 1990 that Krätschmer *et al.*[14] announced their well-known breakthrough that made C_{60} a readily-accessible "chemical." Their work had started years earlier when they examined differently configured carbon structures produced by condensing carbon from the vapor phase. As early as 1982, Krätschmer and Huffman noted that a sample produced by resistive evaporation of graphite in a helium environment showed structure in the ultraviolet-visible, UV-vis, spectrum near 250 nm.[18] This was called the camel spectrum since it had a double "hump," but the hump had no clear explanation at the time. After learning of the proposed C_{60} molecule, its formation conditions, and some of its calculated properties, Krätschmer and Huffman speculated that the camel spectrum might have indicated the presence of C_{60} in their soot. Reproducing the conditions of the early experiment again gave material that exhibited the UV-vis camel spectrum. This time the infrared, IR, absorption was examined, and four strong lines were found. This was exactly the number of lines expected for the icosahedral

[17]F. D. Weiss, J. L. Elkind, S. C. O'Brien, R. F. Curl, and R. E. Smalley, *J. Am. Chem. Soc.* **110**, 4464 (1988).

[18]W. Krätschmer and D. R. Huffman, *Carbon* **30**, 1143 (1992).

[19]J. P. Hare and H. W. Kroto, *Accts. Chem. Res.* **25**, 106 (1992).

[20]R. E. Smalley in "Atomic and Molecular Clusters," ed. E. R. Bernstein (Elsevier, Amsterdam, 1990), p. 1.

[21]S. W. McElvany, M. M. Ross, and J. H. Callahan, *Accts. Chem. Res.* **25**, 162 (1992).

Buckminsterfullerene molecule, and the energy positions were in reasonable agreement with calculations of vibronic excitation for C_{60}.[22] Repeating the experiment with ^{13}C revealed a shift of the IR lines that was consistent with that expected for the all-carbon molecule.

In their landmark paper, Krätschmer et al.[14] showed that fullerenes could be extracted from condensed carbon soot by dissolving them in benzene or subliming them at $\sim 400°C$. This enabled them to measure the first UV-vis and IR absorption spectra of purified C_{60} with small amounts of higher fullerenes. By evaporating the benzene solvent they demonstrated that fullerene crystals could be formed. X-ray and electron diffraction analyses indicated close-packed structures with a nearest neighbor spacing of 10 Å, in agreement with expectations for close-packed arrays of Buckminsterfullerene molecules. Independent work at Sussex, spurred on by the IR results of Krätschmer et al., also found C_{60} and C_{70} to be soluble and, furthermore, chromatographically separable so that nuclear magnetic resonance, NMR, could be used to provide further support for the fullerene structures.[23,24]

The Materials Research Society Meeting in November of 1990 provided the first major forum for fullerene mania. In his address at the special postdeadline session,[25] Donald Huffman described their breakthrough and announced that several laboratories around the world had reproduced it. The fullerene story has continued, punctuated by a postdeadline session at the American Physical Society Meeting in Cincinnati in March 1991,[26] a high-intensity fullerene workshop in Philadelphia in August 1991,[27] and a good many other meetings and workshops worldwide. The substantive progress has been amply documented in the literature.[28]

[22]W. Krätschmer, K. Fostiropoulos, and D. R. Huffman, Chem. Phys. Lett. **170,** 167 (1990).
[23]R. Taylor, J. P. Hare, A. K. Abdul-Sada, and H. W. Kroto, J. Chem. Soc., Chem. Comm. 1423 (1990).
[24]H. W. Kroto in "Clusters and Cluster-Assembled Materials," Materials Research Society Symposium Proceedings, Vol. 206, eds. R. S. Averback, J. Bernholc, and D. L. Nelson (1991), p. 611.
[25]Special Session of the Materials Research Society, Boston, November 1990, organized by D. M. Cox and A. Kiddor.
[26]Special Post Deadline Session of the American Physical Society, Cincinnati, March 1991, organized by J. H. Weaver and J. Bernholc.
[27]Workshop on Fullerites and Solid State Derivatives, University of Pennsylvania, Philadelphia, August 1991, organized by J. E. Fischer.
[28]A Buckminsterfullerene bibliography electronic mail service was created to help researchers stay on top of the tremendous amount of research being reported. Copies of a fullerene publication data base, "The almost (but never quite) Complete Buckminsterfullerene Bibliography," are available via electronic mail by sending the message BIBLIO to the internet address BUCKY@SOL1.LRSM.UPENN.EDU.

3. The Fullerene Family

C_{60} resides at the center of the fullerene family. While smaller fullerenes are produced by gas phase clustering, they have not been isolated in quantity and are probably unstable in air. C_{70}, the next most abundant fullerene, resembles a rugby ball with D_{5h} symmetry, as shown in Fig. 1(b). The effect of adding 10 carbon atoms is to elongate the molecule by about 1 Å. The apparent pinch around the molecular equator was deduced by electron diffraction analyses of McKenzie et al.[11] While larger fullerenes are produced by the laser desorption and K–H techniques, the challenge has been to separate them in pure form. In early studies, the UCLA group headed by Robert Whetten and François Diederich isolated C_{76} and demonstrated the chirality of the molecule,[13] producing structures with both left- and right-handedness. They also succeeded in separating C_{78}, C_{84}, C_{90}, and C_{94}.[12,29,30] The group of Yohji Achiba at Tokyo Metropolitan University isolated several of the higher fullerenes using high performance liquid chromatography, HPLC.[31–33] These studies have led to investigations that included NMR,[30,32] photo-emission,[34–37] optical absorption,[29,33] and scanning tunneling micro-scopy, STM.[38]

One of Euler's theorems shows that a polyhedron with n vertices (n carbon atoms if the polyhedron is a fullerene) derived from hexagons and pentagons will consist of 12 pentagons and $[(n/2) - 10]$ hexagons.[39] The

[29]F. Diederich, R. Ettl, Y. Rubin, R. L. Whetten, R. Beck, M. Alvarez, S. Anz, D. Sensharma, F. Wudl, K. C. Khemani, and A. Koch, *Science* **252**, 548 (1991).

[30]F. Diederich, R. L. Whetten, C. Thilgen, R. Ettl, I. Chao, and M. M. Alvarez, *Science* **254**, 1768 (1991).

[31]K. Kikuchi, N. Nakahara, T. Wakabayashi, M. Honda, H. Matsumiya, T. Moriwaki, S. Suzuki, H. Shiromaru, K. Saito, K. Yamauchi, I. Ikemoto, and Y. Achiba, *Chem. Phys. Lett.* **188**, 177 (1992).

[32]K. Kikuchi, N. Nakahara, T. Wakabayashi, S. Suzuki, H. Shiromaru, Y. Miyake, K. Saito, I. Ikemoto, M. Kainosho, and Y. Achiba, *Nature* **357**, 142 (1992).

[33]K. Kikuchi, N. Nakahara, M. Honda, S. Suzuki, K. Saito, H. Shiromaru, K. Yamauchi, I. Ikemote, T. Kuramochi, S. Hino, and Y. Achiba, *Chem. Lett.*, 1607 (1991).

[34]S. Hino, K. Matsumoto, S. Hasegawa, H. Inokuchi, T. Morikawa, T. Takahashi, K. Seki, K. Kikuchi, S. Suzuki, I. Ikemote and Y. Achiba, *Chem. Phys. Lett.* **197**, 38 (1992).

[35]S. Hino, K. Matsumoto, S. Hasegawa, K. Kamiya, H. Inokuchi, T. Morikawa, T. Takahashi, K. Seki, K. Kikuchi, S. Suzuki, I. Ikemoto, and Y. Achiba, *Chem. Phys. Lett.* **190**, 169 (1992).

[36]D. M. Poirier, J. H. Weaver, K. Kikuchi, and Y. Achiba, *Z. Physik D* **26**, 79 (1993).

[37]S. Hino, K. Matsumoto, S. Hasegawa, K. Iwasaki, K. Yakushi, T. Morikawa, T. Takahashi, and K. Seki, *Phys. Rev. B* **42**, 8418 (1993).

[38]Y. Z. Li, J. C. Patrin, M. Chander, J. H. Weaver, K. Kikuchi, and Y. Achiba, *Phys. Rev. B* **47**, 10867 (1993).

[39]T. G. Schmalz, W. A. Seitz, D. J. Klein, and G. E. Hite, *J. Am. Chem. Soc.* **110**, 1113 (1988).

pentagons on the fullerene surface allow for local curvature, but they also induce local strain since three-coordinated carbon atoms are normally sp^2 bonded and prefer a planar geometry. This leads to a concept known as the isolated pentagon rule. Described by Schmalz et al.[40] and by Kroto[41] in the context of fullerene stability, this concept suggests that the most favorable structures are those in which the strain associated with curvature is spread isotropically; i.e., the pentagons should be isolated to the maximum extent possible. C_{60} is the smallest fullerene in which all 12 pentagons have no common edges. C_{70} is the next smallest, and, of the many possible isomers for higher fullerenes, only those with isolated pentagons are generally considered.

While the isolated pentagon rule is very useful, it is not very restrictive. Depictions of pseudo-icosahedral fullerenes as large as C_{960} have been presented that show the 12 pentagons to be maximally separated at the vertices of largely planar graphitic sheets.[2] Though aesthetically pleasing, these depictions are not unique. Indeed, Manolopoulos and Fowler[42] considered the number of possible isomers that could form, finding a dramatic increase with size. Although C_{60} and C_{70} have unique isolated-pentagon structures, there are 2 possible isomers for C_{76}, 5 for C_{78}, and 24 for C_{84}.[43] These possibilities notwithstanding, the kinetic pathways for fullerene formation and other ultimate stabilities yield a far smaller number of experimentally observed isomers. For example, the NMR measurements of Diederich et al.[29] and Kikuchi et al.[32] indicated that chromatographically purified C_{84} was composed of 2 isomers rather than 24. Kikuchi et al. determined that these isomers had D_2 and D_{2d} symmetry.[32] The four isolated pentagon D_2 structures are shown in Fig. 3.[44] A tight-binding search[45] for the ground state of C_{84} indicated that the $D_2(22)$ form had the lowest energy, though only by ~ 33 meV/molecule relative to $D_{2d}(23)$. Independently, Raghavachari[46] also found that the D_{2d} and D_2 isomers were the most stable. Saito et al.[47] and Wang et al.[48] compared electronic state distributions for C_{84} with photoemission

[40]T. G. Schmalz, W. A. Seitz, D. J. Klein, and G. E. Hite, Chem. Phys. Lett. **130,** 203 (1986).

[41]H. W. Kroto, Nature **329,** 529 (1987).

[42]D. C. Manolopoulos and P. W. Fowler, J. Phys. Chem. **96,** 7603 (1992).

[43]P. W. Fowler, R. C. Batten, and D. E. Manolopoulos, J. Chem. Soc. Faraday Trans. **87,** 3103 (1991).

[44]T. Wakabayashi, H. Shiromaru, K. Kikuchi, and Y. Achiba, Chem. Phys. Lett. **201,** 470 (1993).

[45]B. L. Zhang, C. Z. Wang, and K. M. Ho, J. Chem. Phys. **96,** 7183 (1992).

[46]K. Raghavachari, Chem. Phys. Lett. **190,** 397 (1992).

[47]S. Saito, S. Sawada, and N. Hamada, Phys. Rev. B **45,** 13845 (1992).

[48]X. Q. Wang, C. Z. Wang, B. L. Zhang, and K. M. Ho, Chem. Phys. Lett. **297,** 349 (1993).

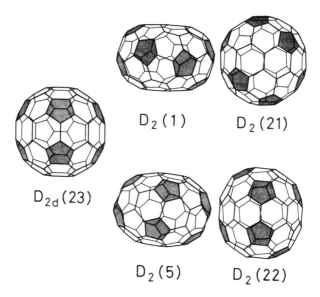

$$D_2(1) \qquad D_2(21)$$

$$D_{2d}(23)$$

$$D_2(5) \qquad D_2(22)$$

FIG. 3. Five of the possible isomeric structures for C_{84}, corresponding to species observed experimentally with D_{2d} and D_2 symmetries. The challenge has been to identify which of the candidates are produced and whether the processing conditions influence the isomeric distribution. From Wakabayashi and Achiba.[44]

results reported by Hino *et al.*[35] in attempts to determine favorable isomeric structures.

An important challenge of research with the larger fullerenes involves preparation of mass-pure and isomer-pure material. For solid state studies, deviation from the spherical shape of C_{60} introduces complexities as far as crystal structures, symmetries, and possible ordering transitions are concerned. For the theorist, the level of complexity also increases, and there will remain ambiguities until the isomeric structures and the appropriate distribution of bond lengths are determined experimentally. While these complexities offer intriguing opportunities, they also hint at difficulties associated with higher fullerene research.

4. FULLERENE FORMATION

The remarkable tendency to form closed-shell structures in abundance upon carbon vapor condensation must indicate that there are multiple pathways leading to the same end product. The delineation of these pathways is, however, far from trivial and there is no consensus as to which dominates.

Early descriptions of fullerene formation suggested that small carbon chains would grow by the sequential addition of carbon radicals.[49] At a certain size (25–35 atoms), it becomes favorable to form graphitic sheets. These planar fragments would have many dangling bonds around the periphery, and some of them could be eliminated by incorporating pentagons. This allows the sheet to curve into a bowl-like form that closes upon itself. This atom-by-atom addition offers one pathway, but others have emphasized growth by accretion of more substantial species. For example, Achiba[50,51] has described pathways involving carbon rings that are abundant in the gas phase. At the high temperatures found in the clustering region, the rearrangement of the constituent atoms was postulated to be quite facile. Achiba has argued that the ring stacking model can account for the abundances and structures of the higher fullerenes.

Several mass spectroscopy experiments have demonstrated that fullerene formation is not a result of "tearing off" of graphitic fragments or ribbons during ablation and the subsequent closing of these fragments. To the contrary, isotope scrambling[52] and a few other techniques[21] have shown that the starting material is single carbon atoms or ions. Fullerene formation from nongraphitic precursors has also been demonstrated[53] as well as considered theoretically.[54–56] Recent ion chromatography experiments by Bowers and Kemper[57] suggested that fullerenes result from annealing of large polycyclic rings that form at high temperature. McElvany et al.[58] also showed that monocyclic rings can be used as

[49]R. E. Smalley, Accts. Chem. Res. 25, 98 (1992).

[50]T. Wakabayashi and Y. Achiba, Chem. Phys. Lett. 190, 465 (1992).

[51]T. Wakabayashi, K. Kikuchi, H. Shiromaru, S. Suzuki, and Y. Achiba, Z. Physik D 26, (1993).

[52]G. Meijer, D. S. Bethune, W. C. Tang, H. J. Rosen, R. D. Johnson, R. J. Wilson, J. J. Chambliss, W. G. Golden, H. Seki, M. S. de Vries, C. A. Brown, J. R. Salem, H. E. Hunziker, and H. R. Wendt in "Clusters and Cluster-Assembled Materials," Materials Research Society Symposium Proceedings, Vol. 206, eds. R. S. Averback, J. Bernholc, and D. L. Nelson (1991), p. 619.

[53]J. B. Howard, Carbon 30, 1183 (1992), and references therein.

[54]J. R. Chelikowsky, Phys. Rev. Lett. 67, 2970 (1991). See also X. Jing and J. R. Chelikowsky, Phys. Rev. B 46, 15503 (1992); X. Jing and J. R. Chelikowsky, Phys. Rev. B 46, 5028 (1992); J. R. Chelikowsky, Phys. Rev. B 45, 12062 (1992).

[55]C. J. Brabec, E. B. Anderson, B. N. Davidson, S. A. Kajihara, Q.-M. Zhang, J. Bernholc, and D. Tománek, Phys. Rev. B 46, 7326 (1992).

[56]C. Z. Wang, C. H. Xu, C. T. Chan, and K. M. Ho, J. Phys. Chem. 96, 3563 (1992).

[57]M. T. Bowers and P. R. Kemper, J. Phys. Chem. 95, 5134 (1991).

[58]S. W. McElvany, M. M. Ross, N. S. Goroff, and F. Diederich, Science 259, 1594 (1993).

precursors and that it may be possible to control the fullerene size distribution by judicious choice of the starting carbon allotrope.

The growth of fullerenes has been described using molecular dynamics simulations of carbon–carbon interactions. Chelikowsky,[54] for example, introduced 60 independent carbon atoms into a constrained volume at $T = 7000$ K and examined the sequence of events followed as the system was cooled over a time of 20 psec. Representative configurations shown in Fig. 4 show that there is dimer formation at very high temperature,

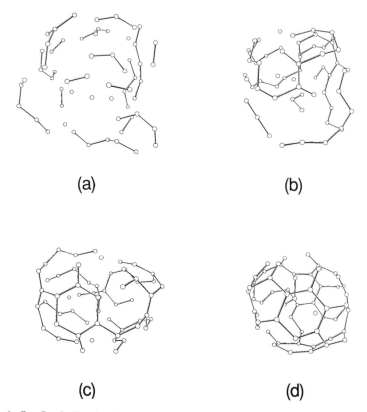

(a)   (b)

(c)   (d)

FIG. 4. Car–Parrinello simulation of C_{60} formation from 60 isolated atoms starting at 7000 K. (a) shows formation of carbon dimers and more extended chains. (b) shows elaborate structures that form and break at high temperature, 5000 K. (c) corresponds to a possible structure after cooling to 4000 K, revealing chains, branched chains, and nascent polycyclic rings. (d) represents a structure obtained after cooling from 7000 to 1000 K in 20 psec, ~1000 times faster than experimentally observed. Though defective, this structure shows five and sixfold ring formation and recognizable pseudospherical shape. From Chelikowsky.[54]

then chain growth, and ultimately polycyclic ring formation. Chelikowsky reported that the C–C bonds were able to open and close at high temperature and that the ring structures were constantly breaking and reforming. Only when the system was cooled to ~3000 K were the structures constrained. While the final structure of Fig. 4 was still rather defective, it had a more-or-less spherical shape and demonstrated pentagon and hexagon formation. Chelikowsky noted that annealing simulations produced more symmetric structures.

Wang et al.[56] simulated C_{60} formation with somewhat different constraints. Whereas the Chelikowsky model had 60 atoms in a fixed volume with the temperature quenched so rapidly that the atoms could not escape, Wang et al. configured their atoms inside a compressible sphere. Under their conditions of compression and cooling, Wang et al. found rapid nucleation of carbon chains. The fullerene cage structure was readily apparent when the radius of the sphere was reduced to ~3.8 Å, but there were defects that were attributed to the simulation constraints.

Yi and Bernholc[59] investigated the energetics of C_{60} isomers and the barriers against concerted exchange involving sequential rotations of pairs of atoms around the bond center. They concluded that transformation between isomers required high temperatures and long annealing times. Fowler et al.[60] discussed at length how isomeric forms of various fullerenes are connected by such bond rotations. Raghavachari and Rohlfing[61] considered defective C_{60} and found that the lowest energy alternative to ideal C_{60} had two pairs of edge sharing pentagons and an energy of ~2 eV higher than the ground state.

Simulations such as these are valuable in distinguishing among possible growth pathways and clarifying the important energetic considerations. With the constraints imposed by today's computers, however, it is not yet feasible to simulate the conditions of the experiments and approach realistic cooling rates.

II. Fullerenes in Gas Phase and Solution

This volume emphasizes fullerenes in the solid state, focusing on developments after the Krätschmer–Huffman breakthrough of 1990.

[59]J.-Y. Yi and J. Bernholc, J. Chem. Phys. 96, 8634 (1992).
[60]P. W. Fowler, D. E. Manolopoulos, and R. P. Ryan, Carbon 30, 1235 (1992).
[61]K. Raghavachari and C. M. Rohlfing, J. Phys. Chem. 96, 2463 (1992).

Well before 1990, however, there had been a great deal of activity within the theoretical chemistry community and by those able to experimentally probe C_{60} in the gas phase. Significantly, calculations for carbon atoms arranged in the icosahedral structure had been performed prior to 1985. For a history of this early work, as well as a thorough review of the experimental and theoretical work up to mid-1991, the interested reader is referred to the review by Kroto et al.[3] and references therein. We will touch on just a few points in this section.

5. FULLERENES IN GAS PHASE

Prior to 1990, it had been determined that there should be only 4 infrared-active vibrational modes for C_{60},[22,62] a prediction that was particularly important because its realization allowed Krätschmer et al. to conclude that they had produced C_{60} in quantity. It had also been determined that there should be 10 Raman-active modes, and their positions had been calculated.[62,63] The optical absorption spectrum was calculated[64] for comparison with the diffuse interstellar lines. A variety of calculational techniques were used to determine the distribution of molecular energy levels and their character.[65] The thermodynamic stability of C_{60} was calculated to be favorable compared with graphitic sheets of 60 atoms, consistent with intuition that dangling bonds should be minimized to maximize stability.[66,67]

Early ultraviolet photoemission spectroscopy studies of fullerene anions[68] indicated that C_{60} and C_{70} were electronically closed shell molecules with appreciable gaps between the highest occupied and lowest unoccupied molecular orbitals, HOMO and LUMO. Such experiments also provided a measure of the electron affinities. The adiabatic ionization potentials were bracketed by examining charge transfer from

[62]R. C. Haddon, L. E. Brus, and K. Raghavachari, Chem. Phys. Lett. **125**, 459 (1986).

[63]F. Negri, G. Orlandi, and F. Zerbetto, Chem. Phys. Lett. **144**, 31 (1988).

[64]S. Larsson, A. Volosov, and A. Rosén, Chem. Phys. Lett. **137**, 501 (1987); M. Braga, S. Larsson, A. Rosén, and A. Volosov, Astron. Astrophys. **245**, 232 (1991).

[65]For an extensive listing of these calculations, see Ref. 3. See also the references cited by Pickett in this volume.

[66]M. D. Newton and R. E. Stanton, J. Am. Chem. Soc. **108**, 469 (1986).

[67]H. P. Lüthi and J. Almlöf, Chem. Phys. Lett. **135**, 357 (1987).

[68]S. H. Yang, C. L. Pettiette, J. Conceicao, O. Cheshnovsky, and R. E. Smalley, Chem. Phys. Lett. **139**, 233 (1987).

fullerene cations to neutral molecules in a Fourier transform mass spectrometer.[21]

The mass spectrometry studies of O'Brien et al. performed in 1987 showed that the fullerenes, under laser irradiation, were fragmented by loss of C_2 units.[69] Before the fullerenes were discovered, Bloomfield et al.[70] had also shown this fragmentation pattern for, coincidentally, C_{60}. Since the apparent threshold for C_{60} photofragmentation was about 18 eV, it was clear that the isolated molecules were highly stable. An elegant model of fragmentation was offered by O'Brien et al.[69] wherein pentagons on the excited fullerene surface were allowed to "migrate" until two were joined. In these regions of high stress, the C_2 unit joining the pentagons evaporated and the fullerene shell closed, leaving a C_{n-2} fullerene. For fragmentation of gas phase molecules, this process could be continued until the cage reached ~32 atoms. Beyond this, the strain involved in maintaining a closed structure was apparently too great, and the structure broke into chains and rings with both even and odd numbers of carbon atoms. Subsequent measurements by Maruyama et al.[71] suggested that the "even-numbered only" cluster distribution extends well into the hundreds of atoms. For the larger fullerenes, C_2 (or C_4 or C_6) loss was still the dominant fragmentation route.

The C_2 loss mechanism was used to demonstrate the existence of endofullerenes, fullerenes with a metal atom, M, trapped inside the cage, denoted $M@C_n$. Weiss et al.[17] reported that laser ablation of composite graphite discs containing La, K, or Cs gave rise to peaks in the mass spectra that corresponded to a fullerene plus the appropriate metal atom. To demonstrate that the metal atoms were inside the cages, "shrink wrapping" experiments of the fullerenes were undertaken that involved successive photoejection of C_2 units. For the endofullerenes, the endpoint of the C_2 removal process occurred not at 32 C atoms, as for empty cages, but at 44 or 48 atoms. Since fullerenes of this size should just accommodate the metal ions in their interior, it was argued that further removal of carbon would result in collapse, as observed. (Smalley and coworkers[72] recently reported the smallest endofullerene, $U@C_{28}$, the stability of which was related to the valence of the U atom and that of

[69]S. C. O'Brien, J. R. Heath, R. F. Curl, and R. E. Smalley, J. Chem. Phys. **88**, 220 (1988).
[70]L. A. Bloomfield, M. E. Geusic, R. R. Freeman, and W. L. Brown, Chem. Phys. Lett. **121**, 33 (1985).
[71]S. Maruyama, M. Y. Lee, R. E. Haufler, Y. Chai, and R. E. Smalley, Z. Phys. D **19**, 409 (1991).
[72]T. Guo, M. D. Diener, Y. Chai, M. J. Alford, R. E. Haufler, S. M. McClure, T. R. Ohno, J. H. Weaver, G. E. Scuseria, and R. E. Smalley, Science **257**, 1661 (1992).

the shell.) Additional studies involved exposure of the endofullerenes to H_2, O_2, NO, and NH_3. Reaction rates comparable with those of the bare carbon cages[73] indicated that the atoms were indeed inside the cage.

While the 1990 K–H breakthrough made possible studies of fullerenes in the solid state, it also ushered in a new era for gas phase investigations. Fullerene films or powders could now be used as the starting material for desorption so that extensive examinations of fundamental properties could be performed. With gas phase molecules produced in this way, the vertical ionization potential was measured, yielding a value of 7.6 eV for C_{60}.[74,75] The electron affinity of solid C_{60} clusters was measured to be 2.65 eV.[76] The leading portion of the energy level spectrum was probed by photoemission for neutral C_{60} molecules,[74] allowing comparison with theoretical predictions. Using gas phase electron diffraction,[77] the C–C bond lengths of C_{60} were measured, giving 1.485 Å for pentagon edges and 1.401 Å for the remaining hexagon edges. Ion collision studies[21] showed that the fragmentation energy is <23 eV for C_2 loss. Gas phase bombardment of C_{60} with energetic He, Ne, Li, and Na ions led to endohedral fullerene formation, and this route to endohedral formation continues to be an active area of research.[21,78–81]

6. FULLERENES IN SOLUTION

Soon after the K–H breakthrough, several groups developed methods for separating fullerenes of different sizes using column

[73]Q. L. Zhang, S. C. O'Brien, J. R. Heath, Y. Liu, R. F. Curl, H. W. Kroto, and R. E. Smalley, *J. Phys. Chem.* **90**, 525 (1986).

[74]D. L. Lichtenberger, K. W. Nebesny, C. D. Ray, D. R. Huffman, and L. D. Lamb, *Chem. Phys. Lett.* **176**, 203 (1991); D. L. Lichtenberger, M. E. Jatcko, K. W. Nebesny, C. D. Ray, D. R. Huffman, and L. D. Lamb in "Clusters and Cluster-Assembled Materials," Materials Research Society Symposium Proceedings, Vol. 206, eds. R. S. Averback, J. Bernholc, and D. L. Nelson (1991), p. 673.

[75]R. K. Yoo, B. Ruscic, and J. Berkowitz, *J. Chem. Phys.* **96**, 911 (1992).

[76]L. S. Wang, J. Conceicao, C. Jin, and R. E. Smalley, *Chem. Phys. Lett.* **182**, 5 (1991).

[77]K. Hedberg, L. Hedberg, D. S. Bethune, C. A. Brown, H. C. Dorn, R. D. Johnson, and M. de Vries, *Science* **254**, 410 (1992).

[78]Z. Wan, J. F. Christian, and S. L. Anderson, *Phys. Rev. Lett.* **69**, 1352 (1992).

[79]M. M. Ross and J. H. Callahan, *J. Phys. Chem.* **95**, 5720 (1991).

[80]Z. Wan, J. C. Christian, and S. L. Anderson, *J. Chem. Phys.* **96**, 3344 (1992).

[81]T. Weiske, D. K. Böhme, J. Hrušák, W. Krätschmer, and H. Schwarz, *Angew. Chem. Int. Ed. Engl.* **30**, 884 (1991).

chromatography.[23,82,83] Taylor *et al.*[23] and Johnson *et al.*[84,85] undertook NMR measurements of C_{60} in solution. They found a single resonance that indicated that all of the C atoms of C_{60} were chemically equivalent, consistent with the proposed icosahedral symmetry. This, considered along with the Raman[86] and IR spectra,[14,83,87] made a strong case for the soccerball structure. The NMR spectrum for C_{70} was found to consist of five lines,[23,82,88] as expected for the D_{5h} structure of Fig. 1.

A wide variety of chemical reactions have been explored in solution chemistry. The photophysical properties of C_{60} in solution were examined by Arbogast *et al.*,[89] who gave estimates of the first singlet and triplet energies [46.1 kcal/mol (2.0 eV) and 37.5 ± 4.5 kcal/mol (1.6 ± 0.2 eV), respectively], as well as the triplet lifetime (40 ± 4 μsec). They also showed that the C_{60} triplet is efficiently quenched by energy transfer to molecular oxygen to form singlet oxygen. In the first demonstration of molecular modification, Haufler *et al.*[90] showed the reversible attachment of 36 hydrogen atoms to the C_{60} cage. Hawkins *et al.*[91] demonstrated the reversible addition of two OsO_4(4-tert-butylpyridine) moieties and subsequently crystallized osmylated C_{60}. The large osmium adduct served to inhibit C_{60} rotation (discussed in the next section). In this way, it was possible to use x-ray diffraction to determine the positions of the individual carbon atoms. This was considered to be the final piece of evidence necessary to confirm the soccer-ball structure. A large number of other chemical additions to fullerene cages have been demonstrated.

[82]H. Ajie, M. M. Alvarez, S. J. Anz, R. D. Beck, F. Diederich, K. Fostiropoulos, D. R. Huffman, W. Krätschmer, Y. Rubin, K. E. Schriver, D. Sensharma, and R. L. Whetten, *J. Phys. Chem.* **94**, 8630 (1990).

[83]D. S. Bethune, G. Meijer, W. C. Tang, H. J. Rosen, W. G. Golden, H. Seki, C. A. Brown, and M. S. de Vries, *Chem. Phys. Lett.* **179**, 181 (1991).

[84]R. D. Johnson, G. Meijer, and D. S. Bethune, *J. Am. Chem. Soc.* **112**, 8983 (1990).

[85]R. D. Johnson, D. S. Bethune, and C. S. Yannoni, *Accts. Chem. Res.* **25**, 169 (1992).

[86]D. S. Bethune, G. Meijer, W. C. Tang, and H. J. Rosen, *Chem. Phys. Lett.* **174**, 219 (1990).

[87]J. P. Hare, T. J. Dennis, H. W. Kroto, R. Taylor, A. W. Allaf, S. P. Balm, and D. R. M. Walton, *J. Chem. Soc., Chem. Comm.* 412 (1991).

[88]R. D. Johnson, G. Meijer, J. R. Salem, and D. S. Bethune, *J. Am. Chem. Soc.* **113**, 3619 (1991).

[89]J. W. Arbogast, A. P. Darmanyan, C. S. Foote, Y. Rubin, F. N. Diederich, M. M. Alvarez, S. J. Anz, and R. L. Whetten, *J. Phys. Chem.* **95**, 11 (1991).

[90]R. E. Haufler, J. Conceicao, L. P. F. Chibante, Y. Chai, N. E. Byrne, S. Flanagan, M. M. Haley, S. C. O'Brien, C. Pan, Z. Xiao, W. E. Billups, M. A. Ciufolini, R. H. Hauge, J. L. Margrave, L. J. Wilson, R. F. Curl, and R. E. Smalley, *J. Phys. Chem.* **94**, 8634 (1990).

[91]J. M. Hawkins, S. Loren, A. Meyer, and R. Nunlist, *J. Am. Chem. Soc.* **113**, 7770 (1991); J. M. Hawkins, A. Meyer, T. A. Lewis, S. Loren, and F. S. Hollander, *Science* **252**, 312 (1991); J. M. Hawkins, *Accts. Chem. Res.* **25**, 150 (1992).

These range from the attachment of individual atoms (such as $C_{60}O$, Ref. 92) to the attachment of large numbers of atoms (such as $Br_{24}C_{60}$, Ref. 93) to the addition of large functional groups (such as $[(C_6H_5)P]_2Pt$, Ref. 94). The reader is referred to excellent review articles by Wudl[95] and by Fagan et al.[96] for more details and in-depth analysis.

III. Fullerenes in the Solid State

Of primary concern for the discussion of a solid state material is the type and the strength of the bonds that hold the material together. For the fullerenes, there are two very different bond strengths, depending on whether one is considering the intermolecular bond or the intramolecular bond. The former is largely van der Waals in character, and the latter is covalent. Krätschmer et al.[14] demonstrated that the fullerenes could be extracted from soot because they were soluble in benzene and were rather weakly bound to the insoluble carbonaceous matrix. They also showed that crystallites could be formed by evaporating the solvent from a fullerene-saturated solution, indicating an important pathway for crystal growth, and they reported that fullerenes could be sublimed intact at temperatures of only a few hundred degrees Celsius. This ability to sublime fullerenes had a profound impact on fullerene research, because it offered an alternate pathway for growth that eliminated the solvent. The ability to sublime the fullerenes also increased the likelihood that they can find applications.

7. SOLID STATE STRUCTURES AND ORDERING TRANSITIONS

Krätschmer et al. reported that C_{60} crystallizes in a close-packed structure with a nearest neighbor spacing of ~ 10 Å. Subsequent studies[97]

[92]K. M. Creegan, J. L. Robbins, W. K. Robbins, J. M. Millar, R. D. Sherwood, P. J. Tindall, D. M. Cox, A. B. Smith III, J. P. McCauley, Jr., D. R. Jones, and R. T. Gallagher, *J. Am. Chem. Soc.* **114**, 1103 (1992).

[93]F. N. Tebbe, R. L. Harlow, D. B. Chase, D. L. Thorn, G. C. Campbell, Jr., J. C. Calabrese, N. Herron, R. J. Young, Jr., and E. Wasserman, *Science* **256**, 822 (1992); P. R. Birkett, P. B. Hitchcock, H. W. Kroto, R. Taylor, and D. R. M. Walton, *Nature*, **357**, 479 (1992).

[94]P. J. Fagan, J. C. Calabrese, and B. Malone, *Science* **252**, 1160 (1991).

[95]F. Wudl, *Accts. Chem. Res.* **25**, 157 (1992), and references therein.

[96]P. J. Fagan, J. C. Calabrese, and B. Malone, *Accts. Chem. Res.* **25**, 134 (1992), and references therein.

[97]R. M. Fleming, T. Siegrist, P. M. Marsh, B. Hessen, A. R. Kortan, D. W. Murphy, R. C. Haddon, R. Tycko, G. Dabbagh, A. M. Mujsce, M. L. Kaplan, and S. M. Zahurak in "Clusters and Cluster-Assembled Materials," Materials Research Society Symposium Proceedings, Vol. 206, eds. R. S. Averback, J. Bernholc, and D. L. Nelson (1991), p. 691.

with solvent-free C_{60} showed that the room temperature crystal structure is fcc (Fm$\bar{3}$m symmetry) with a lattice constant of 14.2 Å. The symmetry of this space group is based on a point group with a fourfold axis. Since this symmetry is not present in the C_{60} molecule (despite the large number of symmetries that are present), the orientation of the molecules on the crystal lattice is then called into question.

The simplest molecular arrangement on an fcc lattice is that depicted in Fig. 1, where every molecule is oriented the same way with the twofold rotation axes of the molecules cut by one of the three $\langle 100 \rangle$ crystallographic directions. This Fm$\bar{3}$ (T_h^3) arrangement does not have the full symmetry of the experimentally determined space group, wherein a fourfold axis should be present along $\langle 100 \rangle$, but it is often employed for calculations of solid state properties that might be somewhat insensitive to the exact molecular orientation. The first such calculation emphasized the electronic structure of solid C_{60}.[98]

The Fm$\bar{3}$m x-ray diffraction pattern was explained by assuming disorder in the C_{60} orientations, and this resulted in a higher apparent symmetry. Fleming et al.[97] showed that a model including static merohedral disorder (wherein half of the molecules were rotated by 90° about [100]) provided a better fit to the observed x-ray powder pattern, although it was not clear if the disorder was static or dynamic. Fischer et al.[99] obtained a good fit to x-ray diffraction data by modeling the C_{60} molecules as spherical shells of charge, thus implying that the molecules were rotating isotropically with no preferred orientation. That model was motivated in part by NMR measurements, to be discussed, that indicated that the C_{60} molecules changed orientations rapidly at room temperature. This model nicely explained the missing ($h00$) reflections for even h in the x-ray powder pattern, reflections that are allowed for an fcc structure.[99–101] The absence of these peaks was reconciled by noting that diffraction peak intensities are proportional to the square of a structure factor. For a spherical shell this structure factor is the zero-order spherical Bessel function, $j_0(Q) = \sin(QR_0)/QR_0$, where the magnitude of the scattering vector Q is $4\pi \sin \theta / \lambda$ and R_0 is the radius of the shell. This function, quite coincidently, has zeros for Q values corresponding to

[98]J. H. Weaver, J. L. Martins, T. Komeda, Y. Chen, T. R. Ohno, G. H. Kroll, R. E. Haufler, and R. E. Smalley, *Phys. Rev. Lett.* **66**, 1741 (1991).

[99]J. E. Fischer, P. A. Heiney, A. R. McGhie, W. J. Romanow, A. M. Denenstein, J. P. McCauley, Jr., and A. B. Smith III, *Science* **252**, 1288 (1991).

[100]S. J. Duclos, K. Brister, R. C. Haddon, A. R. Kortan, and F. A. Thiel, *Nature* **351**, 80 (1991).

[101]P. A. Heiney, J. E. Fischer, A. R. McGhie, W. J. Romanow, A. M. Denenstein, J. P. McCauley, Jr., A. B. Smith III, and D. E. Cox, *Phys. Rev. Lett.* **66**, 2911 (1991).

the missing ($h00$) reflections when $R_0 = 3.5$ Å, the C_{60} radius. Duclos et al.[100] demonstrated that the (200) peak could be moved away from the structure factor zero and was observed if the lattice constant was changed by compression.

The question about whether the disorder was static or dynamic was addressed in early NMR studies by Yannoni et al.[102] and Tycko et al.[103] for fullerene powders. A single sharp resonance, as also observed for solvated C_{60}, indicated that the fullerenes underwent rapid rotational diffusion ($\sim 10^{19}$ Hz by the estimate of Yannoni et al.) that was nearly isotropic at 300 K. Molecular dynamics calculations by Zhang et al.[104] suggested that the C_{60} molecules should rotate at low temperature. Thus emerged the picture of "spinning" balls in the solid state. Rotational correlation times of ~ 12 psec at 300 K and ~ 9 psec at 283 K were reported[105,106] and estimates of the orientational diffusion barrier height of 42 and 60 meV were derived from NMR measurements. These rotational correlation times were only ~ 3 times longer than expected for a free molecule and were actually shorter than observed for C_{60} in solution.[106] Rotational diffusion was subsequently characterized by neutron scattering measurements,[107] and those results were consistent with uncorrelated, thermally activated rotational diffusion with a barrier of 35 meV.

The existence of barriers to reorientation suggested that molecular motion was affected by the potential of the surrounding crystal and that the molecules should have some (weakly) preferred orientation in the solid. A synchrotron radiation x-ray diffraction study by Chow et al.[108] for single crystal C_{60} suggested that the average charge distribution deviated somewhat from spherical. This indicated that the motion was not completely isotropic and that elements of both static and dynamic models were present. Those authors estimated a maximum barrier height of

[102]C. S. Yannoni, R. D. Johnson, G. Meijer, D. S. Bethune, and J. R. Salem, *J. Phys. Chem.* **95**, 9 (1991).

[103]R. Tycko, R. C. Haddon, G. Dabbagh, S. H. Glarum, D. C. Douglass, and A. M. Mujsce, *J. Phys. Chem.* **95**, 518 (1991).

[104]Q. M. Zhang, J.-Y. Yi, and J. Bernholc, *Phys. Rev. Lett.* **66**, 2633 (1991).

[105]R. Tycko, G. Dabbagh, R. M. Fleming, R. C. Haddon, A. V. Makhija, and S. M. Zahurak, *Phys. Rev. Lett.* **67**, 1886 (1991).

[106]R. D. Johnson, C. S. Yannoni, H. C. Dorn, J. R. Salem, and D. S. Bethune, *Science* **255**, 1235 (1992).

[107]D. A. Neumann, J. R. D. Copley, R. L. Cappelletti, W. A. Kamitakahara, R. M. Lindstrom, K. M. Creegan, D. M. Cox, W. J. Romanow, N. Coustel, J. P. McCauley, Jr., N. C. Maliszewskyj, J. E. Fischer, and A. B. Smith III, *Phys. Rev. Lett.* **67**, 3808 (1991).

[108]P. C. Chow, X. Jiang, G. Reiter, P. Wochner, S. C. Moss, J. D. Axe, J. C. Hanson, R. K. McMullan, R. L. Meng, and C. W. Chu, *Phys. Rev. Lett.* **69**, 2943 (1992).

~ 52 meV but pointed out that the barrier height was a function of direction of the reorientation. All the evidence, then, pointed to a spinning molecule that was only slightly hindered relative to a free rotor at 300 K.

The separation between shells of carbon nuclei in the crystalline state can be estimated by taking the nearest neighbor distance of 10 Å and subtracting twice the atomic shell radius of 3.5 Å, giving 3 Å. This is comparable with, though a bit smaller than, the interplanar spacing of ~ 3.35 Å for the planes of graphite. The compressibility of C_{60} was found by Duclos et al.[100] to be consistent with van der Waals bonding of essentially incompressible spheres for pressures up to about 20 GPa. The emergence of the (200) diffraction peak already noted indicated that the molecules themselves were not compressed as their intermolecular spacing was decreased. Fischer et al.[99] found the linear compressibility of C_{60} to be the same, within experimental error, as that of graphite in the interlayer direction for pressures up to 1.2 GPa. The volume compressibility in this regime was 3 times that of graphite and 40 times that of diamond. Changes in optical and vibrational properties also occur with application of pressure, and phase changes could be induced.[109-116] Several authors[114-116] characterized the "collapsed fullerite" material that resulted from compression beyond ~ 20 GPa. While some suggested that it was a new and unique material,[114] Núñez-Regueiro et al.[115,116] found transparent portions of crushed samples that appeared to be microcrystalline diamond.

The high temperature fcc symmetry of C_{60} films is lowered to simple cubic (Pa$\bar{3}$) upon cooling below about 260 K.[117] The transition occurs via orientational ordering of the molecules on their fcc lattice positions. This structure possesses a basis of four molecules that are inequivalent in orientation, as shown in Fig. 5(a). In this configuration, each molecule is

[109]Y. Huang, D. F. R. Gilson, and I. S. Butler, *J. Phys. Chem.* **95**, 5723 (1991).

[110]D. W. Snoke, K. Syassen, and A. Mittelbach, *Phys. Rev. B* **47**, 4146 (1993).

[111]D. W. Snoke, Y. S. Raptis, and K. Syassen, *Phys. Rev. B* **45**, 14419 (1992).

[112]N. Chandrabhas, M. N. Shashikala, D. V. S. Muthu, A. K. Sood, and C. N. R. Rao, *Chem. Phys. Lett.* **197**, 319 (1992).

[113]S. H. Tolbert, A. P. Alivisatos, H. E. Lorenzana, M. B. Kruger, and R. Jeanloz, *Chem. Phys. Lett.* **188**, 163 (1992).

[114]F. Moshary, N. H. Chen, I. F. Silvera, C. A. Brown, H. C. Dorn, M. S. de Vries, and D. S. Bethune, *Phys. Rev. Lett.*, **69**, 466 (1992).

[115]M. Núñez-Regueiro, P. Monceau, and J.-L. Hodeau, *Nature* **355**, 237 (1992).

[116]M. Núñez-Regueiro, L. Abello, G. Lucazeau, and J.-L. Hodeau, *Phys. Rev. B* **46**, 9903 (1992).

[117]P. A. Heiney, *J. Phys. Chem. Solids* **53**, 1333 (1992).

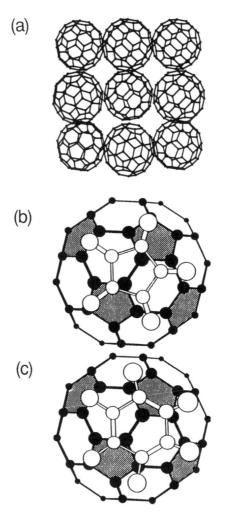

FIG. 5. Molecular orientation for C_{60} in the low temperature simple cubic structure. (a) reveals an array of molecules rotated relative to the orientation of Fig. 1 to achieve the simple cubic $Pa\bar{3}$ structure. The view is from the top of a unit cell with the five smaller molecules forming the bottom corners and the center. The four larger molecules are positioned at face centers of the unit cell sides. (b) and (c) show the relationship between one C_{60} molecule and neighbors in two possible low energy configurations. In (b) the ideal ($Pa\bar{3}$) nearest neighbor configuration is shown in which a pentagon on one C_{60} (only partially shown with open circles) is facing a double bond on its neighbors. In (c) the hexagon to double-bond configuration is shown. From Prassides and Kroto[4] and Copley *et al.*[134]

rotated by an angle of ~22° from the position shown in Fig. 1 around one of the four ⟨111⟩ directions passing through a hexagonal face. The fcc to sc structural transition was observed independently by Dworkin et al.,[118] Heiney et al.,[119] and Tse et al.[120] The correct sc space group was reported by Sachidanandam and Harris[121] based on analysis of the data of Heiney et al. It was verified by the neutron powder diffraction measurements of David et al.[122] An elegant rationalization for the low temperature structure was provided by David et al., who noted that the electron-rich double bonds adjoining hexagons were "nested" in the electron-poor pentagonal faces of neighboring molecules in the Pa$\bar{3}$ arrangement, as depicted in Fig. 5(b). Support for their electrostatic model was given by the failure of van der Waals bonding calculations to predict the correct low temperature structure.[123]

The fcc to sc transition was also observed using differential scanning calorimetry,[117] Raman[124,125] and IR[126,127] spectroscopies, electron diffraction,[128] inelastic neutron scattering,[129] and thermal expansion techniques.[130] A free energy change of ~9.1 J/g (0.4 meV/C_{60}) was estimated from the DSC data.[117] The transition was first order in nature, as can be deduced from the initial and final state symmetries, and it involved a discontinuous change in the lattice constant at the transition,

[118]A. Dworkin, H. Szwarc, S. Leach, J. P. Hare, T. J. S. Dennis, H. W. Kroto, R. Taylor, and D. R. M. Walton, C.R. Acad. Sci. Paris 312 II, 979 (1991).

[119]P. A. Heiney, J. E. Fischer, A. R. McGhie, W. J. Romanow, A. M. Denenstein, J. P. McCauley, A. B. Smith III, and D. E. Cox, Phys. Rev. Lett. 66, 2911 (1991).

[120]J. S. Tse, D. D. Klug, D. A. Wilkinson, and Y. P. Handa, Chem. Phys. Lett. 183, 387 (1991).

[121]R. Sachidanandam and A. B. Harris, Phys. Rev. Lett. 67, 1467 (1991).

[122]W. I. F. David, R. M. Ibberson, J. C. Matthewman, K. Prassides, T. J. S. Dennis, J. P. Hare, H. W. Kroto, R. Taylor, and D. R. M. Walton, Nature 353, 147 (1991).

[123]W. I. F. David, R. M. Ibberson, T. J. S. Dennis, J. P. Hare, and K. Prassides, Europhys. Lett. 18, 219 (1992).

[124]P. H. M. van Loosdrecht, P. J. M. van Bentum, and G. Meijer, Phys. Rev. Lett. 68, 1176 (1992).

[125]M. Matus, T. Pichler, M. Haluska, and H. Kuzmany, Springer Series in Solid State Science 113, 446 (1993).

[126]L. R. Narasimhan, D. N. Stoneback, A. F. Hebard, R. C. Haddon, and C. K. N. Patel, Phys. Rev. B 46, 2591 (1992).

[127]V. S. Babu and M. S. Seehra, Chem. Phys. Lett. 196, 569 (1992).

[128]N. Yao, C. F. Klein, S. K. Behal, M. M. Disko, R. D. Sherwood, K. M. Creegan, and D. M. Cox, Phys. Rev. B 45, 11366 (1992).

[129]D. A. Neumann, J. R. D. Copley, W. A. Kamitakahara, J. J. Rush, R. L. Cappelletti, N. Coustel, J. E. Fischer, J. P. McCauley, Jr., A. B. Smith III, K. M. Creegan, and D. M. Cox, J. Chem. Phys. 96, 8631 (1992).

[130]F. Gugenberger, R. Heid, C. Meingast, P. Adelmann, M. Braun, H. Wühl, M. Haluska, and H. Kuzmany, Phys. Rev. Lett. 69, 3774 (1992).

as demonstrated by Heiney *et al.*[131] and David *et al.*[123] using x-ray and neutron diffraction. Michel[132] reported a calculation for molecular ordering.

Detailed analysis by David *et al.*[123] suggested that there were actually two inequivalent molecular orientations in the low temperature phase. A majority of molecules assumed the "ideal" positions shown in Fig. 5(b), but others were rotated an additional 60° about the ⟨111⟩ direction. In this orientation the molecules presented hexagonal faces, rather than pentagonal faces, to the double bonds on neighboring molecules, as in Fig. 5(c). David *et al.* found a ratio of ~1.7:1 for the two types of orientations just below the fcc–sc transition temperature and reported that this value increased to ~5:1 as the temperature was lowered to 90 K.

The existence of an orientationally ordered structure at first seemed at odds with low temperature NMR data that showed a motionally narrowed line well below the transition temperature. Tycko *et al.*[105] and Johnson *et al.*[106] examined the NMR spin lattice relaxation times as a function of temperature, finding evidence for the 260 K phase transition but suggesting that the narrow NMR peak resulted from continued "ratcheting" of C_{60} molecules between preferred positions in the sc structure. The barrier to reorientation was found to be significantly larger than in the high temperature phase and was estimated to be 250 or 180 meV.

Below ~90 K, it was found that the molecular motion was essentially quenched.[133,134] In this low temperature state, the fullerenes resembled a glass with molecules disordered between two distinguishable orientations. Copley *et al.*[133,134] showed that the diffuse part of the neutron scattering signal could be described by adding small amplitude librational disorder of the molecules to the structural model proposed by David *et al.*[123] The existence of distinguishable molecules suggested possibilities for further long range ordering, and indeed, evidence for a low temperature doubling of the unit cell was found by Van Tendeloo *et al.*[135] in electron diffraction measurements of single crystals. See the chapter by Axe,

[131]P. A. Heiney, G. B. M. Vaughan, J. E. Fischer, N. Coustel, D. E. Cox, J. R. D. Copley, D. A. Neumann, W. A. Kamitakahara, K. M. Creegan, D. M. Cox, J. P. McCauley, Jr., and A. B. Smith III, *Phys. Rev. B* **45**, 4544 (1992).

[132]K. H. Michel, *Chem. Phys. Lett.* **193**, 478 (1992).

[133]J. R. D. Copley, D. A. Neumann, R. L. Cappelletti, W. A. Kamitakahara, E. Prince, N. Coustel, J. P. McCauley, Jr., N. C. Maliszewskyj, J. E. Fischer, A. B. Smith III, K. M. Creegan, and D. M. Cox, *Physica B* **180/181**, 706 (1992).

[134]J. R. D. Copley, D. A. Neumann, R. L. Cappelletti, and W. A. Kamitakahara, *J. Phys. Chem. Solids* **53**, 1353 (1992).

[135]G. Van Tendeloo, S. Amelinckx, M. A. Verheijen, P. H. M. van Loosdrecht, and G. Meijer, *Phys. Rev. Lett.* **69**, 1065 (1992).

Moss, and Neumann for a detailed discussion of structure and dynamics studies. The growth of single crystals or large-grained polycrystalline samples generally involves vapor transport.[136] Meng and coworkers[137] and Kuzmany and coworkers[138] grew crystals several millimeters in size. Xiang *et al.*[139] measured the resistivity as a function of temperature for K_3C_{60} using millimeter-sized doped single crystals. They reported three-dimensional fluctuations near the superconducting transition temperature and T^2 behavior at high temperature. Van Tendeloo *et al.*[135] grew bulk crystals using multiple stages of vapor transport and used them for the electron diffraction studies previously mentioned. Wu *et al.*[140] undertook angle-resolved photoemission studies of C_{60}, reporting that there was no band dispersion for the HOMO band within their experimental uncertainty of 0.05 eV. Pintschovius *et al.*[141] measured the translational phonon band dispersions for C_{60} above and below the fcc–sc ordering transition, and Chow *et al.*[108] investigated orientational order at room temperature. Crystal growth was also reported by Kortan *et al.*,[142] Verheijen *et al.*,[143] Saito *et al.*,[144] and Li *et al.*[145] Electrical conductivity[146,147] and thermal

[136]Special issue of *Applied Physics A*, Vol. **56**, March (1993).

[137]R. L. Meng, D. Ramirez, X. Jiang, P. C. Chow, C. Diaz, K. Matsuishi, S. C. Moss, P. H. Hor, and C. W. Chu, *Appl. Phys. Lett.* **59**, 3401 (1991).

[138]M. Haluska, H. Kuzmany, M. Vybornov, P. Rogl, and P. Fejdi, *Appl. Phys. A* **56**, 161 (1993).

[139]X.-D. Xiang, J. G. Hou, G. Briceño, W. A. Vareka, R. Mostovoy, A. Zettl, V. H. Crespi, and M. L. Cohen, *Science* **256**, 1190 (1992); X.-D. Xiang, J. G. Hou, V. H. Crespi, Z. Zettl, and M. L. Cohen, *Nature* **361**, 54 (1993); V. H. Crespi, J. G. Hou, X.-D. Xiang, M. L. Cohen, and A. Zettl, *Phys. Rev. B* **46**, 12064 (1992).

[140]J. Wu, Z.-X. Shen, D. S. Dessau, R. Cao, D. S. Marshall, P. Pianetta, I. Lindau, X. Yang, J. Terry, D. M. King, B. O. Wells, D. Elloway, H. R. Wendt, C. A. Brown, H. Hunziker, and M. S. de Vries, *Physica C* **197**, 251 (1992).

[141]L. Pintschovius, B. Renker, F. Gompf, R. Heid, S. L. Chaplot, M. Haluska, and H. Kuzmany, *Phys. Rev. Lett.* **69**, 2662 (1992).

[142]A. R. Kortan, N. Kopylov, and F. A. Thiel, *J. Phys. Chem. Solids* **53**, 1683 (1992); E. D. Isaacs, P. M. Platzman, P. Zschack, H. Hamalainen, and A. R. Kortan, *Phys. Rev. B* **46**, 12910 (1992).

[143]M. A. Verheijen, H. Meekes, G. Meijer, E. Raas, and P. Bennema, *Chem. Phys. Lett.* **191**, 339 (1992).

[144]Y. Saito, Y. Ishikawa, A. Ohshita, H. Shinohara, and H. Nagashima, *Phys. Rev. B* **46**, 1846 (1992).

[145]J. Li, S. Komiya, T. Tamura, C. Nagasaki, J. Kihara, K. Kishio, and K. Kitazawa, *Physica C* **195**, 205 (1992).

[146]C. Wen, J. Li, K. Kitazawa, T. Aida, I. Honma, H. Komiyama, and K. Yamada, *Appl. Phys. Lett.* **61**, 2162 (1992).

[147]R.K. Kremer, T. Rabenau, W.K. Maser, M. Kaiser, A. Simon, M. Haluska, and H. Kuzmany, *Appl. Phys. A* **56**, 211 (1993).

conductivity[148] measurements were performed for single crystals and atomic force microscopy, AFM, was used to image the structure of single crystals.[149] Raman spectra[150] and x-ray diffraction[151] measurements were made for single crystals through the fcc–sc phase transition. A variety of reports were made for single crystals grown from solution,[152–155] but those tended to incorporate solvents and were often not fcc. Distilling of the solvent to obtain an fcc structure was reported in at least one case.[154]

The second most abundant fullerene, C_{70}, was the subject of several structural studies in powder or single crystal form.[156–159] At high temperatures, C_{70} forms an orientationally disordered fcc lattice[156] with isotropic tumbling of the rugby ball–shaped molecule. Solvent-free crystals with hcp packing were also observed, but they transformed irreversibly to fcc upon heating.[156] Lucas[160] pointed out that the formation energies for the hcp and fcc structures for spherical or quasi-spherical rotating molecules are nearly identical, although the fcc structure is invariably favored. Following ideas put forward by van de Waal[161] for Lennard–Jones solids, he argued that the dominance of the

[148]N. H. Tea, R.-C. Yu, M. B. Salamon, D. C. Lorents, R. Malhotra, and R. S. Ruoff, *Appl. Phys. A* **56**, 219 (1993).

[149]P. Dietz, P. Hansma, K. Fostiropoulos, and W. Krätschmer, *Appl. Phys. A* **56**, 207 (1993).

[150]M. Matus and H. Kuzmany, *Appl. Phys. A* **56**, 241 (1993).

[151]H. Kasatani, H. Terauchi, Y. Hamanaka, and S. Nakashima, *Phys. Rev. B* **47**, 4022 (1993).

[152]R. M. Fleming, A. R. Kortan, B. Hessen, T. Siegrist, F. A. Thiel, P. Marsh, R. C. Haddon, R. Tycko, G. Dabbagh, M. L. Kaplan, and A. M. Mujsce, *Phys. Rev. B* **44**, 888 (1991).

[153]Y. Saito, N. Suzuki, H. Shinohara, and Y. Ando, *Jpn. J. Appl. Phys.* **39**, 88 (1991).

[154]T. Arai, Y. Murakami, H. Suematsu, K. Kikuchi, Y. Achiba, and I. Ikemoto, *J. Phys. Soc. Jpn.* **61**, 1821 (1992).

[155]Y. Yosida, T. Arai, and H. Suematsu, *Appl. Phys. Lett.* **61**, 1043 (1992).

[156]G. B. M. Vaughan, P. A. Heiney, J. E. Fischer, D. E. Luzzi, D. A. Ricketts-Foot, A. R. McGhie, Y. W. Hui, A. L. Smith, D. E. Cox, W. J. Romanow, B. H. Allen, N. Coustel, J. P. McCauley, Jr., and A. B. Smith III, *Science* **254**, 1350 (1991); G. B. M. Vaughan, P. A. Heiney, D. E. Cox, J. E. Fischer, A. R. McGhie, A. L. Smith, R. M. Strongin, M. A. Chichy, and A. B. Smith III, submitted to *Chem. Phys.*

[157]M. A. Verheijen, H. Meekes, G. Meijer, P. Bennema, J. L. de Boer, S. van Smallen, G. Van Tendeloo, S. Amelinckx, S. Muto, and J. von Landuyt, *Chem. Phys.* **166**, 287 (1992).

[158]V. P. Dravid, X. Lin, H. Zhang, S. Liu and M. M. Kappes, *J. Mater. Res.* **7**, 2440 (1992).

[159]M. A. Green, M. Kurmoo, P. Day, and K. Kikuchi, *J. Chem. Soc. Chem.* **XX**, 1676 (1992).

[160]A. A. Lucas, private communication.

[161]B. W. van de Waal, *Phys. Rev. Lett.* **67**, 3264 (1991).

fcc structure was related to the kinetics of growth, not the thermo-dynamics, and, in part, that accidental nucleation of partial vacancy rows favored ABC stacking over ABAB or random stacking.

Cooling fcc crystals of C_{70} to ~345 K resulted in freezing two degrees of orientational motion, but the molecules continued to spin about long axes that aligned along [111].[156,157] The resulting crystal symmetry is rhombohedral. Further cooling below ~295 K quenched this last degree of freedom[156,157] to yield a monoclinic symmetry. These ordering transitions were observed by DSC,[156,162] Raman spectroscopy,[163,164] IR spectroscopy,[165] electron diffraction,[158,166] neutron scattering,[167] and thermal expansion measurements,[168] and they have been explained through theoretical considerations.[169,170]

While spherical C_{60} and ellipsoidal C_{70} exhibit interesting ordering transitions, there have been no solid state structural studies of the higher fullerenes that we are aware of, mainly due to limitations in quantity. Given their less symmetric form, one can anticipate several ordering transitions. What has been deduced from scanning tunneling microscopy studies[171] of thin films of C_{76}, C_{78}, C_{82}, and C_{84} is that close-packed arrays form at 300 K and that large crystallites can be produced.

8. Thin Films, Thin Film Formation, and Substrate Interactions

The ability to sublime the fullerenes enabled a wide variety of studies using vapor-grown films and crystals. Early studies indicated that

[162]E. Grivei, B. Nysten, M. Cassart, J.-P. Issi, C. Fabre, and A. Rassat, *Phys. Rev. B* **47**, 1705 (1993).

[163]N. Chandrabhas, K. Jayaram, D. V. S. Muthu, A. K. Sood, R. Seshadri, and C. N. R. Rao, *Phys. Rev. B* **47**, 10963 (1993).

[164]P. H. M. van Loosdrecht, M. A. Verheijen, H. Meekes, P. J. M. van Bentum, and G. Meijer, *Phys. Rev. B* **47**, 7610 (1993).

[165]V. Varma, R. Seshadri, A. Govindaraj, A. K. Sood, and C. N. R. Rao, *Chem. Phys. Lett.* **203**, 545 (1993).

[166]M. Tomita, T. Hayashi, P. Gaskell, T. Maruno, and T. Tanaka, *Appl. Phys. Lett.* **61**, 1171 (1992).

[167]B. Renker, F. Gompf, R. Heid, P. Adelmann, A. Heiming, W. Reichardt, G. Roth, H. Schober, and H. Rietschel, *Z. Phys. B* **90**, 325 (1993).

[168]C. Meingast, F. Gugenberger, M. Haluska, H. Kuzmany, and G. Roth, *Appl. Phys. A* **56**, 227 (1993).

[169]M. Sprik, A. Cheng, and M. L. Klein, *Phys. Rev. Lett.* **69**, 1660 (1992).

[170]A. Cheng and M. L. Klein, *Phys. Rev. B* **46**, 4958 (1992).

[171]Y. Z. Li, J. C. Patrin, M. Chander, J. H. Weaver, K. Kikuchi, and Y. Achiba, *Phys. Rev. B* **47**, 10867 (1993).

sublimation at appreciable rates occurred at temperatures of 300–500°C. The first real-space images of the fullerenes[172–174] were obtained with scanning tunneling microscopy, STM, using thin films condensed on gold surfaces and imaged in air. Those images showed more-or-less round objects that were spaced about 10 Å apart and had height variations that could be explained by mixture of C_{70} and C_{60}, as in the starting material. In some cases, internal molecular structure was observed, indicating that the fullerenes were not rotating rapidly on the time scale of STM imaging.[174–176]

In film growth, it is important to understand the character of the overlayer–substrate interaction. Condensation implies physisorption or chemisorption, depending on the strength of the bond. Ohno et al.[177] showed that a condensed layer of C_{60} will be energy-level referenced to the Fermi level of the substrate, producing a dipole at the interface to compensate for the difference in work functions. They reported, for example, that C_{60} condensation on n-type GaAs(110) gave rise to a dipole that represented the transfer in a resonance-like state of 0.02 electrons to each fullerene. For other systems, the amount of charge transfer was larger.

If the surface bonding of the fullerene is strong, then its ability to diffuse on that surface is restricted. Li et al.[178] showed that this was the case for C_{60} deposition onto Si(111) − 7 × 7 at 300 K because isolated individual molecules could be imaged. For this system, the fullerenes could be moved from one position to another using the STM tip, an effect that may be useful in nanotechnology applications. For such strong-bonding systems or for deposition at low temperature where surface diffusion is restricted, continued C_{60} deposition leads to small-grained polycrystalline samples. Such an effect has been reported by Hebard et

[172]R. J. Wilson, G. Meijer, D. S. Bethune, R. D. Johnson, D. D. Chambliss, M. S. de Vries, H. E. Hunziker, and H. R. Wendt, Nature **348**, 621 (1990).
[173]J. L. Wragg, J. E. Chamberlain, H. W. White, W. Krätschmer, and D. R. Huffman, Nature **348**, 623 (1990).
[174]T. Chen, S. Howells, M. Gallagher, L. Yi, D. Sarid, D. L. Lichtenberger, K. W. Nebesny, and C. D. Ray in "Clusters and Cluster-Assembled Materials," Materials Research Society Symposium Proceedings, Vol. 206, eds. R. S. Averback, J. Bernholc, and D. L. Nelson (1990), p. 721.
[175]T. Chen and D. Sarid, Mod. Phys. Lett. **6**, 967 (1992).
[176]T. Chen, S. Howells, M. Gallagher, L. Yi, D. Sarid, D. L. Lichtenberger, K. W. Nebesny, and C. D. Ray, J. Vac. Sci. Technol. B **9**, 2461 (199).
[177]T. R. Ohno, Y. Chen, S. E. Harvey, G. H. Kroll, J. H. Weaver, R. E. Haufler, and R. E. Smalley, Phys. Rev. B **44**, 13747 (1991).
[178]Y. Z. Li, M. Chander, J. C. Patrin, J. H. Weaver, L. P. F. Chibante, and R. E. Smalley, Phys. Rev. B **45**, 13837 (1992).

al.[179] for C_{60} film growth on sapphire. They reported typical grain sizes of ~60 Å, roughly 7–8 molecules wide. Larger grains can be grown, even on strongly interactive surfaces, by increasing the substrate temperature, thereby enhancing thermally activated surface diffusion. There is a limit to the growth temperature, however, because the sticking coefficient of the impinging molecules is also reduced. In any case, the adhesion of the first layer to the substrate generally exceeds that of subsequent layers, an effect that has to be used to obtain single-layer-thick films.[180,181]

The most thoroughly studied fullerene–substrate system is $C_{60}/GaAs(110)$. Li *et al.*[182,183] reported the first ultrahigh vacuum STM studies of fullerenes using films condensed on GaAs(110). They found that the fullerenes were highly mobile, even at 300 K, an effect they related to weak substrate bonding. For growth on stepped GaAs(110) surfaces, they observed fullerene accumulation at step edges. With continued deposition, the fullerene layer grew from the step edges, representing step flow across the terraces. For multilayer films, the consequences of the steps were apparent since fullerene layers were offset by 2 Å, an amount equal to the GaAs step height. This compared with a spacing of 8 Å for close-packed $C_{60}(111)$ layers.

Two-dimensional island formation occurs for C_{60} deposition on large terraces when the molecules cannot diffuse to the steps. Li *et al.*[183] showed that there were two types of bonding sites for C_{60} molecules, namely those positioned over Ga atoms, fourfold coordinated with As, and those positioned over As atoms, fourfold coordinated with Ga. The relaxation of Ga and As surface atoms that is intrinsic to the GaAs(110) surface was unaffected by the growing overlayer. Hence, the two bonding types could be distinguished by their differential vertical displacements, ~0.8 Å. In contrast to most overlayers, C_{60} had the effect of forming an overlayer without changing the reconstruction or relaxation of the substrate, an effect also reported by Hong *et al.*[184] for C_{60} multilayers on

[179] A. F. Hebard, R. C. Haddon, R. M. Fleming, and A. R. Kortan, *Appl. Phys. Lett.* **59**, 2109 (1991).

[180] E. I. Altman and R. J. Colton, *Surf. Sci.* **279**, 49 (1992).

[181] G. Gensterblum, L.-M. Yu, J.-J. Pireaux, P. A. Thiry, R. Caudano, J.-M. Themlin, S. Bouzidi, F. Coletti, and J.-M. Debever, *Appl. Phys. A* **56**, 175 (1993); G. Gensterblum, L.-M. Yu, J.-J. Pireaux, P. A. Thiry, R. Caudano, Ph. Lambin, A. A. Lucas, W. Krätschmer, and J. E. Fischer, *J. Phys. Chem. Solids* **53**, 1427 (1992).

[182] Y. Z. Li, J. C. Patrin, M. Chander, J. H. Weaver, L. P. F. Chibante, and R. E. Smalley, *Science* **252**, 547 (1991).

[183] Y. Z. Li, M. Chander, J. C. Patrin, J. H. Weaver, L. P. F. Chibante, and R. E. Smalley, *Science* **253**, 429 (1991).

[184] H. Wong, W. E. McMahon, P. Zschack, D.-S. Oin, R. D. Aburano, H. Chen, and T.-C. Chiang, *Appl. Phys. Lett.* **61**, 3127 (1992).

Si(111) − 7 × 7. It may be that C_{60} will find applications in surface passivation or protection.

In general, the lattice mesh of $C_{60}(111)$ will differ from that of the substrate on which it is grown. For example, $C_{60}(111)$ is hexagonal and GaAs(110) is rectangular. Therefore, if a pseudomorphic layer of C_{60} is to form, there must be considerable strain that develops in the layer. Although growth at 300 K resulted in a commensurate c(2 × 4) overlayer structure for C_{60}/GaAs(110), Li et al.[182] showed that the interface was strained but contained a large number of defects that served to reduce the strain. Multilayer growth on this lattice-mismatched layer at 300 K produced complicated simultaneous-multilayer-growth patterns with small grains, defects, stacking faults, faceting, and substantial imperfections. Such imperfections are evident in the STM image of Fig. 6(a) for about 3 ML of C_{60} deposited on GaAs(110) at 300 K. It is likely that equally complex nanostructures develop whenever growth conditions are not optimized. Such effects will influence measurements sensitive to order or grain size.

The growth of C_{60} on GaAs(110) or other surfaces can be optimized by deposition at higher temperature. For C_{60}/GaAs(110), the imperfect two-dimensional layer formed at room temperature is replaced by close-packed $C_{60}(111)$ islands that are no longer in registry with the substrate. Hence, an energetically preferred interface is developed in which the overlayer [1$\bar{1}$0] direction is rotated ±3.5° with respect to the substrate [1$\bar{1}$1] direction. Li et al.[183] showed that this produced a a moiré effect as the overlayer lattice moved in and out of registry with the substrate. This modulation had a wavelength of about 45 Å, as shown in Fig. 6(b) for 1 monolayer of C_{60} on GaAs(110). The overall height modulation was ~0.8 Å, corresponding to the relaxation of the substrate itself. Under these growth conditions, flat regions extended for thousands of angströms with defect-free areas limited by the size of the substrate terraces. Such surfaces represent the starting point for studies of surface-related phenomena.

Low energy electron diffraction (LEED) studies of C_{60}/GaAs(110) revealed the splitting of diffraction spots because of the coexistence of two-domain structures. To obtain the LEED pattern shown in Fig. 7, Benning et al.[185] grew a 200-Å C_{60} film at 450 K. When cooled to 40 K, the $C_{60}(111)$ surface underwent a transformation equivalent to that observed for the bulk with an ordering into a simple cubic structure. The doubling of the unit cell introduced half-order spots in the LEED

[185]P. J. Benning, F. Stepniak, and J. H. Weaver, *Phys. Rev. B* **48**, 9086 (1993).

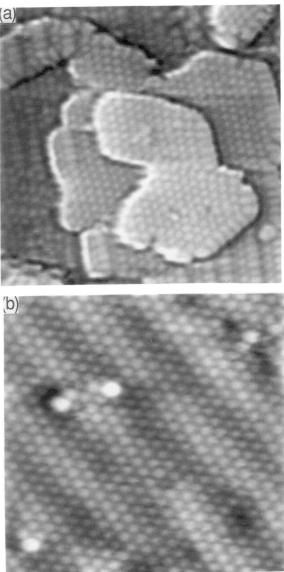

FIG. 6. Scanning tunneling microscope images of a C_{60} layer grown on GaAs(110) at (a) room temperature and (b) 450 K. (a) shows a complex multilayer structure with molecules that are rather disordered in the first layer but more closely packed in layers two and three. Large numbers of defects are evident. The top layer is composed of close-packed molecules, and there is a grain boundary that crosses through its center from left to right. (b) was obtained after 1 monolayer of C_{60} was deposited under conditions that enhanced surface diffusion and led to the formation of a large array of close-packed molecules. These molecules were not in registry with the template of the substrate, resulting in a periodic modulation in height. Molecular rotation precluded detection of intramolecular contrast. From Li et al.[182,183]

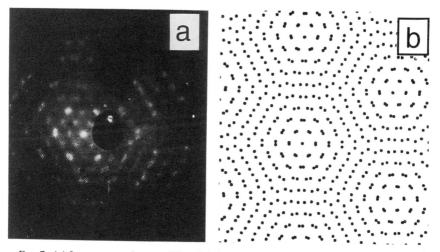

FIG. 7. (a) Low energy electron diffraction (LEED) pattern measured at 40 K for a 200-Å film of C_{60} grown on GaAs(110) at 450 K. LEED shows the hexagon mesh of $C_{60}(111)$ with half-order spots that reflect the low temperature simple cubic structure. The spots are split because of the coexistence of grains rotated 7° relative to one another. (b) shows the superposition of two hexagon arrays rotated by 7° to account for the observed moiré pattern. From Benning et al.[185]

pattern. Part (b) of Fig. 7 depicts the superposition of two hexagonal patterns rotated 7°, thereby accounting for the observed moiré pattern.

In Minnesota, we have sought to understand surface growth structures and optimize overlayer formation. Figure 8 shows a 570×570-Å^2 STM image for 10 monolayers of C_{60} grown on GaAs(110) at 450 K, from Ref. 186. In addition to large regions of perfect order, the image shows a strip of almost-periodic "faults" where molecules on one side appear higher than those on the other. These are produced by the coalescence of two large domains of the type previously described that meet to create a low angle grain boundary. Inspection shows one additional row of molecules on the bright (high) side of each fault compared with the dark (low) side. This image demonstrates that films with a high degree of perfection can be formed and that the imperfections are readily understood from a traditional structural analysis. With such films, it should be possible to undertake detailed studies of surface rotational melting and features associated with the kinetics of growth. For example, it should be possible to search for arrays of partial vacancies of the sort described by van de

[186](a) Y. B. Zhao, D. M. Poirier, and J. H. Weaver, J. Phys. Chem. Solids (in press); (b) Y. B. Zhao, D. M. Poirier, R. J. Pechman, and J. H. Weaver, Appl. Phys. Lett. (in press).

FIG. 8. STM image for a 570×570-$\mathring{A}^2$ surface produced by growing 10 monolayers of C_{60} on GaAs(110) at 450 K. Most of the surface was flat and defect free. This periodic array of dislocations was formed at a 7° grain boundary. Counting shows an extra row of molecules on the bright sides of the faults formed to relieve the build up of strain. From Zhao et al.[186(a)]

Waal[161] and Lucas[160] that would bias the growth of molecular solids in favor of the fcc rather than hcp structure.

Significant insights into fullerene–substrate interactions have come from recent surface science studies. Gensterblum et al.[181] showed that high quality single-domain overlayers of C_{60} could be grown on cleaved GeS(001) and they reported high resolution electron energy loss studies, HREELS,[181] as well as angle-resolved photoemission[187] and inverse photoemission investigations[188] that provided important insight into electronic and vibrational energy levels. Altman and Colton[180] presented a detailed investigation of layer-by-layer formation for $C_{60}/Au(111)$. In contrast to results for Si(111) $- 7 \times 7$ and GaAs(110), they found that C_{60} adsorption lifted the Au surface reconstruction, thus indicating nonnegligible surface interaction. Kuk et al.[189] investigated $C_{60}/Au(100)$ and

[187]G. Gensterblum, J.-J. Pireaux, P. A. Thiry, R. Caudano, T. Buslaps, R. L. Johnson, G. LeLay, V. Aristov, R. Günther, A. Taleb-Ibrahimi, G. Indlekofer, and Y. Petroff, *Phys. Rev. B* **48**, 14756 (1993).

[188]J.-M. Themlin, S. Bouzidi, F. Coletti, J.-M. Debever, G. Gensterblum, L.-M. Yu, J.-J. Pireaux, and P. A. Thiry, *Phys. Rev. B* **46**, 15602 (1992)

[189]Y. Kuk, D. K. Kim, Y. D. Suh, K. H. Park, H. P. Noh, S. J. Oh, and S. K. Kim, *Phys. Rev. Lett.* **70**, 1948 (1993).

found a stressed commensurate monolayer. Hashizume et al.[190,191] examined growth of C_{60} overlayers on Si(100) at room temperature. They
found specific bonding states but little surface diffusion, resulting in
mixed structures of $c(4 \times 3)$ and $c(4 \times 4)$ symmetry. The transition to
ordered close-packed islands was observed for multilayer deposition. For
C_{60}/Si(111) − 7 × 7, Wang et al.[192] reported disordered overlayer formation with preferential molecular adsorption in the faulted half of the
(7 × 7) unit cell. Li et al.[178] used the STM tip to reposition fullerenes on
the Si(111) − 7 × 7 surface. Xu et al.[193] grew a C_{60} overlayer at 475 K and
found distinct island orientation, as for C_{60} on GaAs(110).[182,183]

It was previously noted that C_{60} bonding to the surface can inhibit
molecular rotation enough that internal molecular contrast, IMC, can be
observed using scanning tunneling microscopy. IMC was first reported by
the Arizona group[174–176] but was subsequently observed for C_{60} on
Si(100) and Si(111) − 7 × 7,[190–192] C_{60} on Au(111) at low temperature,[194]
and C_{60} on Cu(111).[195,196] Those authors sought to reconcile the IMCs
with the structure of the HOMO and LUMO orbitals accessed by
electrons tunneling to, or from, the tip. They found reasonable agreement
with the expected molecular structure.

STM was also used to investigate thin films of the higher fullerenes. Li
et al.[171] reported studies of C_{70}, C_{76}, C_{78}, C_{82}, and C_{84} on GaAs(110).
They observed large close-packed islands after deposition at ~450 K with
differences in island growth that they attributed to different activation
energies for surface diffusion. For the large fullerenes, there appeared to
be rows of like-oriented molecules that suggested surface ordering. For
C_{82} and C_{84}, they reported IMCs that they related to the pentagon–
hexagon structure of the molecules. The rotational stabilization for C_{82}
and C_{84} on GaAs(110) was rationalized in terms of enhanced substrate

[190]T. Hashizume, X.-D. Wang, Y. Nishina, H. Shinohara, Y. Saito, Y. Kuk, and T. Sakurai,
Jpn. J. Appl. Phys. 31, L880 (1992).
[191]X.-D. Wang, T. Hashizume, H. Shinohara, Y. Saito, Y. Nishina, and T. Sakurai, Phys.
Rev. B 47 15923 (1993).
[192]X.-D. Wang, T. Hashizume, H. Shinohara, Y. Saito, Y. Nishina, and T. Sakurai, Jpn. J.
Appl. Phys. 31, L983 (1992).
[193]H. Xu, D. M. Chen, and W. N. Creager, Phys. Rev. Lett. 70, 1850 (1993).
[194]S. Behler, H. P. Lang, S. H. Pan, V. Thommen-Geiser, and H.-J. Günterodt, Z. Phys. B
(in press).
[195]T. Hashizume, K. Motai, X.-D. Wang, N. Shinohara, Y. Saito, Y. Maruyama, K. Ohno,
Y. Kawazoe, Y. Nishina, H. W. Pickering, Y. Kuk, and T. Sakurai, Phys. Rev. Lett. 71,
2959 (1993).
[196]K. Motai, T. Hashizume, H. Shinohara, Y. Saito, H. W. Pickering, Y. Nishina, and T.
Sakurai, Jpn. J. Appl. Phys. 32, L450 (1993).

bonding and steric constraints not observed for the more symmetric molecules. Wang et al.[171] reported C_{60} and C_{84} condensation on Si(100), finding disordered adsorption for C_{84} growth at 300 K but a close-packed array for high temperature deposition.

Overlayer formation was investigated for C_{60} on $CaF_2(111)$ by Fölsch et al.[197] They found epitaxial growth for deposition temperatures of 300–575 K and coverage corresponding to 10–500 Å. Fischer et al.[198] discussed film formation on mica with films having in-plane and out-of-plane correlation lengths of 450 and 850 Å, respectively, as determined by grazing-incidence x-ray diffraction. Epitaxial growth on mica had previously been examined by Schmicker et al.[199] using He atom scattering and by Krakow et al.[200] using electron diffraction and microscopy. An x-ray coherence length of ~190 Å was reported by Tong et al.[201] for C_{60} growth on H-terminated Si(111) -1×1. Sakurai et al.[202] reported epitaxial growth on cleaved MoS_2. While $C_{60}(111)$ has been the commonly observed face, Krakow et al.[203] found (100) orientations in NaCl. Ichihashi et al.[204] examined C_{60} films on alkali halide substrates, finding a mixture of hcp and fcc.

9. Electronic Structure

Fullerenes differ from the other forms of carbon in a fundamental way related to the hybridization of the atomic s and p levels. In diamond, sp^3 hybridization assures a bond angle of 109° 28′ and four identical bonds. The filling of the bonding states and the large separation between the sp^3 bonding and antibonding levels results in an insulator (energy gap ~5.5 eV). In graphite, sp^2 hybridization produces three equivalent bonds, a bond angle of 120°, and strong planar bonding. The out-of-plane p orbitals result in π bonds and weak interplanar bonding. Graphite is an

[197]S. Fölsch, T. Maruno, A. Yamashita, and T. Hayashi, Appl. Phys. Lett. 62, 2643 (1993).
[198]J. E. Fischer, E. Werwa, and P. A. Heiney, Appl. Phys. A 56, 193 (1993).
[199]D. Schmicker, S. Schmidt, J. G. Skofronick, J. P. Toennies, and R. Vollmer, Phys. Rev. B 44, 10995 (1991).
[200]W. Krakow, N. M. Rivera, R. A. Roy, R. S. Ruoff, and J. J. Cuomo, J. Mater. Res. 7, 784 (1992).
[201]W. M. Tong, D. A. A. Ohlberg, H. K. You, R. S. Williams, S. J. Anz, M. M. Alvarez, R. L. Whetten, Y. Rubin, and F. Diederich, J. Phys. Chem. 95, 4709 (1991).
[202]M. Sakurai, H. Tada, K. Saiki, and A. Koma, Jpn. J. Appl. Phys. 30, L1892 (1991).
[203]W. Krakow, N. M. Rivera, R. A. Roy, R. S. Ruoff, and J. J. Cuomo, Appl. Phys. A 56, 185 (1993).
[204]T. Ichihashi, K. Tanigaki, T. W. Ebbesen, S. Kuroshima, and S. Iijima, Chem. Phys. Lett. 190, 179 (1992).

accidental metal because the Fermi level is coincident with the top of the bonding π band at the Brillouin zone boundary and that level is not split. Both graphite and diamond have occupied bandwidths of $\sim$25 eV. Graphite and diamond are nearly identical in binding energy, with only about 0.02 eV/carbon atom difference. In comparison, the hybridization of the s and p orbitals in the fullerenes is neither pure sp^3 nor sp^2p, introducing a pyramidalization angle of 11.6° for C_{60}.[15] Theoretical analyses have indicated that C_{60} is less stable than diamond or graphite by about 0.4 eV per atom,[205] a point made by experiment as well.[206] The π bonds are directed radially with a node on the nuclear cage.

The fullerene solids are characterized as molecular crystals. As such, their orbitals are modified, but not too seriously, by condensation to form solids. Several authors[207–209] have pointed out that there are two different energy scales that must be considered when considering the various electronic interactions. The one that describes the overall electronic properties is large and defines on-ball electronic interactions. In particular, the occupied σ and π bandwidth is $\sim$25 eV, and the bonding and antibonding π bands extend over $\sim$20 eV.[210] The other describes the broadening of the molecular orbit upon condensation into the solid state. This is much smaller, and estimates of the widths for the bands derived from the HOMO or LUMO orbitals are $\sim$0.5 eV. The fact that the HOMO and LUMO bands are narrow suggests that the effects of electron correlation will be important, as discussed in the following.

a. *Electronic Energy Levels, Electron Correlation, Optical Absorption, and Excitons*

Measurements of the electronic and optical properties of the fullerenes offer fundamental ways to test descriptions of the electronic energy levels. They also make it possible to quantify changes in those levels induced by condensation on surfaces (as already discussed), isomer formation, radical formation, polymerization, and compound formation. They are essential in discussing superconductivity as well, and we shall consider them in some detail.

[205]N. Troullier and J. L. Martins, *Phys. Rev. B* **46,** 1754 (1992).

[206]H. S. Chen, A. R. Kortan, R. C. Haddon, M. L. Kaplan, C. H. Chen, A. M. Mujsce, H. Chou, and D. A. Fleming, *Appl. Phys. Lett.* **59,** 2956 (1991).

[207]M. Schluter, M. Lannoo, M. Needles, G. A. Baraff, and D. Tománek, *Phys. Rev. Lett.* **68,** 526 (1991).

[208]J. L. Martins and N. Troullier, unpublished preprint.

[209]C. M. Varma, J. Zaanen, and K. Raghavachari, *Science* **254,** 989 (1991).

[210]J. L. Martins, N. Troullier, and J. H. Weaver, *Chem. Phys. Lett.* **180,** 457 (1991).

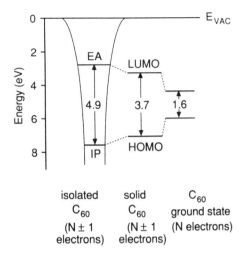

isolated   solid C_{60}
C_{60} C_{60} ground state
(N ± 1   (N ± 1   (N electrons)
electrons)   electrons)

FIG. 9. Energy level picture of isolated C_{60} showing the ionization potential at 7.6 eV and the electron affinity level at 2.6 eV, corresponding to molecular $N - 1$ and $N + 1$ electronic states. The energy separation is reduced by extramolecular screening in the solid, and the HOMO and LUMO band centers are separated by 3.7 eV. For neutral molecules in the ground state, that separation is smaller, about 1.6 eV, as deduced by electron removal from C_{60}^- or A_6C_{60} solids. The latter energy is often taken to be the band gap, but doing so neglects electron correlation. From Ohno et al.[177]

Figure 9 depicts the energy levels of C_{60} in the molecular state referenced to the vacuum level.[177,211] Electron removal from an isolated molecule yields the ionization potential (IP) and addition yields the electron affinity level (EA). Both states are of the ion, not the neutral molecule, and they correspond to the $N - 1$ and $N + 1$ states. For C_{60}, these energies are respectively 7.6 and 2.65 eV.[74–76] Condensation of the molecules introduces extramolecular screening, so that the separation in energy of EA and IP diminishes, as sketched based on results to be described here. In the independent-electron picture of a solid, the wavefunctions describing these molecular states would be delocalized, giving a band of extended states of width W. Electron removal or addition would have a negligible effect on the energy level spectrum. If the wavefunctions remain localized in the solid state, however, then the band is narrow and the energy needed to remove an electron from one molecule and place it on another is appreciable, creating ionized $N - 1$ and $N + 1$ states. This energy corresponds to the Hubbard U. When U is

[211]J. H. Weaver, J. Phys. Chem Solids 53, 1433 (1992).

comparable with W, the system is described as being highly correlated.[212] For an uncorrelated semiconductor, the screened EA and IP would define the conduction band and valence band edges, and IP − EA gives E_g, the energy gap. For correlated systems, the IP − EA includes the energy U, a quantity that is not determined within independent-electron formalisms. A calculation that neglects many-body effects will yield the energy level spectra for the N-electron system, depicted at the right of Fig. 9.

Figure 10(a) depicts a collection of C_{60} molecules illustrated by a photon or electron beam of energy $h\nu$. Photon absorption promotes an electron to a higher energy state and creates a hole, denoted (−) and (+), respectively. For the molecule in the upper right, the electron and the hole are bound, corresponding to an "on-ball," or molecular, excitation. The importance of such on-ball excitations is clear from comparison of optical properties measured for C_{60} in the condensed phase and in solution.[82,179,211,213–217] The optical absorption spectra in Fig. 10(c) were obtained by Ren and coworkers[216] for solid C_{60} and C_{60} in decalin. Other groups observed similar effects, and all show strong absorption features at about 3.6, 4.6, and 5.8 eV, as first reported by Krätschmer et al.[14] Similarities for C_{60} in solution and in the condensed state emphasize the molecular character of these optical structures. Solid state effects contribute to broadening and account for the onset near 2.5 eV.

In solid C_{60}, the electron–hole pairs can hop, behaving like Frenkel excitons. They can also dissociate, producing ionized $N + 1$ and $N - 1$ states if they have sufficient energy. The lowest energy excitation for C_{60} at ~1.6 eV is dipole forbidden, producing an electron in LUMO and a hole in HOMO. This low energy bandgap exciton cannot dissociate directly. For condensed C_{60}, there have been reports of photon-activated transport at energies below the transport gap. In these cases, the nonlinear dependence of conductivity on photon flux suggests a second-order effect, probably the annihilation of two long-lived bandgap triplet

[212]N. F. Mott, "Metal–Insulator Transitions," second edition (Taylor & Francis, New York, 1990).

[213]A. Skumanich, Chem. Phys. Let. 182, 486 (1991).

[214]M. K. Kelly, P. Etchegoin, D. Fuchs, W. Krätschmer, and K. Fostiropoulos, Phys. Rv. B 46, 4963 (1992).

[215]J. P. Hare, H. W. Kroto, and R. Taylor, Chem. Phys. Lett. 177, 394 (1991).

[216]S. L. Ren, Y. Wang, A. M. Rao, E. McRae, J. M. Holden, T. Hager, K.-A. Wang, W.-T. Lee, H. F. Ni, J. Selegue, and P. C. Eklund, Appl. Phys. Lett. 59, 2678 (1991).

[217]S. Leach, M. Vervloet, A. Desprès, E. Bréheret, J. P. Hare, T. J. Dennis, H. W. Kroto, R. Taylor, and D. R. M. Walton, Chem. Phys. 160, 451 (1992).

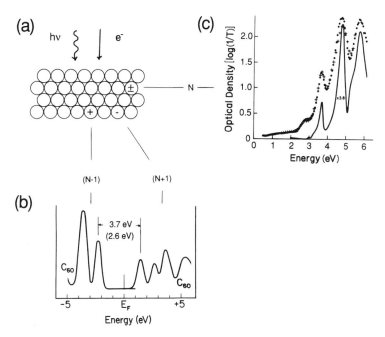

FIG. 10. (a) represents an array of C_{60} molecules exposed to photon and electron beams of variable energy, representing photoemission and inverse photoemission. The lowest energy optical absorption process corresponds to the creation of an electron–hole pair bound by the Coulomb energy on the same molecule, an exciton. Separation requires ionizing radiation that results in isolated electrons and holes, namely the $N-1$ and $N+1$ states of the molecular solid. The energy distribution of $N-1$ and $N+1$ states is measured by photoemission and inverse photoemission. The PES and IPES spectra are reproduced in (b), referenced in energy to the Fermi level. The separation of HOMO and LUMO band centers is 3.7 eV and the extrapolated band edge separation is 2.6 eV. The latter approximates the transport gap. (c) compares optical absorption spectra for C_{60} in decalin (lower curve) with C_{60} in thin film form. The dominant structures reflect dipole-allowed molecular transitions that are only slightly modified in the solid state. From Weaver[211] with (c) adapted from Ren et al.[216]

excitons. Such triplets have been observed in isolated molecules[89] and in the solid state using high resolution electron energy loss spectroscopy.[218] The ionized states achieved by electron–hole separation are equivalent to those reached by photoemission (PES) and inverse photoemission (IPES). The energy distribution of the hole and electron states for the

[218]G. Gensterblum, J.-J. Pireaux, P. A. Thiry, R. Caudano, J. P. Vigneron, Ph. Lambin, A. A. Lucas, and W. Krätschmer, *Phys. Rev. Lett.* **67**, 2171 (1991); A. A. Lucas, G. Gensterblum, J.-J. Pireaux, P. A. Thiry, R. Caudano, J. P. Vigneron, and Ph. Lambin, *Phys. Rev. B* **45**, 13694 (1992).

screened ions then follows directly, as reproduced in Fig. 10(b) and labeled ($N - 1$) and ($N + 1$). Measured with PES–IPES, the HOMO–LUMO gap is greater than the optical absorption (exciton) edge or the gap predicted by various independent-particle calculations. Again, the difference is the energy needed to remove an electron from the N-particle system and create separate $N - 1$ and $N + 1$ system states. This energy difference, U, is ~1.3–1.6 eV.[219] Extrapolation of the band edges to zero emission gives a gap of 2.6 eV, although such extrapolation hides any underlying broadening.

In principle, electron removal or addition will shift the various electronic levels by different amounts because of their differential spatial character. Indeed, Lof et al.[219] reported that U, as determined from comparison of the KVV Auger spectrum to the self-convoluted valence band, depended slightly on the orbital. These differences will not affect the following general comparison with theory but they should be kept in mind. For undoped C_{60}, we will neglect the effects of correlation vis-à-vis the distribution of states within the occupied state or empty state manifolds while emphasizing its role in determining the HOMO–LUMO separation.

b. C_{60} Valence and Conduction Bands

Figure 11 offers an overview of the π electronic levels for an isolated C_{60} molecule and for C_{60} in the solid state. The results for isolated C_{60} were obtained from Hückel calculations performed by Haddon et al.[15,62] The energy level symmetries and degeneracies are as indicated. The highest occupied molecular orbital has h_u character and the lowest unoccupied molecular orbital has t_{1u} character. HOMO to LUMO h_u–t_{1u} transitions are dipole forbidden. These levels are broadened in the solid, as depicted by the right portion of Fig. 11, where the independent-particle energy bands derived from the h_u and t_{1u} levels are shown as a function of crystal momentum in the Brillouin zone of fcc C_{60}. For those calculations, Saito and Oshiyama[9] used the local density approximation, LDA, and assumed the $Fm\bar{3}$ crystal symmetry of Fig. 1. From the calculations, it is evident that the bands are relatively narrow, ~0.4 eV, consistent with the observation that the electronic properties are largely molecular in character. The Brillouin zone is rather small since the lattice constant is large.

Figure 12 shows two representations of the charge distribution for C_{60}

[219]R. W. Lof, M. A. van Veenendaal, B. Koopmans, H. T. Jonkman, and G. A. Sawatzky, Phys. Rev. Lett. **68**, 3924 (1992).

Erratum

Solid State Physics, Vol. 48
Ehrenreich/Spaepen

0-12-607748-7
0-12-606048-7

Regarding Chapter 3 by J. D. Axe, S. C. Moss, and D. A. Neumann, a re-evaluation of the 300 K Bragg data analysis has led to slightly revised coefficients, $C_{l,\gamma}$, in Table 2, page 174. While these corrections do not introduce large changes in the number density maps in Fig. 6, page 175, they have a more profound influence both on the axial density maps in Fig. 7, page 176, and, particularly, on $\bar{V}$ (ω) in Fig. 8, page 180, where a distinct secondary minimum now appears at $0 \cong 38°$. These changes are discussed in detail in a forthcoming paper (see Chow *et al.*, Ref. 18).

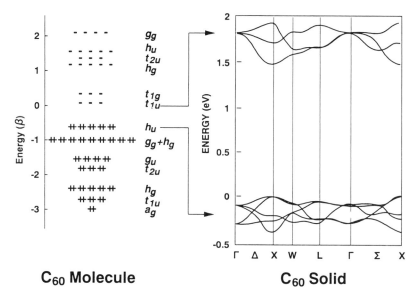

C$_{60}$ Molecule   **C$_{60}$ Solid**

FIG. 11. Electron energy level for C$_{60}$ in molecular and solid state form. The energies deduced from Hückel calculations are shown at the left for the π levels, along with their degeneracy and symmetry. The right portion shows how the degenerate h_u and t_{1u} levels broaden into the valence and conduction bands described in the Brillouin zone of the fcc structure (Fm$\bar{3}$ symmetry). These narrow bands do not overlap with adjacent bands, and the electronic properties retain a high degree of molecular character. From Hebard,[6] adapted from Haddon *et al.*[15,62] and Saito and Oshiyama.[9]

from the work of Martins and Troullier.[220] Figure 12(a) depicts $|\psi^*\psi|$ summed over the five degenerate highest occupied molecular orbitals. The sphere at the center keeps the viewer from seeing structure on the back side of the molecule. The center of the image shows a hexagon defined by double bonds with adjacent pentagons at about 2:00 (two o'clock), 6:00, and 10:00 and hexagons at 4:00, 8:00, and 12:00. The π character of these states is apparent from the node in this charge distribution. The radial character is evident from the greater overlap of the double bond on the inner portion of the nodal surface. Troullier and Martins[205] also plotted charge contours for C$_{60}$ on a planar representation (Mercator projection) to emphasize nodal surfaces and wavefunction character.

Figure 12(b) shows a charge density plot for C$_{60}$(100), again in the Fm$\bar{3}$ structure. This figure emphasizes the ~4-Å void in the center of the molecule at the face-centered position and the ~4-Å diameter

---

[220]J. L. Martins and N. Troullier, unpublished.

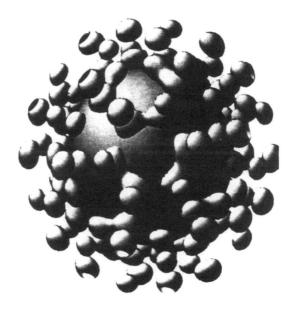

(b)

FIG. 12. Charge distribution representations for (a) an isolated C_{60} molecule and (b) an array of molecules on the (100) face of C_{60} oriented as in Fig. 1. The distribution of the HOMO levels shown in (a) emphasizes the node on the 7-Å nuclear cage and the double-and-single bond alternation. The large sphere at the molecular center serves to eliminate visual confusion arising from structure on the opposite side of the molecule. The hexagon at front-center is surrounded by three hexagons (at about 4 o'clock, 8, and 12) and three pentagons (at about 2, 6, and 10). The total electron density shown in (b) shows very little charge in the center of the molecule and in octahedral lattice sites. The low charge density along close-packed directions emphasizes the slight amount of overlap. From J. L. Martins.[220]

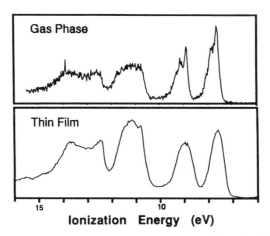

Gas Phase

Thin Film

15   10
Ionization   Energy   (eV)

FIG. 13. Photoionization spectra for C_{60} in the gas phase and in a thin film. The gas phase results show frontier orbitals that exhibit structure because of coupling of electronic and vibronic channels. The broadening of electronic levels in the solid is not very different from that for isolated molecules, but the relative importance of solid state broadening vs. vibronic broadening is not well known. From Lichtenberger et al.[74]

octahedral holes between molecules along $\langle 100 \rangle$. The paucity of charge between molecules along close-packed directions speaks to the small intermolecular overlap.

Lichtenberger et al.[74] provided the first comparison between the distribution of electronic states of C_{60} in the gas phase and C_{60} in the solid. Their results, reproduced in Fig. 13, showed broadening in the gas phase spectra with structure attributed to transitions that coupled electron removal with the vibronic modes of the molecule. As will be discussed, these vibronic modes extend in energy from ~0.03 to ~0.2 eV, although not all modes couple equally. Their results for thin films showed very nearly the same widths as those found for isolated molecules. It is still not certain how the broadening observed in the solid state relates to the band structure (Fig. 12) and how much reflects vibronic coupling, as in the gas phase. The low energy solid state photoemission results of Gensterblum et al.[187] hint at both band dispersion and vibronic coupling, while the electronic structure calculations of Shirley and Louie,[221] which included corrections for many-body effects, indicated a substantially broader band than in Fig. 11. The resolution of this issue is very important, and several groups are undertaking angle-resolved photoemission measurements with single crystals of C_{60} to resolve it.

[221]E. Shirley and S. G. Louie, Phys. Rev. Lett. 71, 133 (1993).

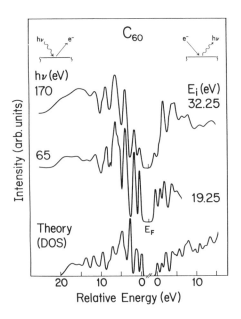

FIG. 14. Photoemission and inverse photoemission spectra obtained at several excitation energies for condensed C_{60}. The spectra are referenced to the Fermi level. The calculated results shown at the bottom are shifted rigidly to reproduce the HOMO–LUMO band separation. The frontier orbitals have dominant π character, while those deep in the valence band have σ character. The calculations underestimate the separation of σ and π states, as is evident from the position of the third peak, but they demonstrate the state sequencing very well. From Weaver.[211]

Figure 14 provides an overview of the distribution of electronic states for solid C_{60}, from Weaver.[211] The PES spectra were obtained with synchrotron radiation with $h\nu = 65$ and 170 eV, and the IPES spectra were obtained with incident electron energies of 19.25 and 32.25 eV. The photoionization cross sections are such that the p-derived states are emphasized at lower excitation energies and the s-derived states dominate at higher energy. The Fermi level lies 2.25 eV above the center of the HOMO-derived band in the $N - 1$ final state and 1.5 eV below the center of the LUMO band in the $N + 1$ final state. The spectra are scaled so that the intensity of the LUMO feature is 3/5 the intensity of the HOMO feature according to the degeneracies of the h_u and t_{1u} orbitals.

The electronic structure of solid C_{60} has been calculated by several groups. Martins et al.[210] calculated the distribution of both occupied and empty states, finding ~250 bands in the energy range from the base of the valence band to 15 eV above the conduction band minimum. The bottom curve of Fig. 14 reproduces their density of states based on their soft pseudopotential LDA method with C_{60} molecules arranged in space group $Fm\bar{3}(T_h^3)$. The appropriately weighted eigenvalues calculated at special $\bar{k}$ points in the Brillouin zone were broadened with a Gaussian having a width that scaled in energy as 0.23 eV $+ 0.02 |E|$ to simulate the experimental energy resolution and lifetime and vibronic broadening. The theoretical results of Fig. 14 were referenced in energy to the HOMO- and LUMO-derived band centers observed experimentally. Analysis showed that the valence band states and leading conduction band states had well-defined angular momentum and molecular σ or π character. Those at higher energy were not "confined" to the molecule and extended throughout the crystal. Martins et al.[210] and Saito and Oshiyama[9] described the electronic states with angular momentum quantum numbers associated with the spherical harmonics. The $l = 0$ state then corresponded to electrons with no nodal plane, the $l = 1$ state had one node (p-like), and so forth. σ states at the bottom of the valence band had $l = 0$, whereas π states at the top had $l = 5$. As the quantum number l increased, the number of angular nodal planes passing through the molecule increased.

The HOMO and HOMO $- 1$ features have π_u and π_g character with fivefold and ninefold degeneracy, respectively.[210] The photoemission results showed that the relative intensities of these bands changed with photon energy because of matrix element effects.[222] States 5–10 eV below E_F were of mixed π and σ character, and those more than $\sim$10 eV below E_F were σ-derived. The overall valence band width was very nearly the same as that for graphite or diamond,[223] but the high molecular symmetry of C_{60} assured sharp spectral features. Agreement of theory and experiment for the electronic state features was good, particularly if account was taken of an underestimate of the binding energy of the σ states relative to the π states in the LDA calculations. Correcting this would shift feature three and those below it to higher binding energy.

In the empty states, the LUMO feature is derived from three bands with t_u symmetry at Γ. The LUMO $+ 1$ arises from three bands with t_g symmetry, and the next band represents the superposition of t_g, t_u, e_g, and a_g states at Γ. The a_g state is localized in the center of the C_{60} molecule and does not correspond to any of the canonical σ or π bonds.

[222]P. J. Benning, D. M. Poirier, N. Troullier, J. L. Martins, J. H. Weaver, R. E. Haufler, L. P. F. Chibante, and R. E. Smalley, Phys. Rev. B **44**, 1962 (1991).
[223]F. R. McFeely, S. P. Kowalczyk, L. Ley, R. G. Cavell, R. A. Pollak, and D. A. Shirley, Phys. Rev. B **9**, 5268 (1974).

The empty states can also be probed by electron energy loss spectroscopy, EELS,[224-227] and by x-ray absorption spectroscopy, XAS,[228-230] wherein a C $1s$ electron is excited and occupies a $\pi*$ or $\sigma*$ state. Distributions of the empty states probed in this way showed sharp features, as observed in the IPES spectra, but with somewhat different spacings due to the interaction of the core hole and the $\pi*$ or $\sigma*$ electrons.

c. *C 1s Satellites and Plasmons for* C_{60}

The fullerenes exhibit a novel variety of spectral features, some of which are observed in simple molecules while others are not. Figure 15 shows the core level spectrum referenced in energy to the center of the C $1s$ main line as measured with x-ray photoemission, XPS.[98] The C $1s$ binding energy was 285.0 eV, and its full width at half maximum, 0.65 eV, was almost certainly resolution limited. The satellite structures at low binding energy were due to on-site and off-site $\pi-\pi*$ and $\sigma-\sigma*$ molecular excitations and to molecular plasmon losses. On-site processes are those that occur on the molecule when the electron is ejected, whereas off-site lines are due to energy loss events suffered by the electron as it propagates through the lattice. The first satellite (labeled 2), located 1.9 eV below the main line, was due to on-ball excitations from HOMO- to LUMO-derived states induced by the creation of the core hole where the electron and hole remained in the excited molecule. This transition is monopole-allowed ($\Delta l = 0$). Although it is dipole-forbidden in the isolated C_{60} molecule, weak absorption is seen in optical absorption[213,216] and high resolution electron energy loss spectroscopy, HREELS.[218] Gensterblum et al.[218] resolved features at 1.55 and 1.8 eV related to triplet and singlet excitons, respectively.

The satellite features at 3.7 5.0, and 6.1 eV originated from $\pi-\pi*$

[224]E. Sohmen, J. Fink, and W. Krätschmer, Z. Phys. B **86,** 87 (1992).
[225]E. Sohmen, J. Fink, and W. Krätschmer, Europhys. Lett. **17,** 51 (1992).
[226]P. L. Hansen, P. J. Fallon, and W. Krätschmer, Chem. Phys. Lett. **181,** 367 (1991).
[227]Y. Saito, H. Shinohara, and A. Oshita, Jpn. J. Appl. Phys. **30,** L1145 (1991).
[228]L. J. Terminello, D. K. Shuh, F. J. Himpsel, D. A. Lapiano-Smith; J. Stöhr, D. S. Bethune, and G. Meijer, Chem. Phys. Lett. **182,** 491 (1991).
[229]H. Shinohara, H. Sato, Y. Saito, K. Tohji, and Y. Udagawa, Jpn. J. Appl. Phys. **30,** L818 (1991).
[230]C. T. Chen, L. H. Tjeng, P. Rudolf, G. Meigs, J. E. Rowe, J. Chen, J. P. McCauley, Jr., A. B. Smith III, A. R. McGhie, W. J. Romanow, and E. W. Plummer, Nature **352,** 603 (1991).

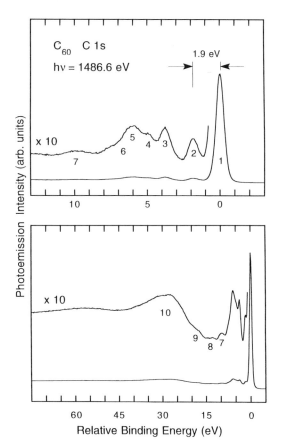

FIG. 15. C 1s-derived satellite structures for C_{60}. The width of the main line is instrument limited. Feature 2 reflects a HOMO–LUMO "on-ball" excitation induced by 1s electron expulsion. Features 3, 4, and 5 are the counterparts of dipole excitations seen in optical absorption. Features 6 and 10 represent on-ball collection oscillation of the π and σ charge distributions, the molecular equivalent of a plasmon.

electron–excited dipole transitions. They correspond to features reported in EELS data from the Karlsruhe group[224,225] and optical measurements.[213,216] The HREELS experiments of Gensterblum et al.[218] indicated that the peak at ~6 eV included contributions from π–π* transitions at 5.5 and 5.8 eV and the excitation of a π plasmon at 6.3 eV. Peaks 7–9 in Fig. 15 were attributed to σ–σ* transitions. Feature 10 was related to a high energy plasmon (~28 eV) that reflected the collective oscillation of the σ and π electrons on the molecule. This molecular plasmon was first discussed by Weaver et al.[98] but was subsequently observed in inverse photoemission[231] and other experiments.[218,224,225,232-234] Krummacher et al.[234] compared C 1s satellite spectra for isolated C_{60} to those for thin films, showing that the plasmon energy shifted depending on which of the multipole polarizabilities were excited. Burose et al.[232] and Hertel et al.[233] examined gas phase C_{60} and C_{70}, finding similar results to the solid state studies except that the high energy plasmon appeared shifted in energy. Collective oscillations were considered by theoretical investigations by Bertsch et al.,[235] Lambin et al.,[236] and Barton and Eberlein.[237]

d. Electronic Structure of the Higher Fullerenes

The electronic structures of C_{60} were investigated by many groups, but much less has been done with the higher fullerenes. This reflects the scarcity of the material. The fact that the molecules are less symmetric and the bond lengths and crystal structures are not as well known has discouraged theoretical investigations.

The addition of 10 carbon atoms around the equator of C_{60} to produce C_{70} (Fig. 1) introduces 10 π electrons and 30 σ electrons. Molecular elongation for C_{70} has little effect on the overall distribution of bands of σ and π parentage, but the lowering of molecular symmetry results in lower degeneracies of the molecular orbitals. States derived from π orbitals should be more affected than those derived from the σ orbitals because of the reduced bond angle for C atoms at the equator of C_{70}.

Jost et al.[238] investigated C_{70} thin films with photoemission and inverse photoemission. Figure 16 reproduces their figure for comparison of C_{60} and C_{70}. The spectra were plotted in arbitrary units, with no implied normalization to the density of states. Energies were referenced to the center of the HOMO features. The measured HOMO to LUMO band center separation was 3.75 eV; a value that is larger than observed for

[231]M. B. Jost, N. Troullier, D. M. Poirier, J. L. Martins, J. H. Weaver, L. P. F. Chibante, and R. E. Smalley, Phys. Rev. B **44**, 1966 (1991).

[232]A. W. Burose, T. Dresch, and A. M. G. Ding, Z. Phys. D (in press?).

[233]I. V. Hertel, H. Steger, J. de Vries, B. Weisser, C. Menzel, B. Kamke, and W. Kamke, Phys. Rev. Lett. **68**, 784 (1992).

[234]S. Krummacher, M. Biermann, M. Neeb, A. Leibsch, and W. Eberhardt, Phys. Rev. B **48**, 8424 (1993).

[235]G. F. Bertsch, A. Bulgac, D. Tománek, and Y. Wang, Phys. Rev. Lett. **67**, 2690 (1991).

[236]Ph. Lambin, A. A. Lucas, and J.-P. Vigneron, Phys. Rev. B **46**, 1794 (1992).

[237]G. Barton and C. Eberlein, J. Chem. Phys. **95**, 1512 (1991).

[238]M. B. Jost, P. J. Benning, D. M. Poirier, J. H. Weaver, L. P. F. Chibante, and R. E. Smalley, Chem. Phys. Lett. **184**, 423 (1991).

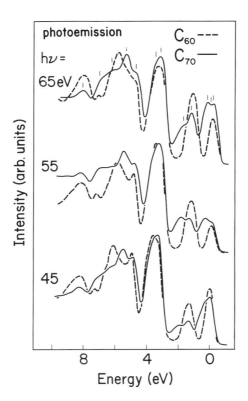

FIG. 16. Photoemission spectra for C_{60} and C_{70} showing extra states and the splitting of the leading valence band features for C_{70}. The spectra are aligned to the center of the leading features and are normalized for visual clarity. From Jost et al.[238]

photoemission from the negative C_{70} ion,[239] ~1.6 eV, for the reasons previously discussed. Comparison with calculated densities of states for C_{70} showed good overall agreement.[240,241]

One of the differences between C_{60} and C_{70} lies in the splitting and broadening of the two leading valence band features. In solid C_{60}, these π-derived bands had no discernible underlying structure when measured with angle-integrated PES, even under conditions that gave ~25 meV total resolution at 9 K. Equivalent measurements for solid C_{70} showed three resolvable features in the first peak and two in the second. Knupfer et

[239]R. E. Haufler, L.-S. Wang L. P. F. Chibante, C. Jin, J. J. Conceicao, Y. Chai, and R. E. Smalley, *Chem. Phys. Lett.* **182**, 491 (1991).
[240]W. Andreoni, F. Gygi, and M. Parrinello, *Chem. Phys. Lett.* **189**, 241 (1992).
[241]S. Saito and A. Oshiyama, *Phys. Rev. B* **44**, 11532 (1991).

al.[242] reported temperature-dependent photoemission studies of C_{70} that indicated changes in these spectral features related to the crystalline ordering transitions previously discussed, revealing solid state effects that were elusive for C_{60}. Lichtenberger *et al.*[243] reported gas phase photoemission results for C_{70} that were similar to those obtained for C_{70} films. Poirier *et al.*[36] compared the valence bands for C_{60}, C_{84}, and graphite using XPS. With an excitation energy of 1486.6 eV, the photoionization cross sections for atomic C 2*s* and C 2*p* electrons differ by a factor of 13 in favor of the C 2*s* orbitals. The results reproduced in Fig. 17 demonstrated the dominance of a broad maximum at ~18 eV due to C 2*s*-derived states. The emission within ~5 eV of E_F was very small in XPS but more evident in ultraviolet phtoemission, as is shown in the

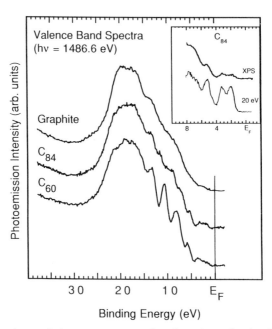

FIG. 17. X-ray photoemission spectra comparing the valence bands of C_{60}, C_{84}, and graphite. While these carbon species have states distributed over the same energy range, the molecular symmetry in the fullerenes produces distinct orbital energies. High energy photoemission emphasizes the deeper states over those close to E_F. The inset compares the XPS spectra with one obtained at lower energy (from Hino *et al.*[35]), showing the enhancement of the π states at low excitation energy. From Poirier *et al.*[36]

[242]M. Knupfer, D. M. Poirier, and J. H. Weaver *Phys. Rev. B* (in press).
[243]D. L. Lichtenberger, M. E. Rempe, and S. B. Gogosha, *Chem. Phys. Lett.* **198**, 454 (1992).

inset of Fig. 17, where a spectrum acquired with $hv = 20$ eV from Hino *et al.*[35] was compared.

The lower molecular symmetry in C_{84} compared with C_{60} or C_{70} reduces the degeneracy of the molecular orbitals still further. Additional broadening is caused by isomeric mixtures, a situation not relevant for C_{60} or C_{70}. Hino *et al.*[34,35,37] reported photoemission spectra for C_{70}, C_{76} C_{82}, and C_{84} that had less distinct features than C_{60}. For graphite, the delocalized π-electron system accounted for significant band dispersion. Mintmire *et al.*,[244] Saito *et al.*,[245] and Wang *et al.*[246] reported electronic calculations for several of the higher fullerenes.

Poirier *et al.*[36] compared the C $1s$ spectra of C_{60}, C_{70}, C_{84}, and graphite, as shown in Fig. 18. Measurements of the main lines of C_{60} and C_{70} were resolution-limited. The increased width for C_{84} indicated contributions from inequivalent carbon atoms. Indeed, there are 21 inequivalent carbon sites for the D_2 structure of C_{84} and 11 for D_{2d}. The results for graphite differed from the fullerenes because it is metallic, and this was reflected in the asymmetry of the main line. In each fullerene, the satellite structure represented $\pi-\pi^*$ transitions within ~6 eV of the main line, a π plasmon at ~6 eV (labeled *P*), $\sigma-\sigma^*$ transitions above ~6 eV, and a higher energy plasmon. While the first $\pi-\pi^*$ feature was quite sharp for C_{60}, the leading feature in C_{70} appeared to be composed of multiple peaks. In C_{84}, the $\pi-\pi^*$ region "filled-in" as a result of an increased number of low energy excitations. The first peak was centered ~1.4 eV from the main line, as determined by peak fitting, and Poirier *et al.* assigned it to the HOMO–LUMO transition. They noted that this excitation energy is ~0.5 eV smaller than in C_{60}. In qualitative agreement, the optical absorption onset for solvated C_{84} was ~0.75 eV lower than for C_{60}.[33]

The C $1s$ satellite region for graphite showed less distinct shakeup structures than the fullerenes, but the plasmon features were still evident. This is consistent with the fact that the electronic density in the π and σ orbitals is not changed significantly when planar graphite is curved to form a fullerene. The difference is that the plasmon is a solid state phenomenon in graphite but it is a collective oscillation of the charge of a single molecule in the fullerenes.

Optical absorption spectra for the higher fullerenes showed an onset shifted to lower photon energy with increasing fullerene size.[33] This was consistent with the reduced difference between the ionization potential

---

[244] J. W. Mintmire, B. I. Dunlap, D. W. Brenner, R. C. Mowrey, and C. T. White, *Phys. Rev. B* **43**, 14281 (1991).
[245] S. Saito, S. Sawada, and N. Hamada, *Phys. Rev. B* **45**, 13845 (1992).
[246] X.-Q. Wang, C. Z. Wang, B. L. Zhang, and K. M. Ho, *Phys. Rev. Lett.* **69**, 69 (1992).

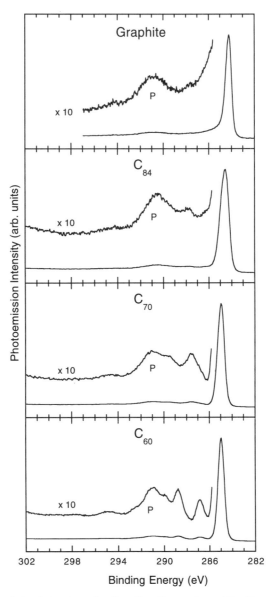

FIG. 18. C 1s-derived structures for C_{60}, C_{70}, C_{84}, and graphite. The main lines for C_{60} and C_{70} are instrument-limited. The greater distribution of inequivalent C atoms accounts for the enhanced main line width for C_{84}. The main line asymmetry for graphite is due to its metallic character. P denotes the π plasmon common to these structures, modified only slightly by the curvature of the carbon network. Adapted from Poirier et al.[36]

and the electron affinity for the higher fullerenes. Ultimately, of course, the gap must vanish as the giant fullerenes converge to graphite. Electron energy loss spectroscopy[224,225,247] results for C_{70} showed structures that can be compared with those of Fig. 18. The high resolution results of Gensterblum et al.[248] showed low energy features associated with band edge excitons, as for C_{60}. Ren et al.[249] measured the dielectric function for $0.5 \leq h\nu \leq 5.3$ eV.

10. VIBRONS AND PHONONS

The C_{60} molecule has 174 normal vibrational modes, but because of its symmetry, many are degenerate and only 46 vibrational lines are expected for the isolated molecule.[62] Of these, 10 are Raman active[62] (8 with H_g and 2 with A_g symmetry) and 4 are IR active[62] (with T_{1u} symmetry). Although condensation to the solid state should break some of the symmetries,[250] the vibrational spectrum of solid C_{60} is described quite well by the molecular levels with little or no apparent solid state symmetry breaking at room temperature. The molecular vibrations span the energy range from about 250 to 1600 cm^{-1} (about 30 to 200 meV), as shown schematically in Fig. 19. Graphite exhibits a comparable range of vibrational modes, which is expected since the C–C bond strength is similar. It was the four strong infrared lines in the soot produced by Krätschmer et al.[22] that suggested the presence of C_{60}. The energies of these modes are about 1428, 1183, 577 and 527 cm^{-1}.

In early studies, Bethune et al.[83,86] investigated the IR and Raman response of K–H soot and chromatographically separated C_{60} and C_{70}. The Raman spectra consisted of 10 lines with 3 particularly strong lines that corresponded to the $H_g(1)$ "squashing" mode at 273 cm^{-1}, the A_g "breathing" mode at 497 cm^{-1}, and the A_g "pentagonal pinch" mode at 1469 cm^{-1}. The symmetries of these lines were confirmed by examining their depolarization ratios. In the breathing mode, the molecule expands and contracts. In the pentagonal pinch mode, the hexagons and pentagons expand and contract out of phase. Both modes maintain the icosahedral symmetry of the molecule, with the breathing mode consisting of radial displacements and the pinch mode of tangential displacements of the carbon nuclei.[251] Other modes are more complicated and

[247]V. P. Dravid, X. Lin, H. Zhang, S. Liu, and M. M. Kappes, *J. Mater. Res.* **7**, 2440 (1992).

[248]A. A. Lucas and G. Gensterblum, private communication.

[249]S.-L. Ren, K. A. Wang, P. Zhou, Y. Wang, A. M. Rao, M. S. Meier, J. P. Selegue, and P. C. Eklund, *Appl. Phys. Lett.* **61**, 124 (1992).

[250]G. Dresselhaus, M. S. Dresselhaus, and P. C. Eklund, *Phys. Rev. B* **45**, 6923 (1992).

[251]R. E. Stanton and M. D. Newton, *J. Phys. Chem.* **92**, 2141 (1988).

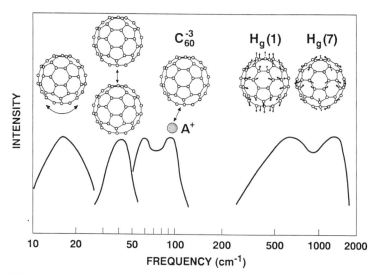

FIG. 19. Energy distribution of the normal modes of C_{60} and A_3C_{60} depicting the low energy librons, the C_{60}–C_{60} phonons, and the C_{60}-ion optic phonons. The two depictions on the right correspond to the vibrons of C_{60} in the solid state. The $H_g(1)$ mode represents largely radial nuclear motion, while the higher energy $H_g(7)$ mode describes tangential motion. From Hebard.[5]

contain elements of both radial and tangential displacements. Additional mode representations appear in Fig. 4 in the chapter by Pickett. In general, the more tangential the character to a mode, the higher its energy. Chase et al.[252] investigated C_{60} monolayers on noble metal surfaces using surface-enhanced Raman scattering. They reported substantial shifts of the high frequency pentagonal pinch mode and attributed the changes to charge transfer from the metal layer.

Inelastic neutron scattering is not subject to the selection rules of optical spectroscopies and is an invaluable probe of vibrational modes that are not IR or Raman active. Neutron scattering experiments by Cappelletti et al.,[253] Prassides et al.,[254] and Colombeau et al.[255] identified

[252]S. J. Chase, W. S. Basca, M. G. Mitch, L. J. Pilione, and J. S. Lannin, *Phys. Rev. B* **46,** 7873 (1992).

[253]R. L. Cappelletti, J. R. D. Copley, W. A. Kamitakahara, F. Li, J. S. Lannin, and D. Ramage, *Phys. Rev. Lett.* **66,** 3261 (1991).

[254]K. Prassides, T. J. S. Dennis, J. P. Hare, J. Tomkinson, H. W. Kroto, R. Taylor, and D. R. M. Walton, *Chem. Phys. Lett.* **187,** 455 (1991).

[255]C. Coulombeau, H. Jobic, P. Bernier, C. Fabre, D. Schültz, and A. Rassat, *J. Phys. Chem.* **96,** 22 (1992).

all of the optically observed modes and several optically "silent" modes for a total of 23 of the expcted 46. Neutron scattering studies have been reviewed by Copley et al.[134] and by Prassides et al.[256] A novel technique for probing the C_{60} vibrational structure was reported by Nissen et al.,[257] who observed 32 modes using singlet oxygen photoluminescence from oxygen molecules intercalated into a C_{60} matrix.

Condensed C_{60} exhibits external vibrations in addition to its molecular modes. Included in this group are librational motions, where the molecules rotationally rock in the potential of the surrounding crystal, and translational C_{60}–C_{60} vibrations, as illustrated in Fig. 19. Neumann et al.[107] used inelastic neutron scattering and a powder sample to show that the librational modes occurred at about 2 to 3 meV ($\sim$16–24 cm^{-1}). This energy was considerably higher than predicted by most models that assumed van der Waals forces,[258] and even than more complicated models where charge was distributed unevenly on the C_{60} sphere.[141,259]

Pintschovius et al.[141] measured the dispersion of several of the translational phonon bands above and below the fcc–sc transition temperature using inelastic neutron scattering from a C_{60} single crystal. These translational modes, which extended to $\sim$6 meV (48 cm^{-1}), were well modeled by rigid spheres interacting through van der Waals interactions. Pintschovius et al. also showed that sharp librational modes exist at temperatures approaching the sc–fcc transition, emphasizing the first-order nature of the transition. The sharpness of these modes indicated that the two molecular orientations present in the sc phase were quite similar in their bonding. Some of the phonon modes were also observed with far infrared spectroscopy.[260]

For C_{70}, there are 204 normal vibrational modes. The lower molecular symmetry of C_{70} leads to a much larger number of expected energy levels, 122. Of these, 31 are IR active and 53 are Raman active. Hare et al.[87] (Bethune et al.[83]) reported positions of 12 (16) IR lines and Dennis et al.[261] (Bethune et al.[83]) reported 27 (21) Raman frequencies for C_{70}.

[256]K. Prassides, H. W. Kroto, R. Taylor, D. R. M. Walton, W. I. F. David, J. Tomkinson, R. C. Haddon, M. J. Rosseinsky, and D. W. Murphy, Carbon 30, 1277 (1992).

[257]M. K. Nissen, S. M. Wilson, and M. L. W. Thewalt, Phys. Rev. Lett. 69, 2423 (1992).

[258]A. Cheng and M. L. Klein, J. Phys. Chem. 92, 2027 (1992).

[259]M. Sprik, A. Cheng, and M. L. Klein, J. Phys. Chem. Solids 96, 2027 (1992).

[260]S. Huant, J. B. Robert, G. Chouteau, P. Bernier, C. Fabre, and A. Rassat, Phys. Rev. Lett. 69, 2666 (1992); H. Bonadeo, E. Halac, and E. Burgos, Phys. Rev. Lett. 70, 3176 (1993); S. A. FitzGerald and A. J. Sievers, Phys. Rev. Lett. 70, 3175 (1993); S. Huant, J. B. Robert, and G. Chouteau, Phys. Rev. Lett. 70, 3177 (1993).

[261]T. J. S. Dennis, J. P. Hare, H. W. Kroto, R. Taylor, D. R. M. Walton, and P. J. Hendra, Spectrochim. Acta 47A, 1289 (1991).

Christides et al.[262] performed inelastic neutron scattering for C_{70}, finding 32 intramolecular modes and good agreement with a calculated density of states.

For C_{84}, intramolecular vibrational spectra have been calculated for various isomers by Zhang et al.[263] For the D_2 isomer, all 246 vibrational modes are nondegenerate and Raman active, while 183 are IR active. For the D_{2d} isomer, there should be 185 lines. Of these, 155 are Raman active and 92 are IR active. Clearly, the vibrational structure increases in complexity for the larger and less symmetric fullerenes.

11. FULLERENE OXIDATION, MODIFICATION, AND POLYMERIZATION

The extraordinary abundance of C_{60} in vaporized graphite products was first explained based on stability arguments.[8] The absence of dangling bonds on the molecular surface was believed to make the molecule impervious to chemical attack or further carbon addition. While explaining the graphite vaporization data, this argument suggested that C_{60} should be a rather robust, long-lived species. On the other hand, evidence for the natural occurrence of fullerenes is quite rare. There was early speculation that helical soot particles and meteor fragments contained fullerenes, but this remains unconfirmed. More recently, C_{60} and C_{70} molecules were detected in two geological samples. The first was a sample of shungite, a carbon-rich Precambrian rock found in Russia.[264] The formation mechanism of the fullerene deposit, and indeed of the shungite itself, was not clear. The second deposit was found in a fulgerite specimen from Sheep Mountain, Colorado.[265] Fulgerite is a glassy rock formed when lightning strikes the earth. Local conditions similar to the intense heat of laser ablation were assumed to accompany a lightning strike, and fullerene formation was considered conceivable.

Perhaps it is not the lack of fullerene producing environments that has limited their abundance, but rather the conditions under which they can be destroyed. The rich variety of chemical modifications reported to date demonstrate that fullerenes are by no means impervious to attack. It has been demonstrated, for example, that fullerenes are produced in certain

[262]C. Christides, A. V. Nikolaev, T. J. S. Dennis, K. Prassides, F. Negri, G. Orlandi, and F. Zerbetto, J. Phys. Chem. **97**, 3641 (1993).

[263]B. L. Zhang, C. Z. Wang, and K. M. Ho, Phys. Rev. B **47**, 1643 (1993).

[264]P. R. Buseck, S. J. Semeon, J. Tsipursky, and R. Hettich, Science **257**, 215 (1992).

[265]T. K. Daly P. R. Buseck, P. Williams, and C. F. Lewis, Science **259**, 1599 (1993).

regions of sooting flames,[53,266,267] but that these hot fullerenes burn in the presence of oxygen if they are not extracted.

Taylor et al.[268] reported degradation of laboratory stock C_{60} under a variety of conditions. In particular, they noted that ultraviolet light and/or oxygen exposure resulted in degradation, and they urged that fullerenes be stored in the dark under vacuum or in a nitrogen environment. Their analysis of reaction products also suggested that ozone might play an important role in decomposition. The intense light emitted from arcs in which the fullerenes are produced might also contribute to their destruction, and Taylor et al. suggested that production efficiencies might be improved in a "low" system.

Milliken et al.[269] demonstrated that C_{60} burns more easily than graphite. Upon heating in air, a C_{60} sample was consumed by 600°C, while combustion of graphite did not begin until ~700°C. In N_2, graphite was stable to 860°C, but sublimation of C_{60} started at ~600°C. Ismail and Rodgers[270] and McKee[271] demonstrated that C_{60} gasifies more readily than other highly reactive forms of disordered carbon (such as carbon black) when heated in air. Chen et al.[206] demonstrated a two-step destruction process upon heating in O_2 with chemisorption starting at ~200°C to form amorphous carbon–oxygen species followed by decomposition by evaporation of CO and CO_2 above about 350°C.

Photoinduced oxidation was examined by Kroll et al.[272] after they condensed O_2 on C_{60} multilayers at 20 K. They found that O_2 did not react at low temperature unless the sample was irradiated with photons. This indicated that there was a kinetic barrier to reaction. CO, CO_2, and carbonyl-like structures were observed following photon irradiation. Kroll et al. suggested that reaction occurred when photoexcited low energy electrons were captured by O_2 to form the more reactive O_2^- species.

Chibante et al.[273] found a temperature-dependent decomposition rate

[266]Ph. Gerhardt, S. Löffler, and K. H. Homann, Chem. Phys. Lett. 137, 306 (1987).

[267]J. B. Howard, J. T. McKinnon, Y. Makarovsky, A. L. Lafleur, and M. E. Johnson, Nature 352, 139 (1991).

[268]R. Taylor, J. P. Parsons, A. G. Avent, S. P. Rannard, T. J. Dennis, J. P. Hare, H. W. Kroto, and D. R. M. Walton, Nature 351, 277 (1991).

[269]J. Milliken, T. M. Keller, A. P. Baronavski, S. W. McElvany, J. H. Callahan, and H. H. Nelson, Chem. Mater. 3, 386 (1991).

[270]I. M. K. Ismail and S. L. Rodgers, Carbon 30, 229 (1992).

[271]D. W. McKee, Carbon 29, 1057 (1991).

[272]G. H. Kroll, P. J. Benning, Y. Chen, T. R. Ohno, J. H. Weaver, L. P. F. Chibante, and R. E. Smalley, Chem. Phys. Lett. 181, 112 (1991).

[273]L. P. F. Chibante, C. Pan, M. L. Pierson, R. E. Haufler, and D. Heymann, Carbon 31, 185 (1993).

for fullerenes heated from ~150 to 250°C in air and in the dark. Each heat treatment rendered a portion of the starting fullerene powder insoluble in toluene, implying that the fullerene form was modified. They estimated that a given C_{60} sample would be rendered completely insoluble after about 2000 years of exposure to air at 25°C. They also estimated that the lifetime of a mixture of 15% C_{70} in C_{60} would be 450 years. From these various oxidation studies, it was clear that naturally occurring C_{60} should be quite rare.

Saunders et al.[274] found that helium can be inserted into the C_{60} cage by heating to ~600°C in a He environment. This implied that the cage opens momentarily and seals again. Those authors noted that some of the fullerene powder was no longer soluble in toluene after heating to ~600°C. Sundar et al.[275] reported degradation of C_{60} upon heating to 700°C for 24 hours, although it was not clear if extrinsic effects such as oxygen contamination might have contributed. They showed the loss of the characteristic x-ray diffraction pattern and absorption spectrum of C_{60} after pressed pellets were heated to 700°C. The resulting material had much lower resistivity and positron lifetime than C_{60}. In contrast, calculations[58,104] suggest that the cage structure should be stable to very high temperatures.

Several authors[276] found that the frequency of the pentagonal pinch mode at $1469 \, cm^{-1}$ shifted to ~$1460 \, cm^{-1}$ during Raman studies. While some controversy has existed regarding the role of oxygen in this mode shift, it is clear that a chemical change was induced by photon irradiation above the C_{60} absorption threshold. Irradiation with photon energies below the absorption threshold had no such effect.

Rao et al.[276] suggested that C_{60} polymerization occurred when fullerene films were irradiated with ultraviolet light, and this was proposed as the explanation for the Raman mode shift. Exposure of C_{60} films to the focused beam from an intense Hg lamp resulted in material that was insoluble in toluene. Laser desorption of the material gave a mass spectrum having broad peaks at approximately integer multiples of 720 amu, the C_{60} molecular mass. Rao et al. presented a model wherein adjacent C_{60} molecules were joined by two C–C bonds formed from erstwhile double bonds of the C_{60} framework. Remarkably, heating to

[274]M. Saunders, H. A. Jiménez-Vásquez, R. J. Cross, and R. J. Poreda, Science 259, 1428 (1993).

[275]C. S. Sundar, A. Bharathi, Y. Hariharan, J. Janaki, V. S. Sastry, and T. S. Radhakrishnan, Solid State Commun. 84, 823 (1992).

[276]A. M. Rao, P. Zhou, K.-A. Wang, G. T. Hager, J. M. Holden, Y. Wang, W.-T. Lee, X.-X. Bi, P. C. Eklund, D. S. Cornett, M. A. Duncan, and I. J. Amster, Science 259, 955 (1993), and references therein.

~170°C restored the fullerenes, and they were again soluble. Zhou *et al.*[277] showed that C_{60} could be hardened (made resistant) to photo-transformation by "contaminating" with oxygen. It was then suggested[276,278] that triplet C_{60} was a precursor to polymerization and the ability of oxygen to quench the C_{60} triplet[89] accounted for the hardening.

Yeretzian *et al.*[279] suggested a different model for fullerene–fullerene bonding. In their work, they used a laser to desorb fullerenes for time-of-flight mass spectrometry. Their results also demonstrated a distribution of masses peaked at near-multiples of C_{60}. They attributed these features to giant fullerenes created by the coalescence of C_{60} molecules.

Zhao *et al.*[16] investigated the effects of electron tunneling between C_{60} and an STM tip in ultrahigh vacuum. Figure 20 shows a series of STM images obtained by those authors. Image (a) represents a 320×320 Å^2 area of a large defect-free surface. After this image was acquired, the authors scanned a smaller area of $\sim 80 \times 80$ Å^2 for ~ 30 min and then acquired the 320×320 Å^2 image shown in (b). Comparison with (a) shows that the small area was modified, appearing "speckled". This procedure was repeated on an 80×80 Å^2 area to the right of the first speckled region, and (c) shows that this area was also modified. Inspection of the speckled region indicated that the close-packed structure was maintained, but also that most of the fullerenes exhibited internal structure, suggesting that they were immobilized. Intermolecular distinction, imaged as dark regions between the fullerenes, was nearly lost in many cases. This suggested increased electronic state density between balls, as would be expected if intermolecular bonds were formed. In addition to these structural changes, the electronic states were also altered, as evidenced by the ability to image the region at tunneling voltages as low as 0.7 V, compared with more than ~ 1.5 V required for pristine C_{60} on GaAs(110). The authors suggested that C_{60}–C_{60} polymerization could explain their results.

Finally, Núñez-Rugueiro[280] succeeded in inducing what appeared to be cross linking by applying unidirectional pressure. It was found that heating to ~ 200°C resulted in fullerene separation. As of this writing, then, it is apparent that C_{60}–C_{60} bonding interactions can be activated by photons,[276] electrons,[186] and pressure.[280] While intriguing and possibly

[277]P. Zhou, A. M. Rao, K.-A. Wang, J. D. Robertson, C. Eloi, M. S. Meier, S. L. Ren, X.-X. Bi, and P. C. Eklund, *Appl. Phys. Lett.* **60**, 2871 (1992).

[278]M. Matus, J. Winter, and H. Kuzmany, Springer Series in Solid State Science, edited by H. Kuzmany and S. Roth (in press).

[279]C. Yeretzian, K. Hansen, F. Diederich, and R. L. Whetten, *Nature* **359**, 44 (1992).

[280]M. Núñez-Rugueiro, private communication.

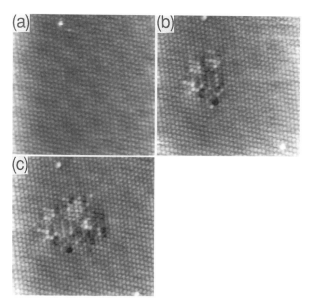

FIG. 20. Scanning tunneling microscope images of a C_{60} thin film showing the effect of concentrated electron flux between the STM tip and the sample. (a) shows the ordered starting surface, (b) shows the effect of rastering the tip over an $80 \times 80 \, \text{Å}^2$ area, and (c) shows that the modification apparent from (b) can be expanded by extensive scanning over an adjacent region. These results demonstrate electron-induced modification that appears to produce polymerized species. From Zhao et al.[186(b)]

useful for thin film applications, the mechanisms for such linking are not well understood.

IV. Doped Fullerenes: Fullerides

The fullerenes exhibit a range of interesting properties, and one of the most captivating is the diversity of their interactions with other species. In some cases, the added species are weakly interactive, as in the mixed molecular solid formed by I_2 uptake.[281] In others, there are strong chemical interactions and charge transfer, as typified by the alkali metal fullerides. What is particularly intriguing is that the doping level x in the fullerides is very high, written A_xC_{60} or AE_xC_{60} where A or AE denotes an alkali or alkaline earth element. While these fullerides are analogous to NaCl or CaF_2, the fact that x can reach 6 for K-, Rb-, or Cs-C_{60} or 10 for Na-C_{60} is remarkable. As for the alkali halides, there is nearly

[281]Q. Zhu, D. E. Cox, J. E. Fischer, K. Kniaz, A. R. McGhie, and O. Zhou, *Nature* **355**, 712 (1992).

complete charge transfer, at least for $x \leq 6$. This demonstrates that the fullerene molecule is a strong electron acceptor, readily becoming C_{60}^{6-} in the solid state. Haddon et al.[15,282] related this to the fact that the LUMO of C_{60} is threefold degenerate, as shown in Fig. 11. Among the properties that attracted the most attention from the physics community was the display of superconductivity by K_3C_{60}, Rb_3C_{60}, and related fullerides.[283–286] Table VII of the chapter by Pickett summarizes many of the properties of the different fullerides. For $Rb_1Cs_2C_{60}$ the superconducting transition temperature T_c reached 33 K.[286]

The first evidence that the fullerenes could be doped was reported by Haddon et al.[282] in early 1991. They had grown thin films of C_{60} after first attaching electrical leads to the glass substrate. When they exposed the film to a flux of alkali metal atoms, they found that the resistivity dropped with exposure time, reached a minimum, then increased and ultimately saturated. They reasoned that the drop was because electrons were transferred from the alkali metal dopants to the fullerene LUMO-derived conduction bands. They argued that saturation corresponded to filling the LUMO level, achieved at A_6C_{60}. At that time, little was known about the doping processes and, in particular, whether distinct phases might be formed. Within weeks, however, Hebard and coworkers[283] reported that K-doped C_{60} underwent a superconducting transition at ~18 K. Shortly thereafter, the UCLA group[284] reported that the diamagnetic susceptibility for well-annealed K_xC_{60} reached a maximum for $x = 3$. They also noted that a single transition temperature was found, regardless of the overall stoichiometry, thus suggesting that K_3C_{60} was a stable line compound. Almost simultaneously, Holczer et al.[284] and Rosseinsky et al.[285] established that Rb_xC_{60} also showed superconductivity, with $T_c = 28$ K.

For packing fraction, an fcc structure derived from spherical objects in contact is 74%. For elemental solids such as Cu, the interstitial spaces

[282]R. C. Haddon, A. F. Hebard, M. J. Rosseinsky, D. W. Murphy, S. J. Duclos, K. B. Lyons, B. Miller, J. M. Rosamilia, R. M. Fleming, A. R. Kortan, S. H. Glarum, A. V. Makhija, A. J. Muller, R. H. Eick, S. M. Zahurak, R. Tycko, G. Dabbagh, and F. A. Thiel, *Nature* **350,** 320 (1991).

[283]A. F. Hebard, M. J. Rosseinsky, R. C. Haddon, D. W. Murphy, S. H. Glarum, T. T. M. Palstra, A. P. Ramirez, and A. R. Kortan, *Nature* **350,** 600 (1991).

[284]K. Holczer, O. Klein, S.-M. Huang, R. B. Kaner, K.-J. Fu, R. L. Whetten, and F. Diederich, *Science* **252,** 1154 (1991).

[285]M. J. Rosseinsky, A. P. Ramirez, S. H. Glarum, D. W. Murphy, R. C. Haddon, A. F. Hebard, T. T. M. Palstra, A. R. Kortan, S. M. Zahurak, and A. V. Makhija, *Phys. Rev. Lett.* **66,** 2380 (1991).

[286]K. Tanigaki, T. W. Ebbesen, S. Saito, J. Mizuki, J. S. Tsai, Y. Kubo, and S. Kuroshima, *Nature* **352,** 222 (1991).

are relatively small, accounting for the fact that the common interstitial compounds are derived from H, B, and C. The octahedral and tetrahedral radii are found from $r_{oct} = 0.5a_o - r_a$ and $r_{tet} = 0.433a_o - r_a$, where r_a is the hard-sphere radius of the host "atom." For C_{60}, r_a is ~5.0 Å and $a_o = 14.2$ Å, so that the octahedral and tetrahedral holes have radii of about 2.1 and 1.1 Å, respectively. As depicted in Fig. 12 for the octahedral sites, these interstices are more than adequate to accommodate guest species.

Size considerations notwithstanding, it is thermochemical arguments that determine whether intercalation is an energy gaining process. For K-C_{60} compounds, Martins and Troullier[287] estimated the exothermic heat of formation to be 1.4–1.7 eV per K ion relative to the standard states of C_{60} and K metal. This exothermic reaction constitutes a strong driving force, and, since diffusion of K ions is relatively facile in C_{60}, fulleride formation can be achieved at low temperature. Mixing of fullerenes has been observed for the alkali metals, the alkaline-earth metals, and some of the rare-earth metals. Mixing has been reported for the higher fullerenes as well, although much less is known about the structures and possible phases formed for these systems. For the larger ions, diffusion is less facile and growth must be done at elevated temperature, particularly to produce bulk samples. Wang et al.[288] considered thermodynamic stability for C_{60} ionic compounds with a wide range of elements, finding the alkali metal and alkaline-earth fullerides to be most stable.

The inability to mix Au, for example, with C_{60} can be understood by noting that the large cohesive energy of Au, 3.81 eV/atom, places a lower limit on the energy that must be recovered upon mixing. Au–C bonding is weak, and there is every reason to expect that thermodynamic attempts to mix Au with C_{60} will produce phase-separated Au and C_{60}, as observed when Au was deposited onto a C_{60} surface.[289] At best, metastable systems might be produced by such nonequilibrium processes as implantation or low temperature codeposition. [We shall return to the Au–C_{60} system when considering onions in Section VI. We also note that the bonding of C_{60} to Au(110) is sufficient to change the Au surface reconstruction.[180]]

Many systems exhibit chemical tendencies that frustrate mixing. For example, Ohno et al.[289] reported that attempts to mix Ti with C_{60} by

[287]J. L. Martins and N. Troullier, *Phys. Rev. B* **46**, 1766 (1992).

[288]Y. Wang, D. Tománek, G. F. Bertsch, and R. S. Ruoff, *Phys. Rev. B* **47**, 6711 (1993); R. S. Ruoff, Y. Wang, and D. Tománek, *Chem. Phys. Lett.* **203**, 438 (1993).

[289]T. R. Ohno, Y. Chen, S. E. Harvey, G. H. Kroll, P. J. Benning, J. H. Weaver, L. P. F. Chibante, and R. E. Smalley, *Phys. Rev. B* **47**, 2389 (1993).

exposure of a C_{60} film to Ti adatoms in vacuo led to disruption of some of the C_{60} molecules, consistent with the tendency of Ti and C to form a carbidic phase. Wang et al.[191] showed that heating of a C_{84} film on Si(100) to 1000°C led to SiC formation. This returns, then, to the theme developed in the section on fullerene modification: The fullerenes are not chemically indestructible.

12. FULLERIDE CRYSTAL STRUCTURES

Figure 21 depicts the crystal structures of the alkali metal fullerides.[290,291] C_{60}, with its fcc structure, is drawn in a way to emphasize the locations and sizes of the octahedral and tetrahedral sites. These structures are also relevant for alkaline-earth fullerides, as will be discussed in Section 16.

Filling of the octahedral sites of the fcc lattice produces the NaCl structure, A_1C_{60}. Evidence for the existence of this phase for $A =$ Rb and Cs was obtained using x-ray photoemission.[292] In their work, Poirier et al.[292] took advantage of the fact that distinct core level binding energies could be associated with ions in the tetrahedral and octahedral sites. They demonstrated only octahedral site occupancy for Rb and Cs for samples doped to give x values less than 1. X-ray diffraction work by the University of Pennsylvania Group[291,293,294] also suggested NaCl phases at high temperature and they proposed rhombohedral structures (distorted NaCl) for Rb_1C_{60} and Cs_1C_{60} at room temperature. Raman studies by Winter and Kuzmany[295,296] of K-C_{60} at elevated temperature indicated a K_1C_{60} phase. For K-C_{60}, detailed XPS studies of the temperature-dependent occupation of the tetrahedral and octahedral sites made it

[290]D. W. Murphy, M. J. Rosseinsky, R. M. Fleming, R. Tycko, A. P. Ramirez, R. C. Haddon, T. Siegrist, G. Dabbagh, J. C. Tully, and R. E. Walstedt, *J. Phys. Chem. Solids* **53**, 1321 (1992).

[291]O. Zhou and D. E. Cox, *J. Phys. Chem. Solids* **53**, 1373 (1992).

[292]D. M. Poirier, T. R. Ohno, G. H. Kroll, P. J. Benning, F. Stepniak, J. H. Weaver, L. P. F. Chibante, and R. E. Smalley, *Phys. Rev. B* **47**, 9870 (1993).

[293]O. Zhou, Q. Zhu, G. B. M. Vaughan, J. E. Fischer, P. A. Heiney, N. Coustel, J. McCauley, Jr., A. B. Smith III, and D. E. Cox in "Novel Forms of Carbon," Materials Research Society Symposium Proceedings, Volume 270, eds. C. L. Renschler, J. J. Pouch, and D. M. Cox (1992), p. 191.

[294]Q. Zhu, O. Zhou, J. E. Fischer, A. R. McGhie, W. J. Romanow, R. M. Strongin, M. A. Chichy, and A. B. Smith III, *Phys. Rev. B* **47**, 13948 (1993).

[295]J. Winter and H. Kuzmany, *Solid State Commun.* **84**, 935 (1992).

[296]J. Winter and H. Kuzmany, Springer Series in Solid State Science, edited by H. Kuzmany and S. Roth (in press).

ALKALI METAL FULLERIDES

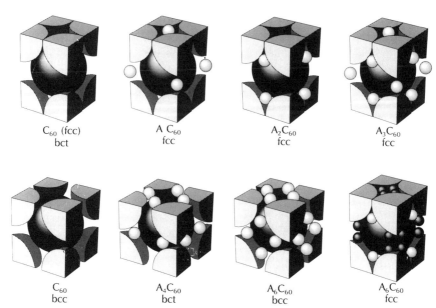

| C_{60} (fcc) bct | A C_{60} fcc | A_2C_{60} fcc | A_3C_{60} fcc |

| C_{60} bcc | A_4C_{60} bct | A_6C_{60} bcc | A_6C_{60} fcc |

FIG. 21. The crystal structures of the fulleride family. The top row shows the fcc structure of undoped C_{60}, the A_1C_{60} structure with octahedral site occupancy, the A_2C_{60} structure with tetrahedral site occupancy, and the A_3C_{60} structure with tetrahedral and octahedral site occupancy. The bottom row shows C_{60} molecules arranged in the bcc structure, the filling of tetrahedral sites producing A_4C_{60} and A_6C_{60}. The bottom-right structure corresponds to A_6C_{60} where there are four ions in the octahedral site and one in each tetrahedral site. A_1C_{60} structures have been reported for K, Rb, and Cs; A_2C_{60} for Na; A_3C_{60} for K and Rb; A_4C_{60} for K, Rb, and Cs; A_6C_{60} (bcc) for K, Rb, and Cs; and A_6C_{60} (fcc) for Na. Ba_4C_{60} also adopts the A_6C_{60} (bcc) structure, while Ca_5C_{60} adopts a modified fcc structure with half of the tetrahedral sites filled and four ions in octahedral sites that maximize the distance from occupied tetrahedral sites. Courtesy D. W. Murphy,[290] AT&T Bell Laboratories.

possible for Poirier and Weaver[297] to propose the phase diagram to be discussed in Section 13.

Filling of the tetrahedral sites produces the CaF_2 structure with 2^- charging of the fullerenes. An A_2C_{60} phase was reported by Rosseinsky *et al.*[298] for Na-C_{60} but not for the larger alkali metals. The tendency to form this structure probably reflects the fact that the Na ions are so small

[297](a) D. M. Poirier and J. H. Weaver, *Phys. Rev. B* **47,** 10959 (1993); J. H. Weaver, D. M. Poirier, and Y. B. Zhao, Springer Series in Solid State Science, edited by H. Kuzmany and S. Roth (in press); (b) M. Knupfer, D. M. Poirier, and J. H. Weaver, *Phys. Rev. B.* (in press).

[298]M. J. Rosseinsky, D. W. Murphy, R. M. Fleming, R. Tycko, A. P. Ramirez, T. Siegrist, G. Dabbagh, and S. E. Barrett, *Nature* **356,** 416 (1992).

that there is a stabilizing energy only for the smaller tetrahedral sites. Conversely, the fact that the octahedral site is large accounts for preferred octahedral occupation by the larger Rb and Cs ions in the A_1C_{60} phases. The situation is particularly intriguing for K, because these ions occupy both tetrahedral and octahedral sites at low temperature, but heating above 425 K destabilizes tetrahedral occupation in favor of the octahedral-only occupation.[294–297]

Figure 21 shows that the occupation of each octahedral and tetrahedral site produces the A_3C_{60} phase with 3^- charging of the fullerenes, corresponding to half-filling of the LUMO-derived levels. Stephens et al.[299] first determined the K_3C_{60} structure with x-ray diffraction. They concluded that the C_{60} molecules were randomly ordered between the position of Fig. 1 and an orientation rotated 90° about [100]. NMR work by Barrett and Tycko[300] indicated that the disorder may be more severe. This A_3C_{60} phase is the important one as far as superconductivity is concerned, because all of the alkali metal fullerides that are superconductors crystallize in this structure.[290] This is true even for those derived from two different alkali metals such as $Rb_1Cs_2C_{60}$. None of the C_{60} fullerides with $x \neq 3$ are metals in the normal state or superconductors at low temperature. The preference of larger ions for the octahedral site was shown for a variety of mixed alkali metal fullerides by x-ray diffraction,[290,298,301–303] NMR,[304,305] and XPS.[292] A recent NMR[305] study of Rb_3C_{60} indicated a possible structural distortion from the ideal fcc lattice that makes the two tetrahedral Rb ions inequivalent.

Doping beyond $x = 3$ requires occupation of some interstitial sites by more than one ion or a structural transformation of the host lattice to provide more interstitial space. For Na doping, the former is observed and the fcc lattice is maintained until $x = 6$, as depicted by the drawing at the lower right of Fig. 21. Yildirim et al.[306] have demonstrated that the large octahedral site will accept up to eight Na ions under proper

[299]P. Stephens, L. Mihaly, P. Lee, R. L. Whetten, S.-M. Huang, R. B. Kaner, F. Diederich, and K. Holczer, Nature 351, 632 (1991).

[300]S. E. Barrett and R. Tycko, Phys. Rev. Lett. 69, 3754 (1992).

[301]K. Tanagaki, I. Hirosawa, T. W. Ebbesen, J. Mizuki, Y. Shimakawa, Y. Kubo, J. S. Tsai, and S. Kuroshima, Nature 356, 419 (1992).

[302]I. Hirosawa, K. Tanagaki, J. Mizuki, T. W. Ebbesen, Y. Shimakawa, Y. Kubo, and S. Kuroshima, Solid State Commun. 82, 979 (1992).

[303]O. Zhou, R. M. Fleming, D. W. Murphy, M. J. Rosseinsky, A. P. Ramirez, R. B. van Dover, and R. C. Haddon, Nature 362, 433 (1993).

[304]Y. Maniwa, K. Mizoguchi, K. Kume, K. Tanigaki, T. W. Ebbesen, S. Saito, J. Mizuki, J. S. Tsai, and Y. Kubo, Solid State Commun. 82, 783 (1992).

[305]R. E. Walstedt, D. W. Murphy, and M. Rosseinsky, Nature 362, 611 (1993).

[306]T. Yildirim, O. Zhou, J. E. Fischer, N. Bykovetz, R. A. Strongin, M. A. Chichy, A. B. Smith III, C. L. Lin, and R. Jelinek, Nature 360, 568 (1992).

reaction conditions to form a cube of Na ions (possibly with another ion in the center of the cube to give the stoichiometry $Na_{11}C_{60}$). Multiple occupancy is observed for doping with the alkaline-earth element Ca, to be discussed in Section 17. For K, Rb, and Cs doping, however, Fleming et al.[307] demonstrated that the fcc lattice transforms to a body centered tetragonal structure, as shown at lower left in Fig. 21. In this structure, the tetrahedral site radius, ~1.5 Å, is larger than for the tetrahedral sites of the fcc structure.

Filling all of the tetrahedral sites results in a transformation to the bcc structure and A_6C_{60}, as first shown by Zhou et al.[308] For A_6C_{60}, each fullerene is coordinated with 24 akali ions, producing a charge state of C_{60}^{6-}. This is the saturated state for K, Rb, and Cs. Interestingly, there has been no evidence of a distinct A_5C_{60} structure Likewise, the $A-15$ structure familiar from the Nb_3Sn family has not been found for alkali metal doping, although it would have the same stoichiometry as the A_3C_{60} fcc phase. The $A-15$ structure does form for Ba-doped C_{60}, as reported by Kortan et al.[309]

Most of the experimental evidence reported to date[230,292,295–297,310–314] points to phase separation into these distinct stoichiometric compounds for global stoichiometries away from integer values. On the other hand, Zhu et al.[315] have argued that there is a large solid solution field extending from $x = 3$ to at least $x = 1.6$ in the range $353 < T < 423$ K for K_xC_{60}, but they have not delineated the phase boundaries for this "lattice-gas." In Na_xC_{60}, one might expect a wide range of disordered solid solution because occupation of the octahedral site could vary continuously from zero to eight (or nine).

[307]R. M. Fleming, M. J. Rosseinsky, A. P. Ramirez, D. W. Murphy, J. C. Tully, R. C. Haddon, T. Siegrist, R. Tycko, S. H. Glarum, P. Marsh, G. Dabbagh, S. M. Zahurak, A. V. Makhija, and C. Hampton, Nature 352, 701 (1991).

[308]O. Zhou, J. E. Fischer, N. Coustel, S. Kycia, Q. Zhu, A. R. McGhie, W. J. Romanow, J. P. McCauley, Jr., A. B. Smith III, and D. E. Cox, Nature 351, 462 (1991).

[309]A. R. Kortan, N. Kopylov, R. M. Fleming, O. Zhou, F. A. Thiel, R. C. Haddon, and K. M. Rabe, Phys. Rev. B 47, 13070 (1993).

[310]P. W. Stephens, L. Mihaly, J. B. Wiley, S.-M. Huang, R. B. Kaner, F. Diederich, R. L. Whetten, and K. Holczer, Phys. Rev. B 45, 543 (1992).

[311]R. Tycko, G. Dabbagh, M. J. Rosseinsky, D. W. Murphy, R. M. Fleming, A. P. Ramirez, and J. C. Tully, Science 253, 884 (1991).

[312]Q. Zhu, O. Zhou, N. Coustel, G. B. M. Vaughan, J. P. McCauley, Jr., W. J. Romanow, J. E. Fischer, and A. B. Smith III, Science 254, 545 (1991).

[313]D. M. Poirier, T. R. Ohno, G. H. Kroll, Y. Chen, P. J. Benning, J. H. Weaver, L. P. F. Chibante, and R. E. Smalley, Science 253, 646 (1991).

[314]T. Pichler, M. Matus, J. Kürti, and H. Kuzmany, Phys. Rev. B 45, 13841 (1992).

[315]Q. Zhu, J. E. Fischer, and D. E. Cox, Springer Series in Solid State Sciences, edited by H. Kuzmany and S. Roth (in press).

13. The K-C_{60} Phase Diagram

There have been a number of important contributions to our understanding of the phase diagrams for the alkali metal fullerides. Early x-ray diffraction work demonstrated $x = 3$, 4, and 6 phases at room temperature[299,307,308] for K-C_{60}, and XPS,[292,313] NMR,[311] and Raman spectroscopy[314] indicated phase separation between C_{60} and K_3C_{60} for $x \leq 3$. Based on studies with the Rb fullerides, Zhu et al.[312] proposed a provisional phase diagram that served as a starting point for discussion of phase equilibria. Winter and Kuzmany[295,296] later presented evidence that an $x = 1$ phase was formed at elevated temperature for K. From x-ray diffraction analyses, Zhu et al.[294] also reported an $x = 1$ NaCl phase for K-C_{60} at high temperature. Poirier and Weaver[297] noted that core level photoemission was able to distinguish K ions in octahedral sites from those in tetrahedral sites because of the large difference in Madelung energies. They measured changes in the site occupancy as a function of temperature for K_xC_{60} films and concluded that a transformation involving the occupancy of tetrahedral and octahedral sites occurred at $T = 425$ K for all $x < 3$.

The phase diagram for K-C_{60} proposed by Poirier and Weaver[297] is shown in Fig. 22. Their data indicated a eutectoid transition, $K_1C_{60} \rightarrow K_3C_{60} + \alpha\text{-}C_{60}$, for $x \leq 3$, and the remainder of the phase diagram was adapted from that of Zhu et al.[312] In Fig. 22, symbols denote stoichiometries studied and identify the eutectoid temperature. The phase fields were assumed to reflect solid solutions at very high temperature, although data were not taken for temperatures above the breaks in the vertical scale. $\alpha\text{-}C_{60}$ denotes the dilute solid solution of K in C_{60} with no attempt to differentiate K in the fcc phase from K in the sc phase. Technically, the doping should change the sc $\rightarrow$ fcc transformation temperature, resulting in another invariant reaction, as has been observed for dilute amounts of K in C_{70}.[297(b)] The phase boundaries around 3 and 4 were drawn vertically because the bounds and vacancy energies are not known. A fairly high vacancy concentration for the A_6C_{60} phase was suggested by Zhu et al.[312] based on results for Rb_xC_{60}. This phase diagram for $0 \leq x \leq 3$ was appealing, because it was very simple and was based on a classic transformation familiar from conventional materials science.[316] It was also consistent with the room temperature NMR results of Tycko et al.[311] and the Raman results of Winter and Kuzmany.[295,296] We note that NMR has been used[317] to show the inequivalence of

---

[316]See, for example, W. D. Callister, Jr., "Materials Science and Engineering: An Introduction" (Wiley, New York, 1985).
[317]K. Holczer and R. L. Whetten, *Carbon* **30**, 1261 (1992).

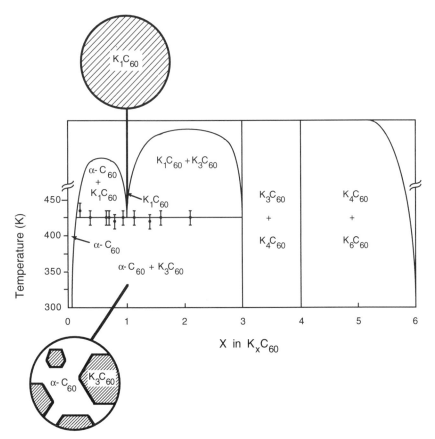

FIG. 22. Phase diagram for K-C_{60}. The horizontal line at 425 K represents a eutectoid temperature, T_E, that defines the transition $K_1C_{60} \rightarrow (\alpha\text{-}C_{60} + K_3C_{60})$. Two-phase fields exist above T_E for $\alpha\text{-}C_{60} + K_1C_{60}$ and for $K_1C_{60} + K_3C_{60}$. Below T_E, the broad field between the $\alpha\text{-}C_{60}$ and K_3C_{60} boundaries represents phase separation. The microstructure depicted for this region is an oversimplification, because it implies large crystallites for $\alpha\text{-}C_{60}$ and K_3C_{60}; a more likely microstructure would display lamellar or nodular grains, depending on kinetics. The lines for $x = 3$ and 4 are drawn vertically because the vacancy fields are not known. The vacancy field for K_6C_{60} is pictorial, as is the behavior above 450 K for any stoichiometry. Adapted from Poirier and Weaver.[297]

octahedral and tetrahedral ions in K_3C_{60} and should prove to be a powerful tool in further defining the equilibrium phase fields.

The invariant temperature defined by the horizontal line at 425 K is recognized to be a eutectoid temperature, T_E, when the transformation occurs from one solid phase into two solid phases.[316] Cooling through T_E converts the high temperature K_1C_{60} phase into phase-separated $\alpha\text{-}C_{60}$

and K_3C_{60}. When $x = 1$, all of the sample undergoes this transformation, depicted in Fig. 22 as the change from a homogeneous material into a two-phase material. The microstructure at lower temperatures is determined by the kinetics of the transformation. Rapid cooling will produce small grains of K_3C_{60} in the majority of α-C_{60} phases, producing lamellar or granular microstructures, as extensively studied for carbon steels (but not yet demonstrated for the fullerides).[316] Rapid cooling or quenching can freeze in nonequilibrium structures.

K_3C_{60} grains will coexist with K_1C_{60} grans for $T > T_E$ when the global stoichiometry is above $x = 1$ in a quantity dictated by the lever rule. Cooling through T_E will have little effect on the K_3C_{60} grains (unless the solubility near $x = 3$ is drastically altered at T_E, perhaps because of rotational motion of the fullerenes), but the K_1C_{60} grains must undergo the eutectoid $K_1C_{60} \rightarrow \alpha$-$C_{60} + K_3C_{60}$ transformation.

In resistivity studies of doped single crystals of C_{60}, Xiang[318] observed a dip at ~400 K, consistent with the K_3C_{60} sample being slightly K-deficient. The resistivity drop will be a result of a eutectoid transformation for the K_1C_{60} component. For fullerides grown from powders with small grains and substantial disorder, the implication is that the sample microstructure may be quite complex when measured at low temperature for intermediate stoichiometry, even if the rate of cooling allows near-equilibrium conditions. Unfortunately, very little single crystal diffraction work has been done to date. This should change as questions regarding phase formation are addressed in detail.

14. ELECTRICAL RESISTIVITY AND GRANULAR METAL BEHAVIOR

The discussion of phase diagrams should make it clear that the properties of a fulleride sample will reflect two-phase coexistence for stoichiometries outside what appear to be rather narrow phase fields. This is a consequence of the Gibbs phase rule,[316] regardless of whether the sample is a thin film or a large crystal and independent of whether the undoped sample was a single crystal or was polycrystalline.

While Haddon et al.[282] first described the resistivity vs. dopant concentration, Kochanski et al.[319] presented the first detailed measurements for K-doped C_{60}. Using Rutherford backscattering, they showed that the minimum resistivity occurred for $x = 3.00 \pm 0.05$. Stepniak et al.[320] subsequently measured the electrical resistivity for all of the alkali

[318]X.-D. Xiang, private communication.
[319]G. P. Kochanski, A. F. Hebard, R. C. Haddon, and A. T. Fiory, *Science* **255**, 184 (1992).
[320]F. Stepniak, P. J. Benning, D. M. Poirier, and J. H. Weaver, *Phys. Rev. B* **48**, 1899 (1993).

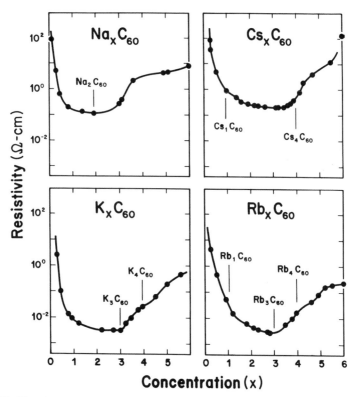

FIG. 23. Electrical resistivity for alkali metal C_{60} films doped showing a drop upon initial doping, a minimum value, and then a return to a more resistive state. The structure evident for intermediate stoichiometries depends on the phases formed. The stoichiometries corresponding to bulk compounds are indicated. For these granular metal systems, only K_3C_{60} and Rb_3C_{60} exhibit metallic temperature coefficients of resistance; all other phases show activated transport. From Stepniak et al.[320]

metal fullerides (except Li), determining x directly with photoemission during doping. Their results are summarized in Fig. 23. While similar to those of Kochanski et al., they also identified structures in $\rho(x)$ that were related to the phases formed, as labeled.

Undoped C_{60} is an insulator with very high resistivity, $>10^{14}$ Ω-cm estimated by Mort et al.[321] Doping from $x = 0$ assured that the sample would enter a two-phase regime when the solubility limit was crossed. For K-C_{60}, this means that K_3C_{60} grains nucleate in the majority α-C_{60}

[321]J. Mort, R. Ziolo, M. Machonkin, D. R. Huffman, and M. I. Ferguson, Appl. Phys. Lett. **61,** 1829 (1992); Chem. Phys. Lett. **186,** 281 (1991) and **186,** 284 (1991).

phase. These grains would be randomly distributed for homogeneous nucleation, but the role of kinetics and the details of diffusion are not well known. In any case, the nuclei are small and the grains are not connected until the percolation threshold is reached. Upon reaching that threshold, the temperature dependence of the resistivity changed from being activated (semiconductor behavior) to being independent of temperature, as discussed by Kochanski et al.[319] and Stepniak et al.[320] Activation reflected the transfer of charge from one metal grain to another, and the percolation network introduced a second conduction path. Ultimately, the temperature dependence reflected metallic character for $K-C_{60}$, and the resistivity reached its minimum. Given the phase diagram of Fig. 22, continued doping must produce insulating K_4C_{60} grains in a background of metallic K_3C_{60}. These grew and ultimately consumed the metallic pathways, resulting in an overall insulating behavior. The transformation to K_6C_{60} does not alter the insulating behavior, since K_6C_{60} is also an isulator. This behavior is very similar to that encountered in studies of granular metal films derived from Ni in SiO_2, Al in Al_2O_3, or Au in Al_2O_3.[322] Perhaps the greatest difference was that the metallic state had a higher resistivity and the insulating state had a lower resistivity than those found in the metal and insulating counterparts in previously studied systems. All of these granular metals showed that the percolation threshold occurred when the volume fraction of the metallic phase reached about 0.55.

The electrical properties of the Na, Rb, and Cs fullerides showed differences compared with $K-C_{60}$ that were related to the phase diagrams for these materials, namely the existence of the $x = 1$ phase for Rb and Cs, the $x = 2$ phase for Na, and the absence of the $x = 3$ phase for Cs. The magnitudes of the resistivities at the minimum for K_3C_{60} and Rb_3C_{60} were about an order of magnitude smaller than those of $Na-C_{60}$ or $Cs-C_{60}$, and all were very close to the conventional Mott limit.[212] Indeed, the temperature-dependent studies of Stepniak et al.[320] showed activated nonmetallic transport for every fulleride except the $x = 3$ phases.

15. FULLERIDE ELECTRONIC STRUCTURES

The work of Haddon et al.[15,282] indicated that doping involved charge transfer to the fullerenes for intermediate stoichiometries and that

[322]B. Abeles, "Granular Metal Films, Applied Solid State Science," 6, 1 (Academic Press, 1976) and references therein; J. E. Morris, A. Mello, and C. J. Adkins, Mat. Res. Soc. Proc. 195, 181 (1990).

saturation occurred when x reached 6. Subsequent spectroscopic studies investigated this behavior in more detail, confirming it in general but finding complexities that reflected on the fundamental physics of materials near a metal–insulator transition and the details of fulleride phase formation. Before reviewing the experimental situation, it is perhaps useful to give an overview of what might be expected, given the foregoing discussion vis-à-vis samples produced by progressive doping. This can be done with the aid of Fig. 24, where we reproduce the densities of electronic states calculated by Erwin[323] for isolated C_{60} (bottom), C_{60} in

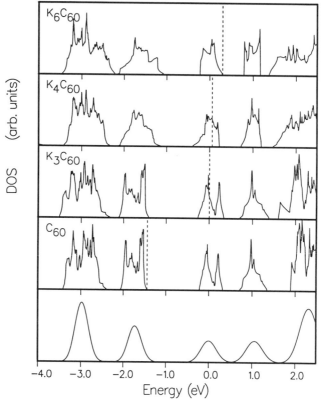

FIG. 24. Densities of states for isolated C_{60} (broadened), solid C_{60}, and K-fullerides showing the movement of the Fermi level (dashed vertical line) through the LUMO band with increased doping. The K atoms transfer their $4s$ charge to the fullerene orbitals and the $4s$ state of the positive K ion appears outside the energy range shown. From Erwin.[323]

[323]S. C. Erwin in "Buckminsterfullerenes," edited by W. E. Billups and M. A. Ciufolini (VCH Publishers, New York, 1993), p. 217.

an fcc lattice with $Fm\bar{3}$ symmetry, K_3C_{60}, K_4C_{60}, and K_6C_{60}. The energy level reference was constrained to be the center of the LUMO band. The Fermi level, identified by the dashed vertical line, moved through LUMO with increased doping.

The calculations for the solid state showed highly structured features centered near the molecular energy levels. While comparison between experiment and density of states calculations for pure C_{60} revealed good agreement as far as the widths and placements of these features are concerned (Fig. 14), experiment did not reproduce the fine structure. Two observations were offered to explain this discrepancy. First, even in the gas phase, the molecular ionizations are broadened by coupling to vibrational modes. Second, the fullerenes studied were dynamically spinning or, at best, merohedrally disordered, as discussed in Section 7. Gelfand and Lu[324] demonstrated that such orientational disorder washes out structure in the density of states.

The fact that doping under thermodynamic equilibrium produces two-phase samples complicates comparison with theory, because two-phase samples reveal features representative of both, weighted accordingly. For example, a photoemission spectrum for a sample with a global stiochiometry of $K_{1.5}C_{60}$ at room temperature would reflect a superposition of features from the dilute α-C_{60} phase and from K_3C_{60}. From the density of states of Fig. 25, this would produce a much-broadened spectrum, with the two spectra offset by the energy necessary to align their Fermi levels.

Figure 25 reproduces photoemission and inverse photoemission curves for K-C_{60} from Benning et al.[325] While several groups[230,326–331] subsequently reported photoemission spectra for K-doped C_{60} with much better resolution, this figure showed the changes in both the occupied

[324]M. P. Gelfand and J. P. Lu, *Phys. Rev. Lett.* **68**, 1050 (1992).

[325]P. J. Benning, D. M. Poirier, T. R. Ohno, Y. Chen, M. B. Jost, F. Stepniak, G. H. Kroll, J. H. Weaver, J. Fure, and R. E. Smalley, *Phys. Rev. B* **45**, 6899 (1992).

[326]P. J. Benning, F. Stepniak, D. M. Poirier, J. L. Martins, J. H. Weaver, L. P. F. Chibante, and R. E. Smalley, *Phys. Rev. B* **47**, 13843 (1993).

[327]J. H. Weaver, P. J. Benning, F. Stepniak, and D. M. Poirier, *J. Phys. Chem. Solids* **53**, 1707 (1992).

[328]M. Merkel, M. Knupfer, M. S. Golden, J. Fink, R. Seeman, and R. L. Johnson, *Phys. Rev. B* **47**, 11470 (1993).

[329]J. Fink, E. Sohmen, M. Merkel, A. Masaki, H. Romberg, A. M. Alexander, M. Knupfer, M. S. Golden, P. Adelmann, and B. Renker in "Fullerenes: Status and Perspectives," eds. C. Taliani, G. Ruani, and R. Zamboni (World Scientific, Singapore, 1992), p. 161.

[330]T. Morikawa and T. Takahashi, *Phys. Rev. B* **48**, 8418 (1993).

[331]M. Knupfer, M. Merkel, M. S. Golden, J. Fink, O. Gunnarsson, and V. P. Antropov, *Phys. Rev. B* **47**, 13944 (1993).

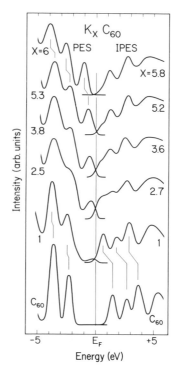

FIG. 25. Photoemission and inverse photoemission results for K-C_{60} as a function of doping concentration x. Electron removal and addition produces $N-1$ and $N+1$ electron states and correlation increases the N-particle energy gap between HOMO and LUMO. The HOMO to LUMO center to center separation is reduced from 3.75 to 1.6 eV when the LUMO level is filled and both are probed by electron removal. While the spectra for intermediate stoichiometries are broadened because of multiple phase coexistence, they show the progressive filling of the LUMO band. From Benning et al.[325]

and empty electronic states. Doping to $x = 1$ broadened the spectra substantially, as expected, and shifted the Fermi level into the region of the LUMO band. The emission at E_F then increased as the amount of the metallic phase increased until the insulating $x = 4$ phase started to form. Ultimately, the fullerene film was saturated as x reached 6. When the LUMO band was filled, E_F fell in the gap between the now-occupied LUMO band and the empty LUMO + 1 band. In this configuration, the system was again a filled-band insulator.

Weaver[211] argued that electron correlation effects were important for pure C_{60}, as has been discussed, and that they were also important for the fully doped fulleride when the gap between the $N-1$ and $N+1$ systems was considered. In particular, when the LUMO band was empty, as in C_{60}, the HOMO–LUMO gap was 3.75 eV, but the center of the LUMO band was only 1.6 eV above HOMO when it was filled. The difference reflected the fact that electrons were removed from both HOMO and LUMO in the fully doped case, while electrons were added to LUMO and removed from HOMO in the undoped case. Comparison of the separation between LUMO and the LUMO + 1 band showed an increase from 1.0 eV for pure C_{60} to ~2.2 eV for K_6C_{60}.

Chen et al.,[230] Fink et al.,[328,329] and Benning et al.[326,327] provided higher resolution views of the electronic states of the K-C_{60} system measured at low temperature. The fullerene films used by Benning et al.[326] were grown at ~450 K on cleaved GaAs(110) and then cooled to 45 K. These conditions produced samples characterized by the LEED pattern of Fig. 7 and the STM image of Fig. 8. (Note that faults of the type shown in Fig. 8 were infrequently observed because the grains were large.) Doping to produce K_3C_{60} yielded results such as shown in Fig. 26, where a low angle (7°) grain boundary of the original C_{60} lattice was imaged and was seen to persist in the $K_3C_{60}(111)$ crystal. The K atoms were not visible because their occupied $3p$ and empty $4s$ states were out of the energy range accessible with STM. This image was obtained with a very low bias, 0.01 V, confirming the metallic character of the sample. Continued doping to saturation required a crystallographic transformation to the bcc phase. This transformation resulted in considerable surface roughening, and large flat areas were rare. While vacancies were seldom observed for an undoped surface, the STM mosaic of Fig. 27 shows that they accounted for 1–2% of the K_6C_{60} surface. Moreover, the vacancies were mobile at the temperature of doping because several appeared as vacancy islands.

The photoemission spectra of Fig. 28 demonstrated that dilute doping induced an initial shift of the C_{60} features as E_F moved from a position 2.2 eV above the HOMO center to a position near the conduction band minimum, as for doping with any alkali metal.[22] Doping to $x = 0.1$ also produced a LUMO-feature that was 1.5 eV wide, a sharp cutoff at E_F, and a shoulder at ~1.6 eV. The LUMO signatures were characteristic of K_3C_{60}, and the shoulder corresponded to the leading portion of HOMO for the K_3C_{60} phase. Benning et al.[326] showed that the shape of the LUMO-feature remained essentially unchanged for $0.1 \le x \le 2.2$, consistent with the presence of only α-C_{60} (which has insignificant emission in this energy window) and K_3C_{60}. The broadening induced by two-phase superposition was particularly evident for $K_{2.2}C_{60}$, where the sample should be ~70% K_3C_{60}. The increase in emission near ~0.5 eV signaled the formation of the K_4C_{60} phase.[327] The $x = 4.2$ spectrum provided the signature of the insulating K_4C_{60} phase.[326] LEED results

GB

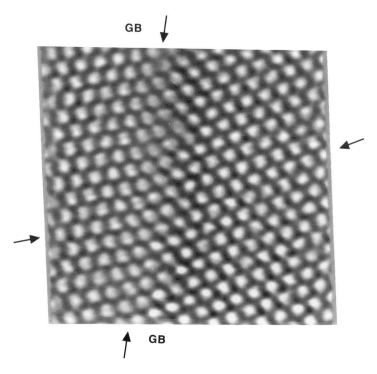

GB

FIG. 26. STM image of a K_3C_{60} surface in the region of a small-angle grain boundary. Viewing along the line defined by the arrows shows the changes in crystallographic orientation. The film was 10 monolayers thick. K ions cannot be detected because their energy states are not accessible to STM. The bias was very small, 0.01 V. From Weaver *et al.*[297]

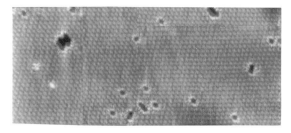

FIG. 27. STM mosaic of a fully doped crystal of K_6C_{60}. The isolated dark objects correspond to single-molecule vacancies that appear after the fcc host lattice transforms to the body centered structure. The formation of vacancy islands indicates that they are mobile at the temperatures used to grow the films. Such vacancies are almost never seen for the fcc structure of pure or doped fcc C_{60}. From Zhao, Poirier, and Weaver (unpublished).

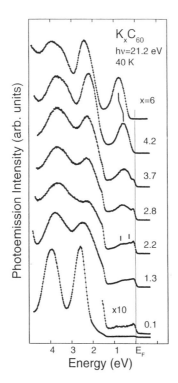

FIG. 28. High resolution photoemission results for K_xC_{60} obtained from samples like those depicted in Figs. 8, 26, and 27. Doping between ~0.1 and ~2.2 introduces emission within 1.5 eV of E_F that is related to filling of the LUMO band of K_3C_{60}. The tick marks highlight features that represent coupling to the vibronic modes (0.3 eV) and to the incoherent part of the spectral function (0.7 eV). The growth of emission at ~0.5 eV reflects the nucleation of the K_4C_{60} phase before x reaches 3. The spectrum for $x = 4.2$ shows the insulating character of the K_4C_{60} phase. Saturation produces a filled-band insulator, K_6C_{60}. From Benning et al.[326]

confirmed the disappearance of the fcc K_3C_{60} pattern at this concentration. Saturation doping to $x = 6$ resulted in complete filling of the LUMO-derived band, a shift of E_F to the LUMO + 1 band edge, and surfaces like that shown in Fig. 27.

Weaver et al.[327] argued that the nucleation of grains of the K_4C_{60} phase occurred for global stoichiometries below $x = 3$ because of growth kinetics. While the Gibbs phase rule of equilibrium thermodynamics precludes three-phase coexistence, kinetic factors were important under vapor phase growth conditions, even at 180°C. Alkali metal diffusion on the surface and into the film is required for fulleride grain growth. One pathway involves diffusion to existing A_3C_{60} grains where incorporation would lead to eventual conversion to a single crystal of A_3C_{60} (the equilibrium pathway, assuming a single crystal host lattice). The other describes the formation of stable seeds of A_4C_{60}, induced by local saturation of the alkali metal concentration on the A_3C_{60} surfaces and enhanced by the relative ease of the fcc→bct transformation at the surface.

The spectral signatures of K_3C_{60} and K_4C_{60} were compared with calculated densities of states by Benning et al.[325] The K_3C_{60} curve of Fig. 29 was obtained by subtracting appropriately scaled contributions for the HOMO and HOMO − 1 features of α-C_{60} from the spectra for $x = 2.2$. The LUMO-feature was much broader than predicted by LDA calculations (1.5 vs 0.3 eV), and the features at 0.3 and 0.7 eV could not be related to band structure effects. Instead, the 0.3 eV feature was

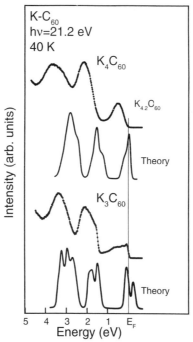

FIG. 29. Comparison of theory and experiment for K_3C_{60} and K_4C_{60}. While the independent-particle calculations do not reproduce the broad LUMO structure of K_3C_{60}, they describe the HOMO and HOMO − 1 bands quite well. For K_4C_{60}, the calculations predict a metal whereas experiment finds an insulator with the LUMO band split into an upper and a lower Hubbard-like band. From Benning et al.[335]

associated with electronic coupling to the vibronic modes. The broad feature at 0.7 eV was discussed in terms of a transfer of weight from the coherent part of the spectral function of E_F to an incoherent part. Analogous electron-correlation-induced broadening was observed in transition metal oxides, where Fujimori et al.[332] related the strength of the on-site Coulomb energy to the relative intensities of a narrow coherent feature at E_F and a broad incoherent feature below E_F.

For K_4C_{60}, the photoemission spectrum was not what would be expected based on the independent-electron picture of filling of the three bands derived from the LUMO levels.[303,326] In particular, K_4C_{60} was an insulator, not a metal. Benning et al. argued that band splitting needed to produce an insulating state again pointed to the importance of electron correlation. Indeed, transport measurements[320] for high quality K-C_{60} films showed that the resistivity was ~25 mΩ-cm for K_4C_{60}, about an order of magnitude higher than the typical limit for metals, and the resistivity increased as temperature decreased (activated transport).[320]

Knupfer et al.[331] suggested an alternate many-body effect that could account for the 0.7 eV feature of K_3C_{60}. They considered coupling of the photoexcitation process to an intrinsic charge carrier plasmon. Previously, Sohmen et al.[225] had demonstrated that the plasmon energy for the electrons of the LUMO band for K_3C_{60} was about 0.5 eV, based on electron energy loss measurements, and additional evidence for the plasmon was obtained in the optical reflectivity measurements of Iwasa et al.[333] Regardless of how the many-body effects are treated, it should be clear that models developed to describe these fullerides as they pass through the metal–insulator transition provide new insight into the physics of such systems.

Spectroscopic studies of Rb_xC_{60} for $0.4 \le x \le 6$ showed commonalities[320,329,334,335] with K-C_{60} but also differences because the Rb_1C_{60} phase was present below room temperature. This required that doping beyond the solubility limit produced Rb_1C_{60}. The presence of emission from Rb_1C_{60} near E_F (discussed in the following) assured a different LUMO appearance than for K-C_{60} for low stoichiometries. For Cs_xC_{60}, Benning et al.[335] showed that doping to $Cs_{0.8}C_{60}$ also revealed emission

[332]A. Fujimori, J. Phys. Chem. Solids **53**, 1595 (1992); A. Fujimori, I. Hase, Y. Tokura, M. Abbate, F. M. F. de Groot, J. C. Fuggle, H. Eisaki, and S. Uchida, Physica B **186–188**, 981 (1993); A. Fujimori, I. Hase, H. Namatame, Y. Fujishima, Y. Tokura, H. Eisaki, S. Uchida, K. Takegahara, and F. M. F. de Groot, Phys. Rev. Lett. **69**, 1796 (1992).

[333]Y. Iwasa, K. Tanaka, T. Yasuda, and T. Koda, Phys. Rev. Lett. **69**, 2284 (1992).

[334]J. Fink, E. Sohmen, and W. Krätschmer, Proc. Int. Workshop on Electronic Properties and Mechanisms in High T_c Superconductors, Tsukuba, July 1991, Physica C (in press).

[335]P. J. Benning, F. Stepniak, and J. H. Weaver, Phys. Rev. B **48**, 9086 (1993).

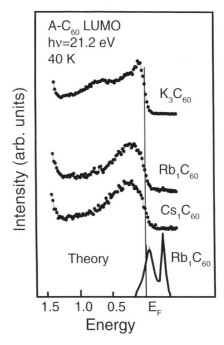

FIG. 30. Comparison of calculated density of states for Rb_1C_{60} (Satpathy et al.[336]) with experimental results for Rb_1C_{60} and Cs_1C_{60}. Experiment shows a broad band centered well below E_F, suggestive of a correlation-induced pseudogap for the $x = 1$ phase. The emission at E_F shows states near the Fermi level, but resistivity measurements indicate activated transport, i.e., Anderson localization and a mobility gap related to merohedral disorder and possible alkali vacancies. The upper curve shows the emission from LUMO for K_3C_{60}, as in Fig. 29, drawn to emphasize its different shape compared with the $x = 1$ phases. The integrated emission intensity under the LUMO of the $x = 3$ phase is about three times that of the $x = 1$ phase. From Benning et al.[335]

at E_F. In contrast to $K-C_{60}$, they reported that Cs incorporation after $x = 1$ results in a reduction of emission at E_F, consistent with the absence of a metallic Cs_3C_{60} phase.

Benning et al.[335] compared the LUMO-derived emission features at 40 K for $K_{2.2}C_{60}$, $Rb_{0.7}C_{60}$, and $Cs_{0.8}C_{60}$, corresponding to stoichiometries at which the LUMO emission was representative of K_3C_{60}, Rb_1C_{60}, and Cs_1C_{60}. The spectra reproduced in Fig. 30 were normalized to the maximum intensity of each curve to emphasize lineshape changes; the integrated LUMO emission for K_3C_{60} was ~3 times that for Rb_1C_{60} and Cs_1C_{60}. The K_3C_{60} spectrum showed the broadening discussed previously with features at 0.3 and 0.7 eV. The A_1C_{60} spectra showed a nearly Gaussian shaped peak ~0.5 eV wide with an asymmetry to higher binding energy. Comparison with the calculated independent particle density of states of Satpathy et $al.$[336] for Rb_1C_{60} showed little agreement, because the calculations predicted a total LUMO bandwidth of ~0.6 eV and an occupied width of ~0.15 eV. The discrepancy was again interpreted in terms of a transfer of spectral weight from E_F into the incoherent part of the spectral feature. Benning et $al.$ noted that the A_1C_{60} phases exhibited emission consistent with a more standard Mott–Hubbard interpretation than did A_3C_{60}, suggesting more pronounced correlation effects. Within the framework of Mott–Hubbard theory, a pseudogap should open when U is comparable with the width of the conduction band,[212] reducing the density of states at E_F and producing a broadened density of states below E_F, as observed for Rb_1C_{60} and Cs_1C_{60}.

The low carrier density of the Rb_1C_{60} and Cs_1C_{60} phases produced resistivities above the classic Mott limit for metals, and measurements of the temperature coefficient of resistivity indicated nonmetallic character for x near 1.[320] From Fig. 30, however, there was emission at E_F, and such emission would normally be associated with metallic character. This suggested Anderson localization of what would ordinarily be the transport states because of disorder that interrupted the periodicity necessary for extended states and created a mobility gap. Benning et $al.$[335] proposed that merohedral disorder and the possibility of vacancies in the octahedral sites of A_1C_{60} could account for the apparent localization, although they noted that little was known about the detailed filling of the octahedral sites for $x < 1$.

Finally, spectroscopic studies of Na_xC_{60} showed that there was no emission at E_F for any doping level and that saturation occurred at $x = 6$ for Na vapor deposition onto C_{60} in vacuo.[335,337] The valence band spectra of Na_6C_{60} were very similar to those of the bcc A_6C_{60} compounds, the differences in crystal structure and site occupancy notwithstanding. This pointed to the overriding importance of the host molecular properties. To date, attempts to increase the Na content beyond $x = 6$ have not been successful under conditions of vapor phase deposition. Benning et $al.$[335] showed that higher alkali doping of the fullerenes could be accomplished, however, by condensing isolated C_{60} molecules onto an alkali metal surface at low temperature. They reported that this resulted in the transfer of eight electrons to the fullerenes based on the observation that all of the LUMO levels and some of the LUMO $-$ 1

[336]S. Satpathy, V. P. Antropov, O. K. Andersen, O. Jepsen, O. Gunnarsson, and A. I. Liechtenstein, $Phys.$ $Rev.$ B **46,** 1773 (1992).
[337]C. Gu, F. Stepniak, D. M. Poirier, M. B. Jost, P. J. Benning, Y. Chen, T. R. Ohno, J. L. Martins, J. H. Weaver, J. Fure, and R. E. Smalley, $Phys.$ $Rev.$ B **45,** 6348 (1992).

levels were occupied. The same effect was observed for Na, K, and Rb. The implication was that the maximum charge states of the fullerenes might be C_{60}^{8-}. Thus, there may not be full charge transfer from the cluster of Na ions in the octahedral hole to the fullerenes in the $Na_{10}C_{60}$ structure.[306]

16. FULLERIDE SUPERCONDUCTIVITY

The observation that the A_3C_{60} fullerides exhibit superconductivity at temperatures as high as 33 K sparked a great deal of interest. Indeed, the transition temperature of $Rb_1Cs_2C_{60}$ is exceeded only by the cuprates. Unlike the cuprates, however, these superconductors are three dimensional in character. While no one considers the fullerides to be serious contenders for superconducting technologies (they are spontaneously combustible; they are expensive), their study is teaching us a great deal about narrowband materials, disordered systems, and relevant energy scales.

Superconductivity in the fullerides is discussed in detail by Lieber in a chapter in this volume devoted specifically to that property and by Pickett in a chapter on electrons and phonons. Here, we offer only a few points that bear repetition.

Fleming et al.[338] first showed that the superconducting transition temperature scaled with the lattice constant for a variety of A_3C_{60} compounds, as in Fig. 31. This suggested that there was a direct relationship between T_c and the density of states at the Fermi level. In particular, it was reasoned that an increase in the lattice constant would reduce the overlap of the LUMO-derived levels, thereby narrowing the LUMO band and increasing $N(E_F)$. Implicit was the assumption that electron coupling with intramolecular phonons was responsible for superconductivity. This coupling would be independent of intermolecular distances and the particular alkali atom mixed with C_{60}. The fact that ternaries such as $Rb_2Cs_1C_{60}$ or $Rb_1Cs_2C_{60}$ could be formed and that they followed the simple depiction of Fig. 31 supported this picture. The observed correlation thus indicated that the alkali atoms dictate the lattice constant but they play little role in superconductivity beyond donating charge to states that are derived from the fullerenes.

[338]R. M. Fleming, A. P. Ramirez, M. J. Rosseinsky, D. W. Murphy, R. C. Haddon, S. M. Zahurak, and A. V. Makhihja, *Nature* **352,** 787 (1991).

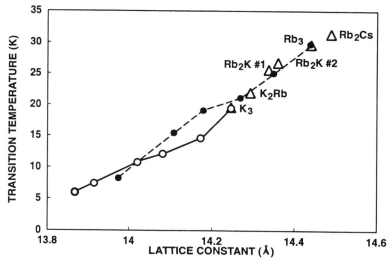

FIG. 31. Correlation of the superconducting transition temperature to the lattice constant for representative alkali metal fullerides. The closed circles connected by the dotted line (open circles connected by the solid line) show that T_c is reduced as the lattice is compressed for Rb_3C_{60} (K_3C_{60}) by applying pressure. From Hebard,[5] as adapted from Fleming et al.[338] and Zhou et al.[340]

These correlations of T_c with lattice constants were supported by pressure-dependent studies. Several authors[339–341] reported that a reduction in the lattice constant induced by pressure had a significant effect on T_c. For the Rb_3C_{60} system, it was found that the lattice could be compressed to the value appropriate for K_3C_{60} and that T_c shifted accordingly. Very recently, Zhou et al.[303] demonstrated that the value of T_c observed for $Na_2Cs_1C_{60}$ could be increased from 10.5 to 29.6 K by the uptake of NH_3 to form $(NH_3)_4Na_2Cs_1C_{60}$. They showed that the NH_3 molecules collected around half of the Na ions to form large $(NH_3)_4Na$ cations. These large ions then exchanged with Cs ions, so that the $(NH_3)_4Na$ ions occupied the octahedral sites and the tetrahedral sites were half-filled with Cs ions and half-filled with bare Na ions. The resulting lattice constant was 14.473 Å, smaller than for $Rb_1Cs_2C_{60}$, and

[339]J. E. Schirber, D. L. Overmyer, H. H. Wang, J. M. Williams, K. D. Carlson, A. M. Kini, U. Welp, and W.-K. Kwok, Physica C 178, 137 (1991).
[340]O. Zhou, G. B. M. Vaughan, Q. Zhu, J. E. Fischer, P. A. Heiney, N. Coustel, J. P. McCauley, Jr., and A. B. Smith III, Science 255, 833 (1991).
[341]G. Sparn, J. D. Thompson, S.-M. Huang, R. B. Kaner, F. Diederich, R. L. Whetten, G. Grüner, and K. Holczer, Science 252, 1829 (1991).

T_c was lower. Note that lattice expansion beyond some critical value must ultimately lead to a saturation or downturn in T_c, as pointed out by Zhou et al.[303] Cs_3C_{60} should have a larger lattice constant, but it has not been synthesized reproducibly.

When superconductivity was discovered in C_{60}-based systems, it was speculated that the fullerides based on the larger or smaller fullerenes might have higher T_c's. Unfortunately, all of the doped higher fullerides (C_{70},[342-344] C_{76},[34] C_{82},[37] C_{84}[35]) appear to be insulating in character, with the possible exception of a K_xC_{70} phase yet to be fully characterized.[345] Presumably, the delicate balance that allows some of the C_{60}-based fullerides to be metallic is not achieved in their larger, less symmetric cousins.

Two theories were proposed to explain superconductivity in the fullerides. One evoked a purely electronic mechanism,[346,347] while the other emphasized electron–phonon coupling.[207-209,348-350] Hebard[6] provided an excellent review of the experimental parameters needed to distinguish between these two models. He and others pointed out that the phonon modes can be distinguished between the high energy intramolecular modes and the low energy modes that include molecular libration and the optic phonon mode sketched in Fig. 19. Electron coupling to the latter would need to be quite large to account for the observed values of T_c, while coupling to the on-ball modes could be much smaller. The different energy scales allowed the two types of modes to be treated independently. The dependence of T_c on a_o evident in Fig. 31 indicated that the optic modes were not strongly coupled, and the absence of any sudden change in the librational modes near T_c (Ref. 351) suggested that those modes were also not significant.

Schluter et al.[207] and Varma et al.[209] considered electron–phonon

[342]T. Takahashi in "Electronic Properties and Mechanisms of High T_c Superconductors," eds. T. Oguchi, K. Kadowaki, and T. Sasaki (Elsevier Science Publishers, 1992), p. 21.

[343]T. Takahashi, T. Morikawa, S. Hasegawa, K. Kamiya, H. Fujimoto, S. Hino, K. Seki, H. Katayama-Yoshida, H. Inokuchi, K. Kikuchi, S. Suzuki, K. Ikemoto, and Y. Achiba, Physica C 190, 205 (1992).

[344]E. Sohmen and J. Fink, Phys. Rev. B 47, 14532 (1993).

[345]K. Imaeda, K. Yakushi, H. Inokuchi, K. Kikuchi, I. Ikemoto, S. Suzuki, and Y. Achiba, Solid State Commun. 84, 1019 (1992).

[346]S. Chakravarty, M. P. Gelfand, and S. Kivelson, Science 254, 970 (1991).

[347]G. Baskaran and E. Tosatti, Curr. Sci. 61, 33 (1991).

[348]V. P. Antropov, O. Gunnarsson, and A. I. Leichtenstein, Phys. Rev. B 48, 7651 (1993).

[349]F. C. Zhang, M. Ogata, and T. M. Rice, Phys. Rev. Lett. 67, 3452 (1991).

[350]J. A. Jishi and M. S. Dresselhaus, Phys. Rev. B 45, 2597 (1992).

[351]C. Christides, D. A. Neumann, K. Prassides, J. R. D. Copley, J. J. Rush, M. R. Rosseinsky, D. W. Murphy, and R. C. Haddon, Phys. Rev. B 46, 12088 (1992).

coupling strengths for the different phonons of A_3C_{60}. They found that the coupling was dominated by the on-ball modes of H_g and A_g symmetry. Coincidently, these were the same modes that were Raman active, as discussed in Section 11, and Raman scattering proved to be an important tool in examining electron–phonon coupling in the fullerides. It was found by Duclos and others[352–354] that coupling was mainly to H_g modes that were considerably broadened for the A_3C_{60} compounds. Similar conclusions were reached via inelastic neutron scattering.[355] Figure 19 depicts representative modes, and we direct the interested reader to the more extensive presentation of A_u, A_g, T_{1u}, and H_g modes given in Fig. 4 of the chapter by Pickett. Schlüter et al.[207] and Varma et al.[209] concluded that the coupling deduced for the molecules is modified only slightly in the solid state. Coupling to such on-ball modes is another of the novel characteristics of C_{60}.

Martins and Troullier[208] and Schluter et al.[207] provided an interesting comparison of the electron–phonon coupling in the fullerides and in intercalated graphite. They noted that the distribution of modes for graphite was comparable with that for the on-ball modes of C_{60}, reflecting strong C–C bonds, but T_c differed by more than a factor of 10. These authors argued that the enhanced T_c for the fullerides reflected the stronger coupling of the electrons to the phonons, an enhancement made possible by the curvature of the fullerenes and a mixing of s and p character in the electronic states of the LUMO-derived band. While the calculation of the coupling constant λ involved complex sums over all vibrational modes and the electronic wavefunctions, the inclusion of new coupling channels for A_3C_{60} was evident from symmetry considerations. In particular, it was argued that matrix elements that were zero because of the symmetry of the electronic states at E_F for graphite were no longer zero for A_3C_{60} because of the $s-p$ admixture. These authors concluded that the coupling strength need not be exceptional to account for the observed high transition temperatures of the fullerides. A similar conclusion was reached by the Stuttgart group.[348]

[352]R. Duclos, R. C. Haddon, S. Glarum, A. F. Hebard, and K. B. Lyons, *Science* **258**, 1625 (1991).
[353]P. Zhou, K.-A. Wang, A. M. Rao, P. C. Eklund, G. Dresselhaus, and M. S. Dresselhaus, *Phys. Rev. B* **45**, 10838 (1992).
[354]M. G. Mitch, S. J. Chase, and J. S. Lannin, *Phys. Rev. Lett.* **68**, 883 (1992); M. G. Mitch, S. J. Chase, and J. S. Lannin, *Phys. Rev. B* **46**, 3696 (1992).
[355]K. Prassides, J. Tomkinson, C. Christides, M. J. Rosseinsky, D. W. Murphy, and R. C. Haddon, *Nature* **354**, 462 (1991); K. Prassides, C. Christides, J. Tomkinson, M. J. Rosseinsky, D. W. Murphy, and R. C. Haddon, *Mat. Res. Soc. Symp. Proc.* Vol. 270, 1992.

After analysis of the various experimental results, Hebard[6] concluded that none of the data were inconsistent with conventional electron–phonon mechanisms of superconductivity. He did, however, point out several experimental inconsistencies and appealed for further refinement. From the theoretical point of view, we note that one of the challenges implicit in studies of superconductivity in the A_3C_{60} phases is that the energy scale of the phonons is the same as the width of the LUMO band, and this pushes the limits of BCS theory. Moreover, the electron energy bands are very narrow and electron correlation effects are large, as is evident from the fact that only the A_3C_{60} phases are metallic.

17. ALKALINE-EARTH FULLERIDES

Early chemical considerations suggested that the alkaline earth metals might mix with fullerenes to form fullerides. Unfortunately, higher temperatures were needed to achieve growth from the vapor phase, diffusion of the alkaline earths into the fullerene crystals was less facile, and, in general, the preparation of these fullerides proved to be more difficult than for the alkali metals.

The earliest report of alkaline earth mixing was that of Chen et al.[356] They exposed fullerene films in vacuo to a flux of Mg, Sr, and Ba and used photoemission to determine whether these materials exhibited metallic character. For Mg, they found limited modification of the electronic states of the fullerene, suggesting minimal interactions. For Sr and Ba, however, they reported metallic character and speculated about the possibility of superconductivity. They also suggested that the LUMO + 1 levels were likely to mix with the alkaline earth s states and that hybrid bonds were possible. Ohno et al.[357] considered mixing with divalent Yb, a close cousin of the alkaline earths, and reported the incorporation of about two Yb atoms per fullerene. They found no evidence for metallic character.

Kortan et al.[358] examined the $Ca-C_{60}$ fullerides and found superconductivity with $T_c = 8.4$ K. The superconducting phase was reported to have a

[356]Y. Chen, F. Stepniak, J. H. Weaver, L. P. F. Chibante, and R. E. Smalley, *Phys. Rev. B* **45**, 8845 (1992).

[357]T. R. Ohno, G. H. Kroll, J. H. Weaver, L. P. F. Chibante, and R. E. Smalley, *Phys. Rev. B* **46**, 10437 (1992).

[358]A. R. Kortan, N. Kopylov, S. Glarum, E. M. Gyorgy, A. P. Ramirez, R. M. Fleming, F. A. Thiel, and R. C. Haddon, *Nature* **355**, 529 (1992).

stoichiometry of Ca_5C_{60} with multiple Ca occupation of the fcc octahedral site. That raised questions pertaining to the filling of LUMO + 1 derived electronic states since Ca tends to form divalent compounds and the LUMO of C_{60} can accommodate only six electrons. Chen *et al.*[359] undertook photoemission and inverse photemission experiments, and Li *et al.*[360] used STM to examine the structure of the Ca fullerides. The STM images showed what appeared to be solid solution formation at intermediate stoichiometries, consistent with the x-ray analysis of Kortan *et al.*[358] They also suggested an ordering of the Ca ions at $x = 5$ such that half of the tetrahedral sites were filled (as in the zincblende structure) and the remaining Ca ions occupied the octahedral site, arranged in a tetrahedral fashion that maximized the Ca–Ca distance from the occupied tetrahedral sites.

LDA calculations by Martins[359] for Ca_3C_{60} in the A_3C_{60} structure (Fig. 21) showed that the Fermi level fell in a small gap above a band derived from the LUMO levels, consistent wih the photoemission and inverse photoemissions results shown in Fig. 32 and the fact that the temperature dependence of the resistivity showed activated transport at this stoichiometry.[361] Saito and Oshiyama[362] also suggested the existence of a gap based on calculations for Ca_3C_{60}, and they showed the effects of placing the ions off center in the octahedral sites. Significantly, LDA calculations[359] for the Ca_5C_{60} structure proposed by Li *et al.*[360] revealed considerable hybridization between the LUMO + 1 level and the Ca s states, as reflected in Fig. 32 by the fact that the distribution of electronic states differed significantly from those of C_{60}. What was clear, then, was that the states responsible for the metallic and superconducting character were no longer simply derived from the LUMO levels of the fullerene, a point supported by subsequent photoemission studies by Wertheim *et al.*[363] Romberg *et al.*[364] reported electron energy loss studies of $Ca-C_{60}$, finding complete occupation of the LUMO level for x near 3 and progressive filling of the LUMO + 1 level thereafter, in good agreement with the conclusions of photoemission and inverse photoemission.

Haddon *et al.*[361] reported temperature-dependent resistivity results for

[359]Y. Chen, D. M. Poirier, M. B. Jost, C. Gu, T. R. Ohno, J. L. Martins, J. H. Weaver, L. P. F. Chibante, and R. E. Smalley, *Phys. Rev. B* **46,** 7961 (1992).

[360]Y. Z. Li, J. C. Patrin, M. Chander, J. H. Weaver, L. P. F. Chibante, and R. E. Smalley, *Phys. Rev. B* **46,** 12914 (1992).

[361]R. C. Haddon, G. P. Kochanski, A. F. Hebard, A. T. Fiory, and R. C. Morris, *Science* **258,** 1636 (1992).

[362]S. Saito and A. Oshiyama, *Solid State Commun.* **83,** 107 (1992).

[363]G. K. Wertheim, D. N. E. Buchanan, and J. E. Rowe, *Science* **258,** 1638 (1992).

[364]H. Romberg, M. Roth, and J. Fink, *Phys. Rev. B* (submitted).

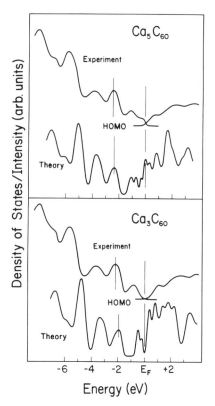

FIG. 32. Comparison of theory and experiment showing the nonmetallic character of Ca_3C_{60} and the metallic character of Ca_5C_{60}. Unlike the alkali metal fullerides, the alkaline-earth fullerides have states that can hybridize with the LUMO + 1 band. From Chen et al.[359]

Ca-C_{60} and Sr-C_{60} that indicated activated conductance over a wide range of stoichiometries with resistivity minima at $x = 2$ and 5 ($\rho_{min} \sim 1\,\Omega$-cm for $x = 2$ and $10^{-2}\,\Omega$-cm for $x = 5$) that suggested phase formation. They associated the increase in $\rho(x)$ near $x = 3$ with the filling of the LUMO band. Using RBS, they estimated their uncertainties in x to be ± 0.3. Their results supported the picture of t_{1g} level occupancy. At both minima in the resistivity, they reported that transport was activated, though with an activation energy of only about 0.02 eV at $x = 5$, and the granularity of the films was thought to have played a role.

For bcc Sr_6C_{60} and Ba_6C_{60}, Saito and Oshiyama[365] reported hybridization and semimetallic character with d–π hybridization. Experimentally, the metallic character of Sr-doped C_{60} was demonstrated by Chen et al.[365] and superconductivity with $T_c = 4$ K was observed by Kortan.[366]

Kortan et al.[309,367] demonstrated two surprises for the Ba-C_{60} system. First, Ba_3C_{60} formed in the $A - 15$ stucture, the only fulleride known to stabilize this structure. Second, Ba_6C_{60} formed in the A_6C_{60} structure of Fig. 21. For the latter, T_c was ~7 K, and Erwin[368] predicted that the Ba d electrons were involved in hybridization with the LUMO + 1 levels. Haddon et al.[369] reported that Ba intercalation into C_{60} films was more difficult than for the other alkaline-earth metals. They found resistivity minima for x between 1 and 2 and at $x = 5$. The activation energy for transport was a minimum at $x = 1$. They reported the lowest ρ_{min} measured for doped films and, further, that the temperature coefficient of resistance was positive for Ba_5C_{60}.

As of this writing, it is clear that the alkaline-earth fullerides offer substantial challenges to theory and experiment, and much remains to be learned. It remains to be seen, for example, exactly what the implications are for the theory of superconductivity since different orbitals are involved. Group theory would indicate the same vibronic coupling for states of t_{1u} and t_{1g} symmetries, i.e., LUMO and LUMO + 1. It may be a coincidence that the value of T_c for Ca_5C_{60} falls on the curve established for T_c vs. a_o for the alkali metal fullerides, but this remains to be explored in detail.

V. Endofullerenes

The widespread fascination with endofullenes reflects, in part, the prospects of producing important new guest-host nanostructured materials. Since all of the elements of the periodic table can fit within the cage of C_{60} or the higher fullerenes, the number of potentially useful materials is very large. The limitation to date has been the production of sufficient quantities that isolation and separation could be undertaken.

[365]S. Saito and A. Oshiyama, Phys. Rev. Lett. **71**, 121 (1993).

[366]A. R. Kortan, private communication.

[367]A. R. Kortan, N. Kopylov, S. Glarum, E. M. Gyorgy, A. P. Ramirez, R. M. Fleming, O. Zhou, F. A. Thiel, P. L. Tervor, and R. C. Haddon, Nature **360**, 566 (1992).

[368]S. C. Erwin and M. R. Pedersen, Phys. Rev. B **47**, 14657 (1993).

[369]R. C. Haddon, G. P. Kochanski, A. F. Hebard, A. T. Fiory, R. C. Morris, and A. S. Perel, Chem. Phys. Lett. **203**, 433 (1993).

While important advances have been made,[10,17,21,72,78–81,370–390] relatively little is known about their solid state properties and their stabilities. Nonetheless, the promise that these materials hold, both scientifically and technologically, drives activity in many laboratories worldwide.

In 1985, Heath et al.[10] reasoned that the central cavity of the C_{60} molecule should be a strong binding site for a wide range of atoms, and they undertook studies of endohedral fullerenes in the gas phase. By using their laser ablation technique with a target of La-impregnated low-density graphite, they were able to produce La@C_{60} as well as the

[370]T. Weiske, D. K. Böhme, J. Housak, W. Krätschmer, and H. Schwarz, *Angew. Chem. Ind. Ed. Engl.* **30,** 884 (1991).

[371]K. A. Caldwell, D. E. Giblin, C. C. Hsu, D. Cox, and M. L. Gross, *J. Am. Chem. Soc.* **113,** 8519 (1991).

[372]E. E. B. Campbell, R. Ehlich, A. Hielscher, J. M. A. Frazao, and I. V. Hertel, *Z. Phys. D* **23,** 1 (1992).

[373]Y. Chai, T. Guo, C. Jin, R. E. Haufler, L. P. F. Chibante, J. Fure, L. Wang, M. J. Alford, and R. E. Smalley, *J. Phys. Chem.* **95,** 7564 (1991).

[374]R. D. Johnson, M. S. de Vries, J. R. Salem, D. S. Bethune, and C. S. Yannoni, *Nature* **355,** 239 (1992).

[375]M. Hoinkis, C. S. Yannoni, D. S. Bethune, J. R. Salem, R. D. Johnson, M. S. Crowder, and M. S. de Vries, *Chem. Phys. Lett.* **198,** 461 (1992).

[376]S. Suzuki, S. Kawata, H. Shiromaru, K. Yamauchi, K. Kikuchi, T. Kato, and Y. Achiba, *J. Phys. Chem.* **96,** 7159 (1992).

[377]J. H. Weaver, Y. Chai, G. H. Kroll, C. Jin, T. R. Ohno, R. E. Haufler, T. Guo, J. M. Alford, J. Conceicao, L. P. F. Chibante, A. Jain, G. Palmer, and R. E. Smalley, *Chem. Phys. Lett.* **190,** 460 (1992).

[378]A. Rosén and B. Wästberg, *J. Am. Chem. Soc.* **110,** 8701 (1988).

[379]B. Wästberg and A. Rosén, *Physica Scripta* **44,** 276 (1991).

[380]A. H. H. Chang, W. C. Ermler, and R. M. Pitzer, *J. Chem. Phys.* **94,** 5004 (1991).

[381]K. Laasonen, W. Andreoni, and M. Parrinello, *Science* **258,** 1916 (1992).

[382]M. M. Alvarez, E. G. Gillan, K. Holczer, R. B. Kaner, K. S. Min, and R. L. Whetten, *J. Phys. Chem.* **95,** 10561 (1091).

[383]H. Shinohara, H. Sato, Y. Saito, M. Ohkochi, and Y. Ando, *J. Phys. Chem.* **96,** 3571 (1992).

[384]H. Shinohara, H. Sato, M. Ohkochi, Y. Ando, T. Kodama, T. Shida, T. Kato, and Y. Saito, *Nature* **357,** 52 (1992).

[385]C. S. Yannoni, M. Hoinkis, M. S. de Vries, D. S. Bethune, J. R. Salem, M. S. Crowder, and R. D. Johnson, *Science* **256,** 1191 (1992).

[386]D. S. Bethune, C. S. Yannoni, M. Hoinkis, M. de Vries, J. R. Salem, M. S. Crowder, and R. D. Johnson, *Z. Phys. D* (in press).

[387]K. Kikuchi, S. Suzuki, Y. Nakao, N. Nakahara, T. Wakabayashi, H. Shiromaru, I. Ikemoto, and Y. Achiba, *Chem. Phys. Lett.* (submitted).

[388]X.-D. Wang, T. Hazhizume, Q. Xue, H. Shinohara, Y. Saito, Y. Nishina, and T. Sakurai, *Jpn. J. Appl. Phys.* **32,** L147 (1993).

[389]T. Pradeep, G. U. Kulkarni, K. R. Kannan, T. N. Guru Row, and C. N. Rao, *J. Am. Chem. Soc.* **114,** 2272 (1992).

[390]L. Soderholm, P. Wurz, K. R. Lykke, and D. H. Parker, *J. Phys. Chem.* **96,** 7153 (1992).

less symmetrical La endofullerenes, La@C_{44} through La@C_{76}. Those results suggested a "superatom" concept in which the caged atom would donate (or accept) charge from the cage, providing a positive (or negative) core. Solids derived from such species offer intriguing possibilities.

Weiss et al.[17] observed that fragments of C_2 could be removed from the carbon cage by laser photodissociation and discussed shrink-wrapping of the cage around the inner metallic core. Wieske et al.[370] demonstrated that He could be trapped in preformed C_{60} and C_{70} and Wan et al.[80] showed Ne incorporation. Experiments by Ross and Callahan,[79] Caldwell et al.,[371] and Campbell et al.[372] have confirmed this ability to insert rare gas atoms into the endohedral cavity during high energy collision with C_{60}^+. Saunders et al.[274] found that He and Ne could be induced to enter the C_{60} shell when the fullerenes were heated, and they showed that these atoms could also be released. Wan et al.[80] investigated Li^+ and Na^+ collisions with C_{60} as a function of collision energy between 0 and 150 eV. They reported energy thresholds for insertion and fragmentation. A rule of thumb developed that said that only those elements with ionization potentials less than 7 eV could form endofullerenes.

The first production of endofullerenes in quantities sufficient for separation was reported in 1991 by Chai et al.[373] They described a modified method for endofullerene production that combined ablation of their lanthana–graphite mixture with a high temperature furnace that enclosed the ablation region. They produced La@C_{82} and demonstrated that it was air-stable and extractable. Other endofullerenes produced by this process proved unstable or nonextractable. Independently, Johnson et al.[374] produced La@C_{82} and suggested that the La atom in the cage was trivalent with the $6s$ and $5d$ electrons transferred to the fullerene cage. They based their conclusion on electron paramagnetic resonance, EPR, studies that showed the presence of an unpaired electron with a small hyperfine splitting constant caused by the La nuclear spin. Hoinkis et al.[375] later reported the presence of two species of La@C_{82}. The work by Suzuki et al.[376] provided additional evidence of La encapsulation based on the ^{13}C hyperfine structures in electron spin resonance, ESR. X-ray photoemission studies by Weaver et al.[377] of a sample containing mixed La endofullerenes and other carbon species showed La $3d$ core level features that were reminiscent of those of La in the La trihalides. It was concluded that the La ion was trivalent with a degree of charge transfer to the carbon cage intermediate between that observed for $LaBr_3$ and LaI_3, both highly ionic materials.

Early theoretical investigations of the endofullerenes by Rosén and Wästberg[378] predicted that a caged La atom in C_{60} would donate charge to the cage, resulting in a structure where the HOMO would have carbon character and the LUMO would have La $5d$–derived character as well. Those authors[379] subsequently calculated the electronic structure of the alkali metal endofullerenes of C_{60}, including $K_2@C_{60}$. They predicted the ionization energies and used them to determine the locations of the alkali metal atoms. Chang *et al.*[380] considered fullerenes containing O, F, K, Ca, Mn, Cs, Ba, La, Eu, and U. They concluded that the cage accepted was one or two electrons from the electropositive element in a formal sense but that the actual charge was generally less, depending on the extent of the *s*-orbital radius.

Laasonen *et al.*[381] recently examined La@C_{82}. Figure 33 shows constant electron density contours for the HOMO state for two different positions of the La atom within the cage. To obtain these results, the authors assumed a molecular symmetry of C_{3v} for C_{82}, while pointing out that the equilibrium structure of La@C_{82} was not known and the C_{3v} isomer of C_{82} was not the most abundant. In doing so, they emphasized one of the challenges in endofullerene theory, because experiments have not yet provided sufficiently refined information concerning the carbon cage bond lengths or the location(s) of the endohedral species.

In calculations, however, the energies for the different sites can be investigated to offer insight into bonding. Part (a) of Fig. 33 reflects the charge distribution for a La atom constrained to the center of the cage, retaining C_{3v} symmetry. In this configuration, two electrons moved to the cage and the third occupied a state localized on the La atom with approximately $d_{3z^2-r^2}$ character. This structure did not agree with the experimental data of Johnson *et al.*[374] or Weaver *et al.*[377] for La@C_{82}. The authors then displaced the La atom to the position indicated in (b). In this location, it was within ~2.86 Å of 10 carbon atoms and was positioned only ~2.5 Å from the cage wall. As a result, the carbon cage relaxed, and there was an energy gain of ~3.5 eV relative to the central position. Significantly, the HOMO was not localized on the La atom, it had no *d* character, and it was diffused on the cage region close to the La atom. Hence, the La atom adopted a formal charge state of $3+$, in agreement with experiment, and the configuration was a doublet with the unpaired electron in a fairly delocalized orbital. As a caution, Laasonen *et al.*[381] emphasized that predictions of special stabilities for these materials are not trivial.

The conclusion that the caged atom would gain energy by moving away from the center suggested that more than one atom could be encapsulated. Indeed, Alvarez *et al.*[382] reported that a soluble dimetallofullerene, $La_2@C_{80}$, could be produced, joining the family of endofullerenes dominated by C_{82}, and they found a series of trimetallofullerenes at

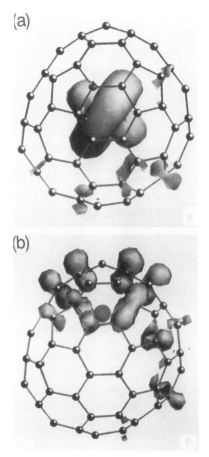

FIG. 33. Constant electron density surfaces corresponding to the HOMO state for La@C_{82}. (a) shows the calculated density with the La atom at the center. Movement off center produces the distribution evident in (b). The La atom is the round object immediately above the center hexagon in (b). Off-center bonding results in a considerable energy gain and delocalization of the La charge on nearby carbon atoms. From Laasonen *et al.*[381]

higher mass. Sc and Y endofullerenes based on C_{82} were produced with one, two, or three caged atoms.[377,383–386] Shinohara *et al.*[384] succeeded in isolating Sc$_3$@C_{82}, and Yannoni *et al.*[385] used solution and solid state electron paramagnetic resonance, EPR, to deduce that the Sc atoms formed an equilateral triangle within the cage wall. For Yz@C_{82} and for Sc$_2$@C_{82}, Sc$_2$@C_{84}, and Sc$_2$@C_{86}, Bethune *et al.*[386] reported no EPR signal and suggested diamagnetic character. Sc solution NMR spectra for those endofullerenes confirmed their proposal.

The smallest endofullerene reported to date[72] is $U@C_{28}$. For this structure, ab initio quantum chemical calculations predicted a tetrahedral cage structure with four sets of triply fused pentagons that would have an open electronic shell. The authors speculated that this effective tetravalence would favor formation of endofullerenes with tetravalent species. Guo *et al.* reported endofullerenes of $U@C_{70}$, $U@C_{60}$, and $U@C_{50}$ in the gas phase. When laser irradiated, these species gave rise to $U@C_{44}$, $U@C_{38}$, and $U@C_{28}$, where the cage of the latter is smaller than any previously observed metallofullerene and below the limit found for empty fullerenes as well, C_{32}.

Guo *et al.*[72] subsequently produced U endofullerenes using the arc discharge method and demonstrated that $U@C_{28}$ was presented in the sublimed film and was able to survive desorption into the gas phase. XPS measurements of the films showed that the atoms were in caged states, immune to oxidation, and had a more covalent bonding than found in UO_2. The photoemission results were consistent with a formal 4^+ valence state. Guo *et al.*[72] also reported $Ti@C_{28}$, $Zr@C_{28}$, and $Hf@C_{28}$ formation and stated that the quantity present decreased for the series U-Hf-Zr-Ti, probably because of the detailed distribution of the valence orbitals.

Achiba and coworkers[387] and Shinohara and coworkers[388] were able to isolate some of the endofullerenes in pure form. The former group reported that $La@C_{82}$ was stable and that the infrared absorption spectrum for it suggested the same cage structure as C_{82} (C_2 symmetry). The presence of near-infrared absorption implied the formation of an open-shell electronic state. The latter group separated $Sc@C_{74}$ and $Sc_2@C_{74}$ and vacuum deposited a fraction of a monolayer of these species onto a clean Si(100) surface. Using scanning tunneling microscopy, they concluded that the charge was at least partially delocalized over the cage. From their analyses, they deduced the endofullerene diameters to be ~ 9.5 Å.

Pradeep *et al.*[389] reported the formation of $Fe@C_{60}$, the only apparent exception to the rule that endofullerenes form only when the ionization potential of the guest is less than about 7 eV. Several groups have tried to reproduce their results, but confirmation has not been forthcoming. In contrast, consistent reproducibility has been found for the more electropositive elements of the periodic table.

In a recent study, Soderholm *et al.*[390] sought to determine the bonding geometry using extended x-ray absorption fine structure spectroscopy, EXAFS. Somewhat surprisingly, they concluded that the Y atom was not inside the cage at all. Instead, they postulated a dimer of the form $C_{82}Y - X - YC_{82}$ where the bridging atom, denoted X, was carbon or oxygen.

At this stage in endofullerene research, there is anticipation that important species will be synthesized and that their study will produce new physical and chemical insights. The challenges, however, should be kept in mind. While endofullerenes can be produced, they are minority species, and extraction and isomeric purification is challenging. Many of the endofullerenes detected in the gas phase are unstable. While some endofullerenes may exhibit properties that are exciting and useful, we must be prepared to find that some of these exotic molecules have rather mundane properties. As is always the case, there can be no certainty in fundamental research.

VI. Tubes, Capsules, and Onions

Tubes, polyhedral capsules, and onions are among the recent additions to the extended family of fullerenes. The tubes[391] can be envisioned as closed cylinders of hexagonal graphite that are capped by structures that contain pentagons, as depicted in Fig. 34(a). They have been discussed in terms of potential applications related to ultrastrong fibers and novel electronic devices. Considerable interest was generated when it was shown that tubes could be prepared in quantity[392] and that they could be filled with other materials,[393] representing a new level of mesoscopic physics (one-dimensional wires of much reduced size). The polyhedral nanocapsules are closed all-carbon structures but are not quite as symmetrical as the tubes. The onions are derived from carbon shells contained within carbon shells, nanoscale versions of Russian dolls.[394,395] The onions appear to be the most stable of the all-carbon family, because they can be formed under extreme conditions of electron beam irradiation from other forms of carbon.[395]

Helical tubules of nanometer diameter composed only of carbon were first reported by Iijima[391] after he inspected the material that grew on the negative electrode in his fullerene-generating machine. Also present were polyhedral particles or capsules with shell structures, 5–20 nm in diameter. Figure 34(b) shows a typical tube grown in this fashion, as imaged

---

[391]S. Iijima, *Nature* **354**, 56 (1991); see also **350**, 123 (1990).
[392]T. W. Ebbesen and P. M. Ajayan, *Nature* **358**, 220 (1992).
[393]P. M. Ajayan and S. Iijima, *Nature* **361**, 333 (1993).
[394]H. Kroto and K. G. McKay, *Nature* **331**, 328 (1988).
[395]D. Ugarte, *Nature* **359**, 707 (1992).

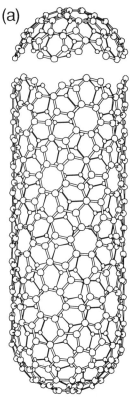

FIG. 34. (a) depicts a chiral carbon fiber with a diameter of 10.36 Å and a chiral angle of 19.11 degrees. This fiber is drawn with hemispherical caps based on an icosahedral C_{140} fullerene. From Saito et al.[419]

by TEM with a 400-keV electron beam.[396] Iijima[391] showed tube structures composed of coaxial arrays of closed graphitic sheets, ranging in number from 2 to about 50. In any given tube, the carbon pentagons were arranged in a helical fashion about the axis with tube-to-tube variation in the pitch angle to allow optimization of the interlayer spacing, 3.4–3.5 Å, slightly more than the spacing of ideal graphite and more characteristic of turbostratic carbon.[397] Iijima found no evidence for a scroll structure and argued that each tube was closed, minimizing the

[396]D. W. Owens, P. G. Kotula, S. K. McKenna, J. H. Weaver, C. B. Carter, L.-S. Wang, and R. E. Smalley (unpublished).
[397]R. Franklin, Acta Cryst. **4**, 253 (1951).

FIG. 34. (b) is a TEM image of a typical buckytube made up of thirteen concentric tubes. The capping of the tubes is accomplished by inserting six pentagons.

number of dangling bonds. The chemical stability of these species strongly supported this picture, as did recent investigations of the burning of the tubes in CO_2[398] or O_2.[399] In the latter studies, it was found that the end cages were etches and layer-by-layer tube removal occurred.

[398]S. C. Tsang, P. J. F. Harris, and M. L. H. Green, *Nature* **362,** 520 (1993); P. J. F. Harris, M. L. H. Green, and S. C. Tsang, *Trans. Faraday Society* (preprint).
[399]P. M. Ajayan, T. W. Ebbesen, T. Ichihashi, S. Iijima, K. Tanigaki, and H. Hiura, *Nature* **362,** 522 (1993).

The tubes in Fig. 34(b) run straight without the addition of incomplete exterior tubes or the sealing off of interior tubes. Inspection reveals 13 concentric tubes around a central void. The tubes are closed at the end, a circumstance associated with the insertion of six pentagons,[400] as for the fullerenes. The tube closure is shown schematically in Fig. 34(a) where hemispherical caps are envisioned for a chiral fiber of diameter 10.36 Å.

Iijima et al.[400,401] described the elongation of tubes at very high temperature on the negative electrode. While the details of tube nucleation were not known, Iijima et al. argued that tube growth occurred by the addition of carbon atoms to the open end. Such elongation was thought to be a consequence of the helical structure of the carbon network, and as long as hexagonal structures were produced, the tube could grow indefinitely. Defect formation in the form of a pentagon introduced positive curvature and cone angles that tended toward convergence. Defects in the form of seven-membered carbon rings, heptagons, also formed and introduced negative curvature.

Figure 34(c) shows a complex structure having both positive and negative curvature. In this case, the structure exhibits 17 concentric tubes at the base, but faults on the left portion of the tubes account for their converging character. The directional change at the left corresponds to the incorporation of a single pentagon. Iijima et al.[400] pointed out that the individual graphite sheets of the multishell structure are continuous, even around the region where defects appear. In Fig. 34(c), 3 of the tubes are seen to close near the base of the tube when the diameter reaches ~20 Å. The center of the image shows that convergence is avoided by changes in direction on both sides of the tube when heptagons are introduced. There is another change toward the end of the tube that ultimately leads to tube closure. Well before this occurs, however, many of the inner tubes have sealed as pairs, of more or less regular intervals. Fully contained within the tube is a closed polyhedral shell three layers in thickness. STM images of such tubes showing carbon details were recently reported by Ge and Sattler.[402]

Ebbesen and Ajayan[392] first described the growth in large quantities of tubes on the negative electrode during arc discharge in a helium atmosphere. It is fascinating to speculate about the growth of

[400]S. Iijima, T. Ichihashi, and Y. Ando, *Nature* **356,** 776 (1992).
[401]S. Iijima, P. M. Ajayan, and T. Ichihashi, *Phys. Rev. Lett.* **69,** 3100 (1992).
[402]M. Ge and K. Sattler, *Science* **260,** 515 (1993).

Fɪɢ. 34. (c) shows a less perfect tube where two of the inner tubes close (bottom center) before the central core converges to a size comparable with a single fullerene. Thereafter, the tubes diverge before faulting and finally closing. Negative curvature is achieved by heptagon insertion, and positive curvature reflects pentagon insertion. There is a closed polyhedral structure that has grown in contact with the outer tubes near the head of the tube. (b) and (c) from Owens *et al.*[396]

tubes.[2,391,394,400–405] Iijima and coworkers[391,400] and Smalley and Curl[2] emphasized carbon addition at an open end. Kroto[406] has argued that carbon addition can occur at the ends of closed tubes in the vicinity of pentagonal faces. Such carbon incorporation is aided by the very high temperature at the electrode. Saito et al.[407] suggested that the difference in density of amorphous carbon and planar graphite accounts for the formation of the large internal voids evident in Fig. 34. They considered fluidlike carbon precursors to the tubes that cooled from the outside, producing graphitic sheets that acted as the templates for outside-to-inside growth, leaving a void as the carbon condenses on the interior surface. At this point, it is impossible to determine whether any of these pathways excludes the others under the extreme conditions of formation.

Dresselhaus[408] observed that the formation of today's nanotubes was preceded by many years by carbon fiber formation, notably in the work in 1960 by Bacon.[409] Today, as then, a dc discharge is struck between carbon electrodes in an inert gas atmosphere. In the early work, however, the pressure was 92 atmospheres, while it is less than one atmosphere for nanotube formation. The whiskers that Bacon grew were much thicker ($\sim$1–5 μm) and longer (up to $\sim$3 cm) than those presently under investigation and were thought to be scroll-like. There is also a mature literature[410–417] that describes fiber formation at high temperature aided

[403]V. P. Dravid, X. Lin, Y. Wang, X. K. Wang, A. Yee, J. B. Ketterson, and R. P. H. Chang, *Science* **259**, 1601 (1993); X. K. Wang, X. W. Lin, V. P. Dravid, J. B. Ketterson, and R. P. H. Chang, *Appl. Phys. Lett.* **62**, 1881 (1993).

[404]X. F. Zhang, X. B. Zhang, G. Van Tendeloo, S. Amelinckx, M. Op de Beeck, and J. van Landuyt, *J. Crystal Growth* (in press).

[405]D. Ugarte, *Chem. Phys. Lett.* **198**, 596 (1992).

[406]H. W. Kroto, private communication.

[407]Y. Saito, T. Yoshikawa, M. Inagaki, M. Tomita, and T. Hayashi, *Chem. Phys. Lett.* **204**, 277 (1993).

[408]M. Dresselhaus, *Nature* **358**, 195 (1992).

[409]R. Bacon, *J. Appl. Phys.* **31**, 283 (1960).

[410]P. L. Walker, Jr., J. F. Rakszawski, and G. R. Imperial, *J. Phys. Chem.* **63**, 133 (1959).

[411]W. R. Ruston, M. Waezee, J. Nennaut, and J. Waty, *Carbon* **7**, 47 (1969).

[412]S. D. Robertson, *Carbon* **8**, 365 (1970).

[413]T. Baird, J. R. Fryer, and B. Grant, *Nature* **233**, 329 (1971).

[414]M. Endo, A. Kathon, T. Sugiura, M. Shiralshi, Proceedings 18th Biennial Conference Carbon, Worcester Polytechnic, Worcester, Massachusetts, 1987, p. 151.

[415]R. T. K. Baker, *Carbon* **27**, 315 (1989).

[416]M. S. Kim, N. M. Rodriguez, and R. T. K. Baker, *J. Catal.* **131**, 60 (1991).

[417]M. José-Yacamán, M. Miki-Yoshida, L. Rendón, and J. G. Saantiesteban, *Appl. Phys. Lett.* **62**, 657 (1993).

by hot metallic particles, as recently applied to nanotube formation by José-Yacamán *et al.*[417]

In 1993, Ajayan and Iijima[393] reported that tubes could be filled by capillary action, realizing a prediction by Pederson and Broughton.[418] They found that heating nanotubes in air in the presence of Pb resulted in the opening of the end caps and the filling of the tubes by Pb (or Pb oxide). They speculated that the tubes could act as molds for nanowires. In their work, they also observed that exposure to molten Pb in vacuum did not induce incorporation, suggesting the critical role of carbon burning.

Nanotube theory has advanced beyond where experiment has been able to go. Theorists have produced perfect single-walled tubes and found that the electronic properties of these tubes are dependent on the diameter and the chirality.[419–422] They have additionally examined doping with nitrogen and boron.[422] It has been predicted that approximately one-third of the possible tubes with diameters close to those observed would be metallic, and the remainder would be semiconducting. This led to speculation concerning possible electronics-type applications using nanotubes of precise size and character. Calculations of the mechanical properties have suggested exceptional stiffness.[423] Other discussions have focused on the energetics of the tubes,[424,425] considering the gains upon closure and weighing those gains against the elastic costs associated with bending planar graphite enough to allow closure. Others[426–431] have considered arrays of tubes in the solid state, some exhibiting toroidal shape, negative curvature, and periodic arrays in two or three dimensions.

[418]M. R. Pederson and J. Q. Broughton, *Phys. Rev. Lett.* **69**, 2689 (1992).

[419]R. Saito, M. Fujita, G. Dresselhaus, and M. S. Dresselhaus, *Phys. Rev. B* **46**, 1804 (1992) and *Appl. Phys. Lett.* **60**, 2204 (1992). See also R. A. Jish and M. S. Dresselhaus, *Phys. Rev. B* **45**, 11305 (1992).

[420]J. W. Mintmire, B. I. Dunlap, and C. T. White, *Phys. Rev. Lett.* **68**, 631 (1992).

[421]N. Hamada, S. Sawada, and A. Oshiyama, *Phys. Rev. Lett.* **68**, 1579 (1992).

[422]J.-Y. Li and J. Bernholc, *Phys. Rev. B* **47**, 1708 (1993).

[423]G. Overney, W. Zhong, and D. Tománek, *Z. Phys. D* (in press).

[424]D. H. Robertson, D. W. Brenner, and J. W. Mintmire, *Phys. Rev. B* **45**, 12592 (1992).

[425]A. A. Lucas, Ph. Lambin, and R. E. Smalley, *J. Phys. Chem. Solids* **54**, 587 (1993).

[426]A. L. Mackay and H. Terrones, *Nature* **352**, 762 (1991).

[427]T. Lenosky, X. Gonze, M. Teter, and V. Elser, *Nature* **355**, 333 (1992).

[428]D. Vanderbilt and J. Tersoff, *Phys. Rev. Lett.* **68**, 511 (1992).

[429]B. I. Dunlap, *Phys. Rev. B* **46**, 1933 (1992).

[430]G. B. Adams, O. F. Sankey, J. B. Page, M. O'Keefe, and D. A. Drabold, *Science* **256**, 1792 (1992).

[431]S. Itoh, S. Ihara, and J. Kitakami, *Phys. Rev. B* **47**, 1703 (1993).

Nanocapsules based on carbon are readily evident in a TEM investigation of the residue grown in the arcing process.[391] Like the tubes, these structures are closed and are derived from multiple layers of graphitic sheets. Figure 35, from Tomita et al.,[432] shows three such objects. In (a), the smaller structure displays a shell four layers thick with a large internal void. Its larger partner and the polyhedral structures shown in (b) and (c) have partially filled interiors. To obtain these objects, Tomita et al.[432] created a discharge using lanthana-impregnated graphite rods. TEM inspection and chemical analysis of the internal objects indicate that they are LaC_2 crystallites of nanoscopic dimension. The perfection in their structure is particularly striking. Indeed, analysis indicates that the interface between them and adjacent C layers is atomically abrupt. It is clear from these results that the growth conditions on the electrode allowed sufficient mobility for the stable carbide phase to form. Further, the stable planar array of the graphite sheet precluded defect incorporation. Ruoff et al.[433] reported independent studies of LaC_2 particles in graphitic nanocapsules. Saito et al.[434] subsequently formed YC_2 nanocrystallites of a few tens of nanometers in size that were also protected by the multilayers of closed graphite nanocapsules, as did Seraphin et al.[435]

An issue that has been much discussed involves the relative stability of the various forms of carbon. It is known, for example, that the cohesive energy of diamond and graphite differ by only about 0.2 eV/atom. Calculations for C_{60} indicate 0.4 eV/atom lower stability than graphite or diamond, and in fact, the measurements of Núñez-Regueiros[115,116] have demonstrated that C_{60} can be converted to diamondlike structures under sufficiently high pressure. What was not known until very recently was that graphite, the fullerenes, and the nanotubes and nanocapsules can all be converted into a more spherical form under sufficiently energetic conditions maintained for long times.

Figure 36 shows a series of TEM images that were obtained by Ugarte.[436] (a) shows a polyhedral structure like that of Fig. 35. To obtain (b), Ugarte subjected this particle to a high flux of energetic electrons ($\sim$150 amps-cm^{-2}, 300 keV). The result was the collapse of the inner shell to eliminate the hollow space. Subsequently, there was regraphitization of several surface layers (b, c), and the particle took on a spherical shape

[432]M. Tomita, Y. Saito, and T. Hayashi, *Jpn. J. Appl. Phys.* **32**, L280 (1993).

[433]R. S. Ruoff, D. C. Lorents, B. Chan, R. Malhotro, and S. Subramoney, *Science* **259**, 346 (1993).

[434]Y. Saito, T. Yoshikawara, M. Okada, M. Ohkohchi, Y. Ando, A. Kasuya, and Y. Nishina, *Chem. Phys. Lett.* **209**, 72 (1993).

[435]S. Seraphin, D. Zhou, J. Jino, J. C. Withers, and R. Loutfy, *Nature* **362**, 503 (1993).

[436]D. Ugarte, *Chem. Phys. Lett.* **207**, 473 (1993).

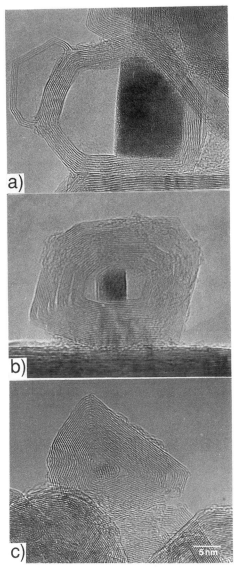

FIG. 35. Polyhedral capsules of graphite that contain crystallites of LaC$_2$. These structures were formed by the arc method using an electrode impregnated with lanthanum oxide. From Tomita *et al.*[432]

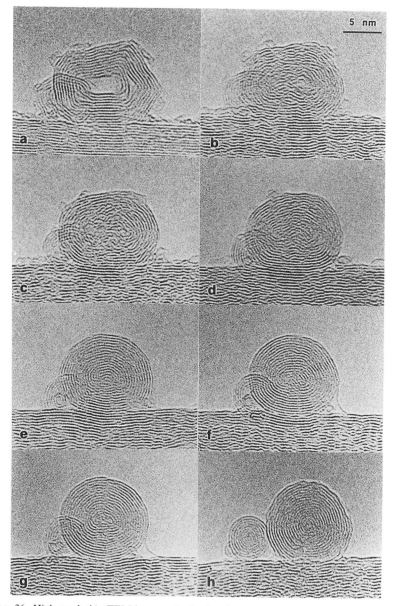

FIG. 36. High resolution TEM images obtained at intervals during intense electron beam irradiation of the polyhedral capsule shown in (a). The internal structure is first lost as the void fills, but then the outer shell becomes more perfect and inner layers grow from outside to inside to produce the nearly perfect onion shown in (b). From Ugarte.[436]

with growth of the inner shells from the outside. The images labeled (f, g) show nearly ideal outer shells but imperfection at the core. The final micrograph displays a nearly spherical onionlike structure in which the central shell has a diameter of 7 Å. Ugarte argued that the growth process that he was able to witness is analogous to that expected for solidification of liquid carbon droplets with surface tension dictating a starting spherical shape and graphitization from the surface toward the center. In this case, the particle is sufficiently hot during formation that atom rearrangement is facile, and in contrast to the capsules or tubes, the empty internal spaces are filled. In other studies, Ugarte[437] has produced micrographs in which tubes and other carbidic structures are modified by electron bombardment heating and onion structures are formed.

Finally, Fig. 37 shows the emptying of a graphitic onion under electron

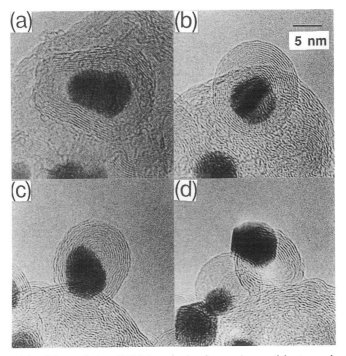

FIG. 37. (a) High resolution TEM image showing a Au particle trapped within a multilayer carbon polyhedron. Under the heating of an intense electron beam, (b)–(d), the carbon structures open, the Au particle is expelled, and the carbon forms an onion. From Ugarte.[438]

[437]D. Ugarte, *Chem. Phys. Lett.* **198,** 594 (1992); W. A. de Heer and D. Ugarte, *Chem. Phys. Lett.* **207,** 480 (1993); D. Ugarte, *Europhys. Lett.* **22,** 45 (1993).

irradiation. In this case, Ugarte[438] started with a cluster of Au that was trapped within a carbon cage. This occurred during discharge experiments in which one of the electrodes contained Au powder. Au was chosen because Au endofullerenes had not been produced previously. The first image shows a gold-filled particle of irregular shape. After 8 minutes of irradiation, the cage was more spherical and exhibited very good crystallization, although the Au cluster was not at the center. In (c), the center of the inner carbon shell was not the center of the overall particle. This, plus the reduction in particle size, indicated that the Au cluster was being expelled. After 44 minutes of irradiation, the Au was completely outside the carbon cage, lying on another spherical particle. It also exhibited faceting that could occur only if it were outside the carbon structure. Ugarte concluded that the carbon capsule that initially trapped the Au particle was opened during expulsion of the Au but then closed again in a more spherical form. Clearly, the expulsion of Au reflects a thermodynamic driving force that favors phase separation rather than mixing, in this case separation into a closed onion of carbon and a faceted crystal of Au.

VII. Summary and Prospects

In this chapter, we have surveyed some of the properties of the fullerenes, the tubes, and the onions in pure form and mixed with other species. Other chapters will provide more details and a different perspective into these fascinating materials. At this stage, it is clear that the fullerenes have captured the attention of a large group of investigators. Their study has provided a great deal of very specific information related to them, as we have tried to show, and it is likely to have a broader impact on the community in general. For example, to understand the electronic interactions of the alkali metal fullerides, we must extend our present models for materials at the verge of a metal–insulator transition. To understand the formation of these carbon structures in the first place, we must develop more effective simulations for existing computers and develop algorithms for the next-generation computers. By investigating the growth kinetics, crystal structures, and ordering transitions, we have unique opportunities to observe phenomena that may be present in other systems but have been more elusive.

A question that seems to come up whenever fullerenes are discussed, is "What are they good for?" Harry Kroto tells a delightful story of what

---

[438]D. Ugarte, *Chem. Phys. Lett.* (in press).

transpired in the British House of Lords when that question was asked. The Lords did not formulate an answer, but one was heard to remark, "Could it be said that it does nothing in particular and it does it very well?" The technical community has a more elaborate answer. Many possibilities have been mentioned, ranging from lubricants to batteries to biological magic bullets, but these remain largely possibilities. A few more realistic possibilities have been demonstrated. C_{70} has been shown to be a seed for diamond growth,[439] presumably because of the C–C motif. Self-supporting membranes of fullerenes have been produced,[440] and fullerenes have been used as passivating layers[184] and optical limiters.[441] Polymerization has been accomplished by light[276] and by an electron beam,[186] and a large family of fullerene derivatives has been synthesized.[95,96]

The cost of the fullerenes today is still very high, roughly $400 per gram (the density of C_{60} is 1.65 gm-cm^{-3}). At about 40 times the cost of Au, C_{60} is still a long way from commercialization, and the high fullerenes are far more dear. This noted, scale-up processes for fullerene production can certainly be achieved once a market is identified. It remains to identify a market where the value-added character will justify the expense. Given their cost, it is unlikely that fullerenes will find applications as property enhancers when added to other materials, as has been suggested, for example, in xerography where quantities added to polyvinylcarbazole enhance photoconductivity.[442]

As of this writing, there are no commercially viable products except, of course, the fullerenes themselves for scientific study, T-shirts depicting the C_{60} structure, and models for instructional use. The same could have been said for the cuprate superconductors at the equivalent point in their development. We should therefore take a longer view and recognize that the fullerenes have been "generally" available only since late 1990 and that the possibilities for fullerene derivatives, endofullerenes, and tubes are extremely diverse.

ACKNOWLEDGEMENTS

This work was supported by the National Science Foundation and the Office of Naval Research. We gratefully acknowledge stimulating

[439]R. Meilunas, R. P. H. Chiang, S. Liu, and M. Kappes, *Nature* **354,** 271 (1991).
[440]C. B. Eom, A. F. Hebard, L. E. Trimble, G. K. Celler, and R. C. Haddon, *Science* **259,** 1887 (1993).
[441]L. W. Tutt and A. Kost, *Nature* **346,** 225 (1992).
[442]R. Wang, *Nature* **356,** 585 (1992).

discussions and interactions with members of the extended fullerene community, and we apologize to those whose work we have been unable to discuss. We are particularly pleased to recognize our colleagues at the University of Minnesota, notably J. L. Martins, N. Troullier, J. R. Chelikowsky, P. J. Benning, F. Stepniak, T. R. Ohno, Y. Z. Li, J. C. Patrin, Y. B. Zhao, M. B. Jost, C. Gu, T. Komeda, Y. Chen, and G. H. Kroll; L. P. F. Chibante and R. E. Smalley at Rice University; K. Kikuchi and Y. Achiba at Tokyo Metropolitan University; and M. Knupfer, A. A. Lucas, and A. F. Hebard. DMP thanks his wife, Jessica, who postponed the delivery of their son, Zachary, until three days after this chapter was submitted. Both authors are indebted to M. J. Weaver for her dedication and patience.

SOLID STATE PHYSICS, VOLUME 48

Preparation of Fullerenes and Fullerene-Based Materials

CHARLES M. LIEBER
CHIA-CHUN CHEN

Division of Applied Sciences and Department of Chemistry
Harvard University
Cambridge, Massachusetts

I. Introduction

Revolutionary rather than evolutionary advances in condensed matter research often arise from the discovery of new materials. The discovery of the first high temperature copper oxide superconductor, $La_{2-x}Ba_xCuO_4$, by Bednorz and Müller in 1986[1] is an excellent example. This discovery initiated an explosion of research activity that has resulted in the development of many new classes of copper oxide superconductors with critical temperatures (T_c) approaching 130 K. Most advances since

---

[1] J. G. Bednorz and K. A. Müller, *Z. Phys. B* **64**, 187 (1986).

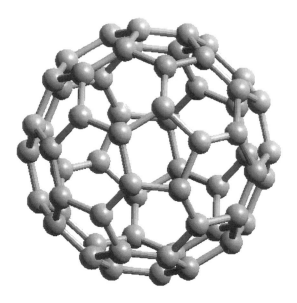

FIG. 1. Molecular structure of Buckminsterfullerene.

this revolutionary period of rapidly increasing T_c's have been evolution-ary, however, in the sense that our understanding of the high temperature copper oxide superconductors has occurred in small steps as a result of increasingly refined measurements on higher and higher quality materials. More recently, the discovery of methods to prepare and isolate macro-scopic quantities of pure carbon clusters such as C_{60} has resulted in another revolution in materials physics and chemistry research.[2] It is this rapdily growing area of research on which we will focus in this chapter.

Historically, Smalley and coworkers first detected C_{60} as an unusually abundant species in mass spectra recorded on carbon clusters produced by laser vaporization of graphite in a helium flow.[3] To account for the unusually high abundance and apparent stability of the C_{60} species they proposed that the 60 carbon atoms were arranged in the shape of the truncated icosohedron or soccer ball, where each of the 60 carbon atoms has an identical environment (Fig. 1). This proposed structure, which consists of 20 hexagonal and 12 pentagonal carbon rings was inspired in

[2]W. Kratschmer, L. D. Lamb, K. Fostiropoulos, and D. R. Huffman, *Nature* **347,** 354 (1990).
[3]H. W. Kroto, J. R. Heath, S. C. O'Brien, R. F. Curl, and R. E. Smalley, *Nature* **318,** 162 (1985).

part by earlier architectural structures, geodesic domes, designed by R. Buckminster Fuller.[4] It is this inspiration that led to the naming of C_{60} as Buckminster-fullerene or Bucky-Ball for short. Interestingly, Euler proved in the 18th century that a closed cage or cluster made up of only hexagons and pentagons must always contain 12 pentagons, but can have any number of hexagons greater than or equal to 2. The general class of carbon clusters described by this rule are now called the fullerenes.

Smalley and coworkers[3,5–7] and others[8,9] provided significant experimental and theoretical evidence supporting the proposed soccer-ball structure of C_{60}. Prior to 1990, however, C_{60} was not a molecule you could collect in a bottle and use to make materials. The laser vaporization techniques that had been used to study C_{60} and other fullerenes produced only on the order of 10^4 clusters, a far cry from the $>10^{18}$ clusters needed for a 1-mg sample of solid. Hence, the discovery by Krätschmer, Huffman, and coworkers[2] of a method to prepare macroscopic quantities of C_{60} and other fullerenes represents a remarkable advance that has provided the raw material fueling an explosion of materials research by physicists, chemists, and materials scientists.[10–16] For example, a number of fullerene-based conductors and superconductors have been prepared and intensively studied (see the chapter on fullerene superconductivity).[12] In addition, C_{60}-based solids that exhibit interesting optical and magnetic properties are beginning to emerge from this work.[17–21] Clearly, the

[4]R. B. Fuller, "Inventions: The Patented Works of Buckminster Fuller." St. Martin's Press, New York, 1983.
[5]S. C. O'Brien, J. R. Heath, R. F. Curl, and R. E. Smalley, *J. Chem. Phys.* **88,** 220 (1988).
[6]J. R. Heath, R. F. Curl, and R. E. Smalley, *J. Chem. Phys.* **87,** 4236 (1987).
[7]R. F. Curl and R. E. Smalley, *Science* **242,** 1017 (1988).
[8]E. A. Rohlfing, D. M. Cox, and A. Kaldor, *J. Chem. Phys.* **81,** 3322 (1984).
[9]H. Kroto, *Science* **242,** 1139 (1988).
[10]R. E. Smalley, *Acc. Chem. Res.* **25,** 98 (1992).
[11]J. E. Ficher, P. A. Heiney, and A. B. Smith, *Acc. Chem. Res.* **25,** 115 (1992).
[12]A. F. Hebard, *Phys. Today* **45,** 26 (1992).
[13]F. Diederich and R. L. Whetten, *Acc. Chem. Res.* **25,** 119 (1992).
[14]R. C. Haddon, *Acc. Chem. Res.* **25,** 127 (1992).
[15]P. J. Fogan, J. C. Calabrese, and B. Malone, *Acc. Chem. Res.* **25,** 134 (1992).
[16]J. H. Weaver, *Acc. Chem. Res.* **25,** 143 (1992).
[17]H. Yonehara and C. Pac, *Appl. Phys. Lett.* **61,** 575 (1992).
[18]Y. Wang, *Nature* **356,** 585 (1992).
[19]B. Miller, J. M. Rosamilia, G. Dabbagh, R. Tycko, R. C. Haddon, A. J. Muller, W. Wilson, D. W. Murphy and A. F. Hebard, *J. Am. Chem. Soc.* **113,** 6291 (1991).
[20]Y. Wang and L. T. Cheng, *J. Phys. Chem.* **96,** 1530 (1992).
[21]P. M. Allemand, K. C. Khemani, A. Koch, F. Wudl, K. Holczer, S. Donovan, G. Grüner, J. D. Thompson, *Science* **253,** 301 (1991).

preparation of macroscopic quantities of fullerenes appears to have provided a revolutionary advance in solid state research.

In this chapter, we will provide the background needed to prepare, isolate, and characterize fullerene clusters, as well as solids based on these clusters. Our goal is to provide physicists and chemists with the information needed to carry out experimental research starting only from carbon rods. First, several techniques that are used to prepare crude mixtures of fullerenes, and isolate and characterize pure C_n clusters will be reviewed. Approaches for preparing isotopically substituted clusters, which are materials essential for physical studies, will also be discussed. We will then review the properties of solid C_{60} and discuss in detail doping of solid C_{60} with different metal species. Lastly, the status and prospects of several new fullerene building blocks, including endohedral metal clusters and carbon nanotubes, will be reviewed.

II. Fullerene Clusters

To carry out well-defined studies of fullerene-based materials requires at the very least macroscopic quantities of pure fullerene clusters. In this section we will first review methods that can be used to generate mixtures of fullerene clusters and the relative merits of these different methodologies. Effective techniques for the isolation of pure C_n ($n = 60, 70, \ldots$) clusters and the characterization of these large molecular species will then be discussed. Finally, we will review techniques that can be used to prepare isotopically substituted fullerenes.

1. Formation of Fullerenes

a. *Arc Vaporization of Graphite*

The fullerene field was revolutionized by the discovery that simple resistive vaporization of graphite rods could produce fullerenes in substantial yield.[2] This procedure, which is often termed the Krätschmer–Huffman method, is a straightforward and low cost method for generating large quantities of fullerene-containing carbon soot. A typical example of an arc vaporization fullerene generating apparatus is shown in Fig. 2. Key components of the fullerene generating apparatus include (1) high-

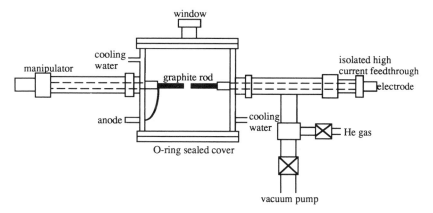

FIG. 2. Schematic illustration of the arc vaporization apparatus for generating fullerene-containing soot.

current electrical feedthroughs; (2) at least one movable electrode; and (3) a water-cooled surface to collect the vaporized carbon products. These components are housed within a vacuum chamber that is connected to a pumping/gas handling system that can evacuate the chamber to $\leq 10^{-3}$ torr and can control the pressure between 10–500 torr. Lastly, a high current (100–200 A), low voltage AC or DC power supply is needed to drive the evaporation; an arc welder is an inexpensive but effective power supply for this purpose.

In a typical preparative experiment, two high purity graphite rods are clamped to the high current feedthroughs, and the chamber is pumped down to $\leq 10^{-3}$ torr and refilled with He gas to a pressure of 150–250 torr. Because both oxygen and water significantly inhibit the formation of fullerenes, it is important to evacuate the chamber carefully and refill it using purified helium. The electrodes are positioned so that the carbon rods are just touching, and then the vaporization is initiated by passing a high current through the rods. For 6.25-mm-diameter rods, currents between 100 and 200 A lead to efficient fullerene formation. Under these conditions, the 6.25-mm rods are consumed at a rate of about 5–10 mm/min. The crude carbon product or soot produced by this vaporization collects on the water-cooled inner surface of the fullerene apparatus (Fig. 2) and is readily removed from the walls and collected using a stiff brush. This soot contains a variety of carbon products including C_{60} and larger fullerenes. The isolation and purification of these fullerene clusters is discussed in Section 2.

There are several experimental factors that can be varied to maximize the yield of fullerene clusters in the soot produced from the vaporization of graphite electrodes. These factors include the vaporization current density and helium partial pressure. The importance of these two experimental variables in determining the yield of fullerenes can be understood by considering the environment in which the clusters form. Specifically, the vaporization of the carbon rods is not driven by ohmic heating but rather by an arc discharge between the ends of the graphite rods. These latter conditions produce a plasma, and hence, control of temperature and cooling of the carbon plasma will determine the yield and size distribution of the fullerene clusters. Qualitatively, increases in the current density will increase the plasma temperature, while increases in the buffer gas (He) pressure will increase the cooling and growth rates of carbon species. A particularly dramatic example of the effects of He pressure is found in comparing the carbon products produced from dc discharges in 150 and 500 torr of helium. Using the former conditions fullerene clusters are observed in high yield, while at higher He pressures carbon nanotubes (see Section 11) are the principle product. Unfortunately, however, systematic studies of how the current density and the buffer gas pressure affect the yield and cluster size distribution have not been carried out. We believe that future investigations addressing these points are worthwhile, since they should provide the data needed to prepare rationally specific cluster sizes.

b. *Laser Ablation*

A second very useful and powerful technique for producing fullerene clusters involves laser ablation of graphite in a helium atmosphere. Historically, laser ablation was the first technique used to generate fullerene clusters in the gas phase,[3,5-7] although initial attempts to prepare isolable quantities of C_{60} were unsuccessful with this technique. The early attempts to prepare fullerenes by laser ablation relied on ablation in an inert gas buffer at room temperature. The failure of these early attempts to yield macroscopic quantities of fullerenes is almost certainly due to the fact that the dense carbon plasma created during laser vaporization cools too quickly. The rapidly cooling plasma does not provide sufficient time for the growing carbon fragments to rearrange into the stable, closed fullerene structures. In support of this scenario, Smalley and coworkers first showed that laser ablation of graphite carried out at elevated temperatures could in fact lead to the efficient production of macroscopic quantities of C_{60} and other fullerenes.[9,22]

A typical laser ablation apparatus that has been used to prepare

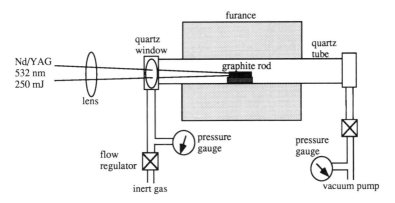

FIG. 3. Schematic view of a laser ablation apparatus for the preparation of C_{60} and other fullerenes. In this apparatus graphite is ablated inside a quartz tube held at 1200°C using the green (532 nm) line from a pulsed Nd–YAG laser. Because this procedure is carried out at elevated temperature, C_{60} and C_{70} condense at the cool end of the tube.

macroscopic quantities of fullerenes in our laboratory is shown in Fig. 3. A key feature of this apparatus is that graphite is ablated within a high temperature furnace. By carrying out the ablation at an elevated temperature, the plasma cools more slowly and growing carbon clusters have sufficient time to rearrange (anneal) into stable fullerenes. We have found that fullerene production is most efficient at 1200°C, the upper limit of our furnace. At lower temperatures, the yield of fullerenes is reduced.[9,22] It is possible, however, that ablation at higher temperatures (i.e., >1200°C) could lead to further enhancements in the yields of fullerenes produced by this technique.

There are several features of the laser ablation method that make it an excellent technique for the preparation of fullerenes. First, it is possible to systematically vary the characteristics of the carbon plasma through variations in the laser pulse energy and wavelength.[23] Second, by controlling the buffer gas pressure and the furnace temperature it is also possible to control systematically the growth of clusters from the carbon plasma. Control of the growth and annealing should increase the efficiency and selectivity of the fullerenes produced. To date, these experimental variables have not been carefully explored, although such investigations could advance significantly our understanding of how to grow rationally new clusters. Finally, since the laser light is easily

[22]R. E. Haufler, Y. Chai, L. P. F. Chibante, J. Conceicao, C. Jin, L. S. Wang, S. Maruyama, and R. E. Smalley, *Mater. Res. Soc. Symp. Proc* **206,** 627 (1991).
[23]J. T. Cheung and H. Sankur, *CRC Critical Rev. in Solid State and Mat. Sci.* **15,** 63 (1988).

positioned and no external electrical connections to the sample are needed, it is possible to produce fullerenes from very small amounts of carbon precursors. We have exploited this feature previously to prepare efficiently ^{13}C-substituted fullerenes,[24,25] while Smalley and co-workers have used it to prepare metal-encapsulated fullerenes (see sections 4 and 10 below).

c. *Other Methods of Fullerene Production*

Several other techniques, including hydrocarbon combustion, low-pressure helium sputtering, electron beam evaporation, and inductively coupled RF evaporation of graphite targets have been used to prepare fullerene containing carbon products.[26-28] Of these, we believe only the combustion technique is sufficiently developed to be considered as a general method for the production of macroscopic quantities of fullerenes.

The production of fullerenes from hydrocarbon combustion was first reported by Howard and co-workers in 1991.[26] Specifically, combustion of benzene-oxygen mixtures in laminar flow flames produced significant quantities of C_{60} and C_{70}. A maximum yield of 0.3% C_{60} and C_{70} was observed when the benzene: oxygen ratio was nearly 1 and these gases were diluted with 10% argon. Although this product yield is an order of magnitude lower than that observed for the arc-discharge and laser ablation methods of fullerene production, there are several attractive features of the flame production process. First, hydrocarbon combustion can be easily scaled-up and operated as a continuous process. These characteristics could make hydrocarbon combustion a key techique for large-scale commercial production of fullerenes. Secondly, variations in the properties of the combustion flame may be used to control systematically the size distribution fullerene clusters.[26] This could be useful for solid state studies of larger fullerene clusters since large clusters are only produced in low yield by the arc-discharge method. Lastly, we believe that there is great potential to investigate systematic doping of fullerene cluster using the flame method since dopants can be easily introduced into the benzene-oxygen mixture prior to combustion.

[24]C.-C. Chen and C. M. Lieber, *J. Am. Chem. Soc.* **114,** 3141 (1992).

[25]C.-C. Chen and C. M. Lieber, *Science* **259,** 665 (1993).

[26]J. B. Howard, J. T. McKinnon, Y. Makarovsky, A. L. Lafleur, and M. E. Johnson, *Nature* **352,** 139 (1991).

[27]R. F. Bunshah, S. Jou, S. Prakash, H. J. Doerr, L. Isaacs, A. Wehrsig, C. Yeretzian, H. Cynn, and F. Diederich, *J. Phys. Chem.* **96,** 6866 (1992).

[28]G. Peters and M. Jansen, *Angew. Chem.* **104,** 240 (1992).

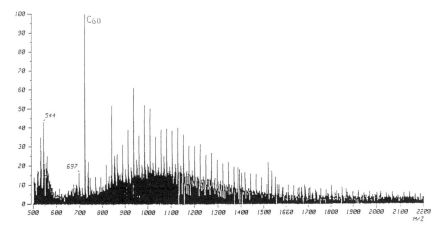

FIG. 4. Fast atom bombardment mass spectrum of the soot produced by arc vaporization of graphite rods. This spectrum illustrates the large number of different carbon products (i.e., different M/Z peaks) produced by arc vaporization.

2. SEPARATION OF PURE CLUSTERS

The methods of fullerene production that have been described all produce the macroscopic quantities of fullerenes needed for solid state research. In all cases, however, the fullerenes are produced as a crude mixture containing C_{60} and other C_n ($n > 60$), as well as conventional hydrocarbon species (Fig. 4). In this crude soot state, the fullerenes are unsuitable for materials research. Individual fullerene clusters must be isolated from the crude soot to carry out meaningful solid state research. In this section we review several methods by which pure C_{60} and other fullerenes can be isolated from the crude soot. An overall flow chart outlining the purification procedures to be discussed is shown in Fig. 5.

The first step in most purification schemes involves the extraction of soluble fullerenes from the crude carbon soot. In the original work of Krätschmer and Huffman, benzene was used to extract C_{60} and larger fullerenes from the carbon soot.[2] For safety reasons, most researchers now employ toluene to extract fullerenes from soot. The extraction may be efficiently carried out using several methods. First, soot can be extracted continuously using boiling toluene in a Soxhlet solid extraction apparatus. The glassware needed for this procedure can usually be found in an organic chemistry laboratory. This method is convenient, since it can be run continuously by itself, and furthermore, completion of the procedure is readily determined when the toluene condensed in the Soxhlet extractor remains colorless. A second method that can used to

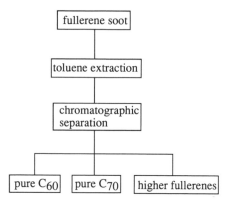

FIG. 5. Flow chart outlining the basic purification steps required to obtain pure C_{60} and other fullerenes from the crude soot produced by arc vaporization.

extract fullerenes from soot involves sonication of toluene–soot mixture at room temperature. In this procedure, a flask containing a toluene suspension of soot is placed in an ultrasonic cleaning bath and sonicated for 30 to 60 minutes. The insoluble carbon products are then removed from the toluene solution of fullerenes by filtration.

The fullerene extract obtained using either of the foregoing methods amounts to approximately 15–20% by weight of the original carbon soot. Although a similar benzene extract was used by Krätschmer et al. to obtain the first measurements of the solid state properties of C_{60}, we now know that this extract contains a number of different species that must be separated to obtain pure C_{60}. The heterogeneous composition of the fullerene extract is clearly illustrated in the mass spectrum shown in Fig. 6. This spectrum shows that the dominant species in the extract are C_{60} and C_{70}; however, small quantities of larger fullerenes and low molecular weight hydrocarbons are also present in the extract. In what follows we describe several successful approaches for separating these fullerene and hydrocarbon species.

Hydrocarbon impurities can be conveniently separated from the extract prior to the isolation of the individual fullerenes. These impurities are believed to be primarily linear hydrocarbon chains that are known to form in carbon plasmas.[29,30] The removal of the hydrocarbons is based on the large difference in solubility of the hydrocarbons versus the fullerenes in ether: the hydrocarbon impurities are soluble in diethyl ether, while the fullerenes are essentially insoluble. Hence, to remove these

[29]G. V. Helden, N. G. Gotts and M. T. Bowers, *Nature* **363**, 60 (1993).
[30]R. F. Curl, *Nature* **363**, 14 (1993).

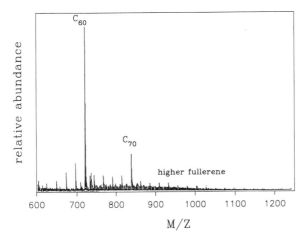

FIG. 6. Fast atom bombardment mass spectrum of the toleuene soluble extract from the soot. The primary products in this extract are C_{60} and C_{70}.

impurities the toluene–fullerene extract is evaporated to dryness using a rotary evaporator, and then the resulting black solid, which contains primarily fullerenes, is washed extensively with diethyl ether to selectively remove the hydrocarbons. The solid that remains after washing contains only fullerenes.

In general, all successful schemes used to isolate pure C_{60} and other individual fullerenes utilize a chromatographic separation. The most widely used chromatographic separation employs a neutral alumina (Al_2O_3) stationary phase and simple gravity feed.[31–34] Typically, a crude mixture of fullerenes is loaded onto the top of the alumina column and then eluted with $95:5$ hexane : toluene mixture. An advantage of this chromatographic separation is that the progress can be monitored visually. In a good separation, at least three distinct bands can be observed during the 5% toluene elution. A purple band corresponding to C_{60} has the highest mobility and can be isolated first from the fullerene mixture. The pure C_{60} band will separate completely from other fullerene bands in a good column, and it can be eluted using the 5% toluene

[31]A. S. Koch, K. C. Khemani, and F. Wudl, *J. Org. Chem.* **56,** 4543 (1991).
[32]F. Diederich, R. Ettl, Y. Rubin, R. L. Whetten, R. Beck, M. Alvarez, S. Anz, D. Sensharma, F. Wudl, K. C. Khemani, and A. Koch, *Science* **252,** 548 (1991).
[33]H. Ajie, M. M. Alvarez, S. J. Anz, R. D. Beck, F. Diederich, K. Fostiorpoulous, D. R. Huffman, W. Kratschmer, Y. Rubin, K. E. Schriver, D. Sensharma, and R. L. Whetten, *J. Phys. Chem* **94,** 8630 (1990).
[34]P. Bhyrappa, A. Penicaud, M. Kawamoto, and C. A. Reed, *J. Chem. Soc., Chem. Commun.* 936 (1992).

solution. A deep red band that corresponds to C_{70} is the second band that separates using the 5% toluene solution. To elute the C_{70} from the alumina column efficiently, however, requires an increase in the toluene:hexane ratio to approximately 1:1. The third yellow band, which has the lowest mobility on the alumina column, corresponds to a mixture of C_{76}, C_{84}, C_{90}, and C_{94} fullerenes.[32] Since these larger fullerenes constitute only a small fraction (<10%) of the crude fullerene mixture, we will not discuss the isolation of these fullerenes here. Increases in the fraction of large fullerenes are needed before these clusters will be useful for general solid state studies.

Chromatography on alumina provides a straightforward method for isolating pure C_{60} and C_{70}; however, it also has several drawbacks. First, the separation procedure is very time consuming and requires at least several days to produce gram quantities of pure C_{60}. Second, this separation procedure requires significant quantities of organic solvents and alumina, and it produces large quantities of organic wastes. For example, it has been estimated that the purification of 1 g of C_{60} requires approximately 10 kg of alumina and 50 L of solvent.[35] To reduce the consumption of costly solvents and alumina, several groups have developed alumina columns within Soxhlet extractors.[36,37] This modified procedure has the advantage of continuously recirculating the hexane solvent; however, it only separates about 50% of the total C_{60} from the crude fullerene extract.

A newer and perhaps more attractive separation procedure has recently been reported by Tour and coworkers.[35-38] In this chromatographic separation, the stationary phase is activated charcoal supported on silica gel and the mobile phase is pure toluene. An important feature of this system is that material can be eluted from column under positive pressure. The positive pressure elution enables much higher solvent flows than possible using gravity feed, and thus this separation procedure is considerably faster than those based on the Al_2O_3 stationary phase. Notably, with this method it is possible to obtain 1 g of pure C_{60} in less than one hour.

A final chromatography technique that can be used to separate the fullerene mixture is high pressure liquid chromatography (HPLC). HPLC instruments can be usually found in organic chemistry laboratories, and typically they have much higher separation efficiencies than the gravity

[35]W. A. Scrivens, P. V. Bedworth, and J. M. Tour, *J. Am. Chem. Soc.* **114**, 7917 (1992).
[36]K. Chatterjee, D. H. Parker, P. Wurz, K. R. Lyke, D. Gruen, and L. M. Stock, *J. Org. Chem.* **57**, 3253 (1992).
[37]K. C. Khemani, M. Prato, and F. Wudl, *J. Org. Chem.* **57**, 3254 (1992).
[38]W. A. Scrivens and J. M. Tour, *J. Org. Chem.* **57**, 6932 (1992).

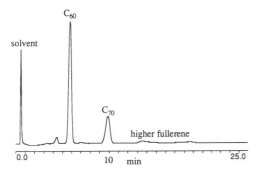

FIG. 7. HPLC chromatograph of the toluene soluble extract from the soot. The chromatograph, which corresponds to the optical density at 310 nm as a function of time, illustrates that C_{60}, C_{70}, and higher (C_n, $n > 70$) fullerenes can be cleanly separated using this experimental technique.

and low pressure techniques already described. Typically, HPLCs also incorporate in-line absorption monitors so that it is possible to quantify the separation process as it proceeds. The optical absorption of the eluant as a function of time for a HPLC separation of a fullerene mixture is shown in Fig. 7. Clearly, the peaks corresponding to C_{60} and C_{70} are well separated under these experimental conditions. If the solution volumes corresponding to the absorption peaks were collected from successive runs, it would then be possible to isolate macroscopic quantities of C_{60} and C_{70}. A drawback of the HPLC technique is, however, that most laboratory research instruments are only equipped for milligram-scale separations. Nevertheless, it is a very convenient technique for assaying the purity of C_{60} and C_{70} samples.

3. FULLERENE CHARACTERIZATION

Methods used to isolate large quantities of C_{60} and C_{70} were reviewed in Section 2. These techniques can provide pure C_{60} or C_{70} that is ideal for solid state studies. It is also important, however, to verify the purity of individual samples, since impurities (e.g., C_{70} or other hydrocarbons in C_{60}) can significantly affect solid state properties. Here we review several techniques that can be used to verify the identity and to ascertain the purity of fullerene samples; these include (1) nuclear magnetic resonance (NMR) spectroscopy; (2) mass spectroscopy; (3) optical spectroscopy; (4) HPLC.

a. *NMR Spectroscopy*

Solution phase carbon-13 NMR spectroscopy is a powerful probe of fullerene purity and structure.[39–42] For example, all 60 carbon nuclei are equivalent in C_{60}, and thus, a ^{13}C-NMR spectrum containing purified C_{60} should exhibit only a single resonance corresponding to the pure fullerene. Indeed, ^{13}C spectra recorded on C_{60} isolated using the procedure we have described show only a single resonance, at 143.2 ppm (Fig. 8).

In contrast, NMR spectra recorded on solutions of purified C_{70} exhibit five distinct resonance at 130.8, 144.4, 147.8, 148.3, and 150.8. These five resonances reflect the lower symmetry of C_{70} versus C_{60}.[42] The significant differences in these spectra indicate that NMR could be used to determine readily the presence of C_{70} in C_{60} samples. Consideration of the sensitivity of typical NMR instruments indicates, however, that impurities can only be detected reliably at the 1% level. Hence, observation of a single resonance in spectra of C_{60} solutions implies $\geq 99\%$ sample purity.

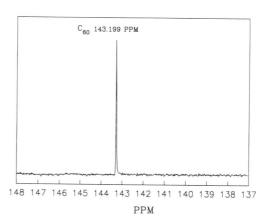

FIG. 8. Carbon-13 NMR spectrum obtained on a sample of purified C_{60} dissolved in C_6D_6. The single peak at 143.2 ppm corresponds to the 60 equivalent carbon atoms in the C_{60} cluster; the absence of other resonances in the spectrum shows that the sample contains only C_{60} at the detection limit of the NRM.

[39] R. Taylor, J. P. Hare, A. K. Abudl-sada, and H. W. Kroto, *J. Chem. Soc., Chem. Commun.* 1423 (1990).

[40] R. D. Johnson, G. Meijer, and D. S. Bethune, *J. Am. Chem. Soc.* **112,** 8983 (1990).

[41] C. S. Yannoni, P. P. Bernier, D. S. Bethune, G. Meijer, and J. R. Salem, *J. Am. Chem. Soc.* **113,** 3190 (1991).

[42] R. D. Johnson, G. Meijer, J. R. Salem, and D. S. Bethune, *J. Am. Chem. Soc.* **113,** 3619 (1991).

b. *Mass Spectroscopy*

Another technique commonly used to determine the composition and purity of fullerene samples is mass spectroscopy. As discussed earlier, mass spectroscopy can provide a quick survey of the different fullerene clusters in carbon soot (Fig. 4). There are several important points, however, that must be considered when using mass spectroscopy as an analytical probe of sample purity. First, the evaporation and ionization process must be carefully controlled to prevent generation of new carbon species from a pure sample. For example, electron impact desorption and ionization typically results in extensive fragmentation of C_{60}. Mass spectra obtained in this way contain a host of peaks with $m/z < 720$, which could incorrectly indicate the presence of hydrocarbon impurities. Desorption and ionization techniques that have been used to determine reliably fullerene sample purity include field desorption and fast atom bombardment (FAB).[43,44] Analysis of purified C_{60} and C_{70} samples using these methods yield spectra corresponding predominantly to C_{60}^+ and C_{70}^+, respectively. Another technique that has been used by several groups to produce molecular ions without significant fragmentation is laser desorption.[43,45] Hence, these last three techniques can be used to analyze the purity or composition of fullerene solid samples. The sensitivity of mass spectrometers to impurities is, however, dependent on a number of experimental and instrumental factors. For this reason, spectra recorded on commercial instruments generally cannot be used to assure purity to better than 99%.

c. *Optical Spectroscopy*

Optical spectroscopy, covering the ultraviolet through infrared regions of the electromagnetic spectrum, represents a third powerful tool for characterizing fullerene clusters. Fullerenes exhibit size- and structure-dependent ultraviolet-visible (UV-VIS) absorption spectra that are due to electronic transitions within the clusters. Qualitatively, pure C_{60} solutions exhibit a deep purple color, C_{70} solutions show an intense wine red color, and higher fullerenes (i.e., C_{76} to C_{94}) exhibit yellow to greenish

[43]S. W. McElvany and M. M. Ross, *J. Am. Soc. Mass. Spectrom.* **3**, 268 (1992).
[44]D. M. Cox, S. Behal, M. Disko, S. M. Gorun, M. Greaney, C. S. Hsu, E. B. Kollin, J. Millar, J. Robbins, W. Robbins, R. D. Sherwood, and P. Tindall, *J. Am. Chem. Soc.* **113**, 2940 (1991).
[45]D. H. Parker, P. Wurz, K. Chatterjee, K. R. Lykke, J. E. Hunt, M. J. Pellin, J. C. Hemminger, D. M. Gruen, and L. M. Stock, *J. Am. Chem. Soc.* **113**, 7499 (1991).

colours.[46–48] These variations can be attributed to the fact that the optical transition between the highest occupied molecular orbital and lowest unoccupied molecular orbital becomes progressively smaller with increasing size, and thus the larger fullerenes absorb light at increasingly longer wavelengths.

In the specific cases of C_{60} and C_{70}, detailed features in the absorption spectra provide an unambiguous method for distinguishing between C_{60} and C_{70} or for detecting C_{70} impurities within a C_{60} sample. Representative spectra for C_{60} and C_{70} are shown in Fig. 9. The UV-VIS absorption spectrum of C_{60} exhibits strong peaks at 213, 257, and 329 nm, and weak, broad absorption between 450 and 600 nm. On the other hand, C_{70} exhibits well-defined absorption peaks at 214, 236, 331, 360, 378, and 468 nm. In comparing the C_{60} and C_{70} spectra, it is apparent that the visible absorption at 468 nm in C_{70} is one of the most clear features by which to distinguish C_{70} from C_{60} or to detect the presence of C_{70} in a C_{60} sample. Using a reference pure C_{60} spectrum, it is possible (by focusing

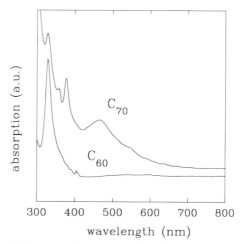

FIG. 9. Ultraviolet-visible absorption spectra of purified samples of C_{60} (bottom) and C_{70} (top) dissolved in hexane solvent. These spectra illustrate the large difference in optical density at 460 nm between C_{60} and C_{70}. This difference can be used to assess the relative purity of samples.

[46]R. Ettl, I. Chao, F. Diederich, and R. L. Whetten, *Nature* **353,** 149 (1991).

[47]L. D. Lamb, D. R. Huffman, R. K. Workman, S. Howells, T. Chen, D. Sarid, and R. F. Ziolo, *Science* **255,** 1413 (1992).

[48]K. Kikuchi, N. Nakahara, T. Wakabayashi, M. Honda, H. Matsumiya, T. Moriwaki, S. Suzuki, H. Shiromaru, K. Yamauchi, I. Ikemato, and Y. Achiba, *Chem. Phys. Lett.* **188,** 177 (1992).

on the region around 468 nm) to detect C_{70} impurity in an unknown C_{60} sample at the 0.1% level. Hence, optical spectroscopy is one of the most sensitive methods for determining the purity of C_{60} samples.

Infrared (IR) absorption spectroscopy is also a useful method for characterizing C_{60} and other fullerene samples. In the case of C_{60}, group theoretical analysis of the truncated icosohedron shows that there should be only 4 IR active modes for C_{60}. These four bands are experimentally observed at 1428, 1183, 577, and 527 cm^{-1}. The observation of additional bands within this region thus indicates the presence of impurities in a C_{60} sample. Since strong C–H bending modes occur within this spectral region, the observation of more than four absorptions in a spectra of C_{60} typically indicates the presence of hydrocarbon impurities.

d. *HPLC*

In Section 2 we discussed the use of chromatography for isolating pure fullerene clusters. While HPLC can be used to separate macroscopic quantities of fullerenes, it is also very useful as an analytical technique for characterizing the purity of fullerene samples. For fixed experimental conditions (i.e., stationary phase, solvent, and flow rate) C_{60}, C_{70}, and other materials elute from the column at specific and reproducible times (e.g., Fig. 6). Hence, measurements of the optical absorption as a function of time represent a straightforward assay for the presence of C_{70} (and other impurities) in a C_{60} sample. Because the intense UV bands are monitored in the HPLC experiment, it is possible to detect at least 0.1% impurities. Analytical HPLC thus represents one of the most sensitive techniques available to assess fullerence sample purity.

4. ISOTOPICALLY SUBSTITUTED FULLERENES

In the foregoing sections we have reviewed techniques needed to prepare, separate, and characterize fullerene carbon clusters. Before reviewing the preparation of solid materials using these pure clusters, we will discuss how the foregoing procedures can be generalized to produce modified fullerene clusters. Specifically, we discuss procedures that can be used to prepare C_{60} substituted with carbon-13. Because isotopic substitution is a powerful probe of solid state processes (see the chapter on fullerene superconductors), it is an essential topic to cover in the preparation of fullerene clusters. Here we concentrate on the preparation and characterization of ^{13}C substituted C_{60}.

A number of techniques have been reported for preparing ^{13}C-

substituted C_{60}, $(^{13}C_x {}^{12}C_{1-x})_{60}$.[24,25,49–52] First, holes in dense ^{12}C rods have been filled with mixtures of ^{13}C powder and a binder. These rods can then be vaporized using the standard arc discharge method. Since the rods contain an inhomogeneous mix of ^{13}C in a ^{12}C binder, and relatively large regions of pure ^{12}C rod, it is not surprising that the C_{60} obtained from these rods contains a poorly controlled mass distribution as well as pure $^{12}C_{60}$.[49–51] Such inhomogeneous samples are not useful for quantitative studies of solid state properties. A second approach that appears to enable more control of ^{13}C substitution was reported by Ramirez and coworkers.[52] In their procedure, ^{13}C powder is mixed with a binder containing 90% ^{13}C, the resulting viscous material is formed into a rod, and then the rod is graphitized at 2800°C. The isotopic composition of the C_{60} material obtained after arc discharge vaporization of the rods differed, however, from the calculated starting isotopic composition and contained some pure $^{12}C_{60}$. It is believed that the ^{12}C impurities observed in the final product probably were incorporated into the rods during the high temperature graphitization step.

In contrast to the shortcomings of the foregoing procedures, we have developed a general approach that is capable of yielding reproducible $(^{13}C_x {}^{12}C_{1-x})_{60}$ samples for all values of x.[24,25] In this procedure a mixture of ^{13}C and ^{12}C powder in the desired stoichiometry (x) is extensively ground, placed in a quartz tube between two tantalum electrodes, and then the $^{13}C_x {}^{12}C_{1-x}$ mixture is converted to dense carbon rods by resistive heating under pressure. A schematic view of the apparatus used to form these homogeneous $^{13}C_x {}^{12}C_{1-x}$ rods is shown in Fig. 10.

The dense rods produced by this procedure can be vaporized using the arc discharge[24] or laser ablation[25] techniques, although we have found that laser ablation is more efficient for the small rods produced by this method.

Because this procedure simply relies on the mixing of pure ^{13}C and ^{12}C powders, it is very reproducible and readily yields controlled ^{13}C substitution in C_{60}. For example, the mass spectra recorded on C_{60} samples prepared from $^{13}C_x {}^{12}C_{1-x}$ rods with $x = 1$ and $x = 0.5$ show that the C_{60} has virtually the same mass as the starting carbon rod (Fig. 11).

[49] R. D. Johnson, G. Meijer, J. R. Salem, and D. S. Bethune, *J. Am. Chem. Soc.* **113**, 3619 (1991).

[50] J. M. Hawkins, S. Loren, A. Meyer, and R. Nunlist, *J. Am. Chem. Soc.* **113**, 7770 (1991).

[51] T. W. Ebbesen, J. S. Tsai, K. Tanigaki, J. Tabuchi, Y. Shimakawa, Y. Kubo, I. Hirosawa, and J. Mizuki, *Nature* **355**, 620 (1992).

[52] A. P. Ramirez, A. R. Kortan, M. J. Rosseinsky, S. J. Duclos, A. M. Mujsce, R. C. Haddon, D. W. Murphy, A. V. Makhija, C. M. Zahurak, and K. B. Lyons, *Phys. Rev. Lett.* **68**, 1058 (1992).

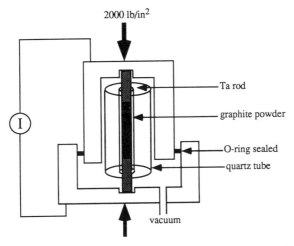

FIG. 10. Schematic illustration of the apparatus used to make dense $^{13}C_x\,^{12}C_{1-x}$ rods using mixtures of only ^{13}C and ^{12}C carbon powders.

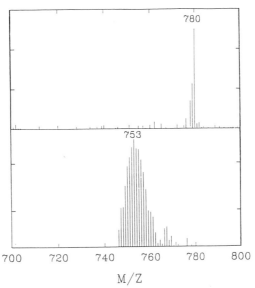

FIG. 11. Fast atom bombardment mass spectra of purified samples of $^{13}C_{60}$ (top) and $(^{13}C_{0.55}\,^{12}C_{0.45})_{60}$ (bottom). The matrix used for the fast atom bombardment mass spectroscopy experiments was 3-nitrobenzyl alcohol.

The use of these isotopically substituted fullerenes will be discussed in detail in the chapter on fullerene superconductivity.

III. Metal-Doped C_{60} Materials

Now that the methods needed to prepare and isolate macroscopic quantities of C_{60} have been reviewed, we turn to the preparation materials based on this carbon cluster building block. In this section we will first discuss the preparation of bulk and thin film C_{60} solids. We will then review the procedures and compounds that can be prepared via metal doping reactions, and finally we will discuss new research directions.

5. C_{60} Solids and Thin Films

a. *Polycrystalline Powders*

Polycrystalline C_{60} is readily obtained by evaporating the solvent from C_{60} solutions. These solutions are obtained from the chromatographic purifications already described (Section 2). This material obtained directly from solvent evaporation is not suitable for solid state studies, however, since significant solvent can be trapped within the lattice. Because solvent molecules can affect the cluster packing and because these molecules may react with dopant species, it is essential to remove solvent impurities from the lattice. The most effective method for removing trapped solvent molecules from solid C_{60} involves heating the solid under vacuum. Typical conditions of temperature and pressure used for this process are 200 to 250°C and $<10^{-3}$ torr. Importantly, C_{60} should not be dried in air, since it may be oxidized in the presence of oxygen at elevated temperature.[53,54]

The vacuum-dried C_{60} powder is an excellent starting material for the preparation of metal-doped solids, since it is very pure. The crystallite grain size of these powders can, however, be quite small ($<100\,Å$). Because the grain size can significantly affect solid state properties, it is important to consider methods by which the grain size can be maximized. First, the rate of solvent evaporation from the starting C_{60} solution will

[53]J. Milliken, T. M. Keller, A. P. Baronavski, S. W. McElvany, J. H. Callahan, and H. H. Nelson, *Chem. Mater.* **3**, 387 (1991).
[54]H. S. Chen, A. R. Kortan, R. C. Haddon, M. L. Kaplan, C. H. Chen, A. M. Mujsce, H. CHou, and D. A. Fleming, *Appl. Phys. Lett.* **59**, 2956 (1991).

affect the crystal grain size; slow solvent removal promotes the growth of large crystallites. Second, extended vacuum annealing at elevated temperature promotes the growth of crystal grains. Taking into account these steps can yield polycrystalline C_{60} solids with approximately 1000 Å sized grains; however, poorly prepared samples may have crystallites as small as 60 Å. In this latter case, the granularity of the solid can severely affect the observed results (see following chapter on fullerene superconductivity).[55]

b. *Single Crystals*

There has also been considerable effort directed toward the growth of high quality single crystals of C_{60}. This work represents an important area of research, since measurements made on single crystals often provide the deepest insight into physical properties.

In general, two approaches have been taken to grow C_{60} single crystals. The first, which is based on the original work of Krätschmer and Huffman, involves growth of the crystals from a supersaturated solution.[2,56,57] Using this method it is possible to obtain crystals with dimensions in excess of 1 mm. Unfortunately, these crystals generally consist of C_{60} co-crystallized with the solvent from which they were grown. Co-crystallized solvent may change the crystal structure of the solid C_{60},[2,11,56] and furthermore, it may react with metal dopants. Hence, we believe that solution phase growth is not a useful method for preparing pure C_{60} crystals that can be used in detailed physical studies.

A second technique better suited to preparation of high quality, pure C_{60} crystals involves vapor phase growth.[58-61] Briefly, purified and vacuum-dried C_{60} powder is sealed in a quartz tube, either under vacuum or with an inert gas such as argon, and then this tube is placed in a gradient furnace with the source end at approximately 600°C and the

[55]T. T. M. Palstra, R. C. Haddon, A. F. Hebard, and J. Zaanen, *Phys. Rev. Lett.* **68,** 1054 (1992).

[56]R. M. Fleming, A. R. Kortan, B. Hessen, T. Siegrist, F. A. Thiel, P. Marsh, R. C. Haddon, R. Tycko, G. Dabbagh, M. L. Kaplan, and A. M. Mujsce, *Phys. Rev. B* **44,** 888 (1991).

[57]Y. Yoshida, T. Arai, and H. Suematsu, *Appl. Phys. Lett.* **61,** 1043 (1992).

[58]R. L. Meng, D. Ramirez, X. Jiang, P. C. Chow, C. Diaz, K. Matsuishi, S. C. Moss, P. H. Hor, and C. W. Chu, *Appl. Phys. Lett.* **59,** 3402 (1992).

[59]X. D. Xiang, J. G. Hou, G. Briceno, W. A. Vareka, R. Mostovoy, A. Zettl, V. H. Crespi, and M. L. Cohen, *Science* **256,** 1190 (1990).

[60]C. Wen, J. Li, K. Kitazawa, T. Aida, I. Honma, H. Komiyama, and K. Yamada, Appl. Phys. Lett. **61,** 2162 (1992).

[61]J. Li, S. Komiya, T. Tamura, C. Nagasaki, J. Kihara, K. Kishio, and K. Kitazawa, *Physics C* **195,** 205 (1992).

growth end at about 500°C. After one week of growth under these conditions, it is possible to isolate pure single crystals with dimensions of up to 1 mm on edge.

c. *Thin films*

There has also been considerable effort directed toward the growth of high quality C_{60} thin films since such samples are ideally suited for many physical measurements.[62–68] Because C_{60} sublimes at relatively low temperatures, it is straightforward to deposit thin films of this material. Early thin film growth studies utilized crude C_{60}/C_{70} mixtures as a source material under the assumption that the more volatile C_{60} would be preferentially deposited during evaporation.[2,62,68] Subsequent scanning tunneling microscopy (STM) studies indicate, however, that C_{70} is incorporated into the films.[62] Since C_{70} "impurities" may significantly affect transport and other properties, this method of film preparation is unsuitable for high quality physical studies.

More recently, C_{60} films have been deposited using pure C_{60} source material in high and ultrahigh vacuum.[65–67] Crystalline films have been grown on a variety of substrates (e.g., KBr, glass, or Si) with source temperatures between 573 and 773 K and the substrate at ambient (300 K) temperature. X-ray diffraction and STM structural analyses of films grown at 300 K indicate that these materials consist of randomly oriented crystalline domains with sizes ranging from 50 to 150 Å. The granular character of these films has important consequences in the interpretation of transport measurements (see the chapter on fullerene superconductivity).

Several strategies have been investigated to reduce the granularity of C_{60} thin films. First, depositions have been carried out with the substrates at elevated temperatures.[67,68] Structural studies of films deposited on

[62] R. J. Wilson, G. Meijer, D. S. Bethune, R. D. Johsnon, D. D. Chambliss, M. S. de Vries, H. E. Hunziker, and H. R. Wendt, *Nature* **348,** 621 (1990).

[63] A. F. Hebard, R. C. Haddon, R. M. Fleming, and A. R. Kortan, *Appl. Phys. Lett.* **59,** 2109 (1991).

[64] P. Dietz, K. Fostiropoulos, W. Krätschmer, and P. K. Hansma, *Appl. Phys. Lett.* **60,** 62 (1992).

[65] G. P. Kochanski, A. F. Hebard, R. C. Haddon, and A. T. Fiory, *Science* **255,** 184 (1992).

[66] Y. Z. Li, J. C. Patrin, M. Chander, J. H. Weaver, L. P. F. Chibante, and R. E. Smalley, *Science* **252,** 547 (1991).

[67] Y. Z. Li, M. Chander, J. C. Patrin, J. H. Weaver, L. P. F. Chibante, and R. E. Smalley, *Science* **253,** 429 (1991).

[68] D. Schmicker, S. Schmidt, J. G. Skofronick, J. P. Toennies, and R. Vollmer, *Phys. Rev. B* **44** 10995 (1991).

substrates heated between 400 and 500 K show that elevated temperature growth enhances grain size. Because the upper temperature limit for growth is limited by the relatively low sublimation temperature of C_{60}, it is unclear how much film quality will be improved using this strategy. A second approach to improving film quality involves growth on a lattice-matched substrate. For example, Schmicher et al. have reported epitaxial growth of C_{60} monolayers on cleaved mica surfaces at 300 K[68]; the tensile strain in the monolayer was 1% based on the bulk C_{60} lattice parameters. Although this result is promising, it is uncertain how well epitaxial growth can be sustained in multilayer films of this van der Waals solid. Further studies of the growth of C_{60} films on lattice-matched substrates at different growth temperatures are clearly needed to advance this area.

d. C_{60} Structure

The structural properties of C_{60} solids will be reviewed in detail in the chapter by Axe, Moss, and Neumann. Here we briefly introduce key features of solid C_{60} that are important to understand metal doping of this solid.

On the basis of x-ray, neutron, and electron diffraction studies it is now well established that C_{60} crystallizes in a face-centered cubic (fcc) lattice with a lattice constant of 14.17 Å at room temperature (Fig. 12).[11,69–74] In this structure the separation between nearest neighbor C_{60} clusters is 10 Å. The distance between adjacent 7.1-Å-diameter C_{60} molecules is thus 2.9 Å. Interestingly, comparison of this intercluster separation with the 3.35-Å interplanar separation in graphite indicates that the C_{60} clusters may be more strongly coupled than the carbon planes in graphite after normalizing for the contact area.

An important feature of an fcc structure made up from these relatively large, spherical clusters is that there are sizable, well-defined intercluster holes within the lattice. These holes constitute 26% of the total cell

[69]S. Liu, Y.-J. Lu, M. M. Kappes, and J. A. Ibers, *Science* **254**, 408 (1991).

[70]P. A. Heiney, J. E. Fisher, A. R. McGhie, W. J. Romanow, A. M. Denenstein, J. P. McCauley, and A. B. Smith, *Phys. Rev. Lett.* **66**, 2911 (1991).

[71]J. E. Fisher, P. A. Heiney, A. R. McGhie, W. J. Romanow, A. M. Denenstein, J. P. McCauley, and I. A. B. Smith, *Science* **252**, 1288 (1991).

[72]W. I. F. David, R. M. Ibberson, J. C. Matthewman, K. Prassides, T. J. S. Dennis, J. P. Hare, H. W. Kroto, R. Taylor, and D. R. M. Walton, *Nature* **353**, 147 (1991).

[73]D. A. Neumann, J. R. D. Copley, R. L. Cappelletti, W. A. Kamitakhara, R. M. Lindstrom, K. M. Creegan, D. M. Cox, W. J. Romanow, N. Coustel, J. P. McCauley, Jr., N. C. Maliszewskyj, J. E. Fischer, and A. B. Smith III, *Phys. Rev. Lett.* **67**, 3808 (1991).

[74]J. Q. Li, Z. X. Zhao, D. B. Zhu, Z. Z. Gan, and D. L. Yin, *Appl. Phys. Lett.* **59**, 3108 (1991).

FIG. 12. Model illustrating the packing of individual clusters in the fcc lattice solid C_{60}. The C_{60} clusters are represented by gray shaded spheres in this model.

volume. On a per C_{60} basis, there is one octahedral hole and two tetrahedral holes (Fig. 13). These octahedral and tetrahedral holes have radii of 2.06 and 1.12 Å, respectively, and thus these spaces can in principle accommodate a large variety of species without a change in the basic C_{60} structure. In the next section we discuss general considerations for the preparation of metal-doped C_{60} solids and review specific compounds formed using alkali and alkaline earth metal dopants.

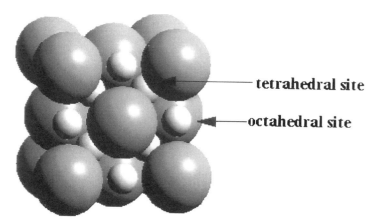

tetrahedral site

octahedral site

FIG. 13. Model illustrating the packing in the alkali metal–doped M_3C_{60} fcc solid. The alkali metal dopants, which are represented by light gray shaded spheres, fill the octahedral and tetrahedral holes that exist in the fcc C_{60} lattice.

6. Considerations for Metal Doping

Both the octahedral and tetrahedral sites can in principle accommodate metal dopant species. The process of adding metals to these lattice holes in solid C_{60} is similar in many respects to the intercalation of metals and other species into the interlayer region of graphite. Hence, it is reasonable to utilize procedures developed for the preparation of graphite intercalation compounds as a starting point for the preparation of metal-doped C_{60} solids.

One of the most common procedures for preparing graphite intercalation compounds involves vapor phase diffusion of a volatile dopant species into the interplanar regions of the solid. Not surprisingly, vapor phase diffusion also constitutes one of the most successful methods used to prepare high quality metal-doped C_{60} samples.

In general, there are several important issues to consider in the vapor phase doping process. First, successful intercalation into the holes in the C_{60} lattice requires that the metal dopant exhibit significant vapor pressure at a temperature below which the metal reacts with C_{60} and below which C_{60} decomposes. Second, the diffusion of dopants within crystallites of a solid C_{60} is much slower than previously found in graphite intercalation compounds. This behavior is due to the fact that the dopant ions in C_{60} must hop between well-separated hole sites in the lattice versus a simple two-dimensional diffusion through interlayer regions in graphite. A schematic view of how the metal doping process is believed to proceed in polycrystalline C_{60} samples is illustrated in Fig. 14. In this general model, vapor phase metal initially condenses around the grains in the solid. Extensive annealing is then needed to achieve an equilibrium distribution of metal dopant ions within the lattice. Notably, this model suggests that it may be very difficult to achieve uniform doping in single

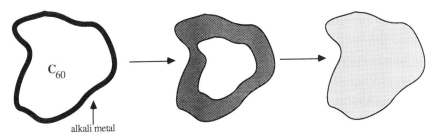

FIG. 14. Schematic view of the "vapor phase" doping process. Initially, alkali metal vapor deposits as a layer on C_{60} grains, and then this metal diffuses into the solid.

crystal samples. A final consideration in working with metal-doped C_{60} solids is that they are air sensitive. This air sensitivity arises from the fact that metal intercalation involves charge transfer from metal to C_{60}, and thus the doped solids are highly reduced and can react with oxygen and water vapor in the atmosphere. Thus, it is essential to always handle the metal-doped C_{60} solids in an inert gas or vacuum environment. In Sections 7 and 8 we will elaborate on these points for the specific cases of alkali and alkaline earth metal–doped C_{60} solids.

7. ALKALI METAL–DOPED FULLERENES

Much of the initial explosion of activity in fullerene research was fueled by the reports from Bell Laboratories of conductivity and then supercon-ductivity in potassium- and rubidium-doped C_{60}.[12,14,75–78] Extensive studies carried out since these initial reports have resulted in the characterization of many of the properties of the superconducting phase. These properties will be discussed in detail in the chapter on fullerene superconductivity. Here we briefly review only those properties relevant to understanding the preparation of these solids.

It is now well established that the superconducting phase of alkai metal–doped C_{60} solids is derived from the fcc C_{60} lattice by filling the single octahedral and two tetrahedral holes with metal ions. This filling yields an overall stoichiometry M_3C_{60}, where M is an alkali metal. A general structural model for the M_3C_{60} structure that highlights the octahedral and tetrahedral sites in the lattice was shown in Fig. 13. comparison of this structure with that for pure C_{60} (Fig. 12) demonstrates that doping to a $3:1\ M:C_{60}$ ratio only fills the holes in the lattice and does not result in gross structural changes.

In addition, it is possible to prepare C_{60} solids in which the $M:C_{60}$ ratio is greater than $3:1$. Specifically, compounds with stoichiometries of

[75]R. C. Haddon, A. F. Hebard, M. J. Rosseinsky, D. W. Murphy, S. J. Duclos, K. B. Lyons, B. Miller, J. M. Rosamila, R. M. Fleming, A. R. Kortan, S. H. Glarum, A. V. Makhija, A. J. Muller, R. H. Eick, S. M. Zahurak, R. Tycko, G. Dabbagh, and F. A. Thiel, *Nature* **350**, 320 (1991).
[76]A. F. Hebard, M. J. Rosseinsky, R. C. Haddon, D. W. Murphy, S. H. Glarum, T. T. M. Palstra, A. P. Ramirez, and A. R. Kortan, *Nature* **350**, 600 (1991).
[77]M. J. Rosseinsky, A. P. Ramirez, S. H. Glarum, D. W. Murphy, R. C. Haddon, A. F. Hebard, T. T. M. Palstra, A. R. Kortan, S. M. Zahurak, and A. V. Maikhija, *Phys. Rev. Lett.* **66**, 2830 (1991).
[78]P. W. Stephens, L. Mihaly, P. L. Lee, R. L. Whetten, S.-M. Huang, R. Kaner, F. Deiderich, and K. Holczer, *Nature* **351**, 632 (1991).

M_4C_{60} and M_6C_{60} have been prepared and structurally characterized.[12,79,80] Neither of these metal-doped materials has the fcc structure of solid C_{60}, but rather both are derived from more open body-centered cells. X-ray diffraction investigations have shown that M_4C_{60} has a body-centered tetragonal (bct) cell, while M_6C_{60} adopts a body-centered cubic (bcc) structure. The unit cells for these two solids are illustrated in Fig. 15. Although the M_4C_{60} and M_6C_{60} compounds are nonsuperconducting, we mention them here since they represent well-defined compounds in the alkali-metal/C_{60} phase diagram.

A number of strategies have been proposed for the preparation of alkali metal-doped C_{60}. These are reviewed in the following subsections with an emphasis on procedures that can be used to prepare high quality M_3C_{60} superconductors.

a. *Vapor Phase Doping*

As has been indicated, vapor phase doping of C_{60} has been a popular and successful technique. The high volatility of all of the alkali metals except Li at low temperature enables doping to be carried out under mild conditions where C_{60} remains stable. A general protocol for doping is as follows: First, polycrystalline C_{60} solid is dried in vacuum at elevated temperature and then transferred to an inert atmosphere glove box. A weighed sample of C_{60} (10–30 mg) is then placed in a pyrex or quartz reaction tube, and then a stoichiometric amount of alkali metal is added to the tube. A convenient and relatively accurate method of adding the metal dopant is to use measured lengths of a glass capillary tube filled with the alkai metal. The reaction tube containing C_{60} and alkali metal are sealed with a high vacuum valve, and then are removed from the inert atmosphere chamber, evacuated, and sealed under high vacuum. The sealed reaction tubes are heated between 200 and 400°C to intercalate the alkali metal into the C_{60} lattice. Although the M_3C_{60} superconducting phase can be detected by magnetization measurements in less than 12 hours, most of the sample remains undoped, as illustrated in Fig. 14. Inhomogeneous doping is an even greater problem for single crystal samples since the surface-to-volume ratio is so small, while in thin films it appears that homogeneous doping can be achieved in a matter of minutes.[65]

[79]R. M. Fleming, M. J. Rosseinsky, A. P. Ramirez, D. W. Murphy, J. C. Tully, R. C. Haddon, T. Siegrist, R. Tycko, S. H. Glarum, P. Marsh, G. Dabbagh, S. M. Zahurak, A. V. Makhija, and C. Hampton, *Nature* **352**, 701 (1991).

[80]O. Zhou, J. E. Fischer, N. Coustel, S. Kycia, Q. Zhu, A. R. McGhie, W. J. Romanow, J. P. McCauley, A. B. Smith, and D. E. Cox, *Nature* **351**, 461 (991).

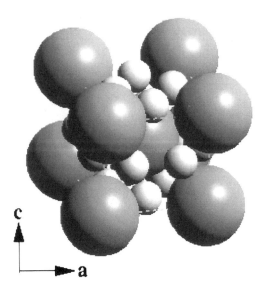

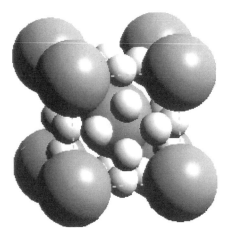

FIG. 15. Models illustrating the packing of C_{60} clusters (gray shaded spheres) alkali metal dopants (light gray shaded spheres) in the body-centered tetragonal M_4C_{60} (top) and body-centered cubic M_6C_{60} (bottom) solids.

Since preparation of samples with homogeneous dopant distributions is essential for physical studies, it is important to consider methods for improving the doping homogeneity in the vapor phase technique. A straightforward approach that has been used by several groups involves annealing at increasingly higher temperatures ranging from 250 to 400°C.[81] The effectiveness of this annealing technique is limited, however, by the poor contact between crystalline grains in powder samples, the stability of other nonsuperconducting phases (e.g., M_4C_{60} and M_6C_{60}), and sample decomposition. The contact between grains can be improved by using pressed pellet samples; however, the other limitations are intrinsic to the C_{60} system. Alternatively, vapor phase doping can be carried out with more than a six to one excess of rubidium to prepare first Rb_6C_{60}.[82] Because no more than six Rb ions can be intercalated into the C_{60} lattice, it is possible to synthesize Rb_6C_{60} as a well-defined, homogeneous compound. Extensive mixing and subsequent heating of equal molar ratios of Rb_6C_{60} and C_{60} can then yield good samples of Rb_3C_{60}.[82]

b. *Metal Alloys and Metal-Doped C_{60}*

To enhance the doping process, we proposed early on that metal alloys be used to intercalate solid C_{60}.[83,84] In these intercalation reactions a metal alloy such as KHg or RbHg is mixed with C_{60} and then heated in a sealed tube as described in Section 7.a. There are several features of this alloy method that make it an attractive approach for preparing metal-doped C_{60}. First, alkali-metal/Hg alloys are hard solids in comparison with pure alkali metals. It is thus possible to grind the MHg alloys into a fine powder and intimately mix this powder with the C_{60} prior to heating. Second, the increased molar mass and greater stability of the alloys (i.e., they are less reactive toward O_2 and H_2O impurities) enable the stoichiometry to be controlled much more accurately than when using pure alkali metal dopants. Third, we have found that intercalation generally proceeds more rapidly at low temperature than when using pure alkali metals.[84] This enhanced reactivity is probably due to the intimate C_{60}/alloy mixture, since alkali metal diffusion is required within

[81]K. Holczer, O. Klein, S.-M. Huang, R. B. Kaner, K.-J. Fu, R. L. Whetten, and F. Diederich, *Science* **252,** 1154 (1991).

[82]J. P. McCauley, Q. Zhu, N. Coustel, O. Zhou, G. Vaughn, S. H. J. Idziak, J. E. Fischer, S. W. Tozer, D. M. Groski, N. Bykovetz, C. L. Lin, A. R. McGhie, B. H. Allen, W. J. Romanow, A. Denenstein, and A. B. Smith, *J. Am. Chem. Soc.* **113,** 8537 (1991).

[83]S. P. Kelty, C.-C. Chen, and C. M. Lieber, *Nature* **352,** 223 (1991).

[84]C.-C. Chen, S. P. Kelty, and C. M. Lieber, *Science* **253,** 886 (1991).

C_{60} grains as in the case of vapor phase doping. One potential limitation of this alloy approach to doping is that components of the alloy other than the alkali metal (e.g., Hg) may be incorporated as an impurity in the C_{60} lattice. In the case of MHg alloys this does not appear to be a major limitation, since Hg can be distilled from the solid.

In addition to the formation of K_3-, Rb_3-, and $K_{3-x}Rb_xC_{60}$ solids using K-, Rb-, and $K_{1-x}Rb_xHg$ alloys,[84] respectively, this general idea has been used to prepare other alkali metal–doped C_{60} solids. For example, Rosseinsky and coworkers have used Na_5Hg_2 and NaH to prepare a variety of Na_xC_{60} and Na_2MC_{60} (M = Cs, Rb) solids.[85] They found that it was possible to achieve much greater control of the sodium stoichiometry in these solids using the alloys versus the pure Na metal. More generally, we believe that low melting point alloys may represent a useful approach to explore for the preparation of C_{60} solids doped with nonvolatile metals.

c. *Solution Phase Doping*

A final approach that has been used successfully to prepare alkali-metal/C_{60} superconductors involves reaction of C_{60} and alkali metal in an organic solvent.[86–88] The motivation behind solution phase doping is that stoichiometric reduction of a homogeneous C_{60} solution and subsequent precipitation should in principle lead to a homogeneously doped solid. Initial studies by Wang and coworkers showed that superconducting K_3C_{60} and Rb_3C_{60} could be obtained from the direct reduction of C_{60} in toluene using an excess of K or Rb metal, respectively.[86,87] The fraction of the superconducting phase observed in these studies, was, however, <10%. This low yield of the superconducting phase is not surprising since C_{60}^{1-} species are expected to precipitate as they are formed from the nonpolar solvent (toluene) used in this work. Indeed, more recent work demonstrates the MC_{60} is the dominant phase obtained from reduction in toluene.[88]

[85]M. J. Rosseinsky, D. W. Murphy, R. M. Fleming, R. Tycko, A. P. Ramirez, T. Siegrist, G. Dabbagh, and S. E. Barrett, *Nature* **356,** 416 (1992).

[86]H. H. Wang, A. M. Kini, B. M. Savall, K. D. Carlson, J. M. Williams, K. R. Lykke, P. Wurz, D. H. Parker, M. J. Pellin, D. M. Gruen, U. Welp, W.-K. Kowk, S. Fleshler, and G. W. Crabtree, *Inorg. Chem.* **30,** 2838 (1991).

[87]H. H. Wang, A. M. Kini, B. M. Savall, K. D. Carlson, J. M. Williams, M. W. Lathrop, K. R. Lykke, D. H. Parker, P. Wurz, M. J. Pellin, D. M. Gruen, U. Welp, W.-K. Kwok, S. Fleshler, G. W. Crabtree, J. E. Schirber, and D. L. Overmyer, *Inorg. Chem.* **30,** 2962 (1991).

[88]J. A. Schlueter, H. H. Wang, M. W. Lathrop, U. Geiser, K. D. Carlson, J. D. Dudek, G. A. Yaconi, and J. M. Williams, *Chem. Mat.* **5,** 720 (1993).

To increase the ratio of $M : C_{60}$ in the solid thus requires a solvent that can dissolve both C_{60} and C_{60}^{n-} ions. This solvation can be accomplished with toluene/benzonitrile mixtures, since benzonitrile can effectively solvate C_{60}^{n-} anions in solution. Hence, reduction of C_{60} in toluene/benzonitrile mixtures using either K or Rb metal and subsequent precipitation of the M–C_{60} solid yield materials with superconducting fractions in excess of 50%. Advantages of the solution phase approach to doping include (1) doping can be carried out on a large scale with little influence on the product homogeneity, and (2) the process can be carried out at significantly lower temperatures ($<110°$C) than possible with either the vapor phase or alloy methods of doping. Such low temperature doping may be utilized in the future to prepare new metastable phases. There is also a significant disadvantage to this approach, since it is known that solution phase C_{60} crystal growth traps solvent in the lattice. Hence, the presence and possible effects of trapped solvent molecules in the metal-doped C_{60} samples produced in solution must be addressed.

d. *Alkali Metal–Doped C_{60} Superconductors*

As discussed in Section 5, the superconducting phase of all of the known alkali metal–doped C_{60} superconductors has a stoichiometry of M_3C_{60} and a fcc structure with the metal ions (M) occupying the one octahedral and two tetrahedral holes in the lattice. The known superconducting phases reported to date with alkali metal dopants and the transition temperatures (T_c) of these materials are summarized in Table I; these compounds can be prepared in high yield using either the vapor phase or alloy doping methods.

Examination of Table I shows that in addition to the pure compounds K_3C_{60} and Rb_3C_{60}, there are a number of mixed metal and/or solid solution superconductors. The total alkali metal ion to C_{60} ratio in the

TABLE I. ALKALI METAL–DOPED
C_{60} SUPERCONDUCTORS

Material	T_c (K)
$N_{a2}CsC_{60}$	10–11
K_3C_{60}	19.2
$K_2R_bC_{60}$	22–23
$K_2C_sC_{60}$	24
$KR_{b2}C_{60}$	26–27
$R_{b3}C_{60}$	29.4
$R_{b2}C_sC_{60}$	31

solid solution compounds is, however, always $3:1$. There are also several notable absences from this table. First, the pure fcc phase Cs_3C_{60} has not yet been clearly observed, although bcc Cs_6C_{60} is a well-characterized solid.[80] It is apparent that Cs_3C_{60} is unstable relative to Cs_6C_{60} under conventional doping conditions (e.g., $>200°C$) since heating $3:1$ mixtures of $Cs:C_{60}$ yields products containing C_{60} and $Cs_{60}C_{60}$. It may be possible, however, to trap the Cs_3C_{60} phase by doping at much lower temperatures. Since Cs_3C_{60} is expected to have a higher T_c than any known fullerene superconductor, this will be an interesting area to investigate in the future. Second, the phase Na_3C_{60}, while it exists at $300 K$, is not superconducting.[85] It is now believed that the absence of superconductivity in this material is due to a structural phase transition below $250 K$ in which the material disproportionates into Na_2C_{60} and Na_6C_{60} domains. These results for the Cs and Na-C_{60} phases suggest that dopant size must be carefully considered when attempting to prepare new materials. If the ions are too large for the tetrahedral holes (i.e., Cs) or if they are too small for the octahedral holes (e.g., Na), then the desired fcc structure may be unstable.

8. ALKALINE EARTH METAL–DOPED C_{60}

For the alkali metals (Li, Na, K, Rb, Cs) only M_3C_{60} fcc solids have exhibited superconductivity (Table I). The discovery of superconductivity in calcium and barium metal–doped C_{60}[89–91] represents, however, structurally and compositionally distinct materials in which to explore fullerene superconductivity. The alkaline earth metal–doped C_{60} superconductors are, however, considerably more difficult to prepare than the alkali metal–doped materials already discussed. This difficulty is due in large part to the low vapor pressure of the alkaline earth metals at convenient reaction temperatures. As a consequence of their low vapor pressure, solid state diffusion is the dominant mechanism by which alkaline earth metal intercalation or doping occurs in bulk C_{60}. In contrast, vapor phase diffusion accounts for the transport of alkali metals to grains in the C_{60} powder, and solid state diffusion is only required within individual crystal grains. Hence, methods that enhance solid state

[89]A. R. Kortan, N. Kopylov, S. Glarum, E. M. Gyorgy, A. P. Ramirez, R. M. Fleming, F. A. Thiel, and R. C. Haddon, *Nature* **355,** 529 (1992).

[90]A. R. Kortan, N. Kopylov, S. Glarum, E. M. Gyorgy, A. P. Ramirez, R. M. Fleming, O. Zhou, F. A. Thiel, P. L. Trevor and R. C. Haddon, *Nature* **360,** 566 (1992).

[91]R. C. Haddon, G. P. Kochanski, A. F. Hebard, A. T. Fiory, and R. C. Morris, *Science* **258,** 1636 (1992).

diffusion are critical to the preparation of high quality alkaline earth metal–doped C_{60} materials. Here we discuss approaches that have been successfully used to prepare Ca- and Ba-doped C_{60} superconductors.

The first report of an alkaline earth metal–doped C_{60} superconductor was made by Kortan and coworkers, who found a single superconducting phase with a stoichiometry of Ca_5C_{60}.[89] This material was prepared by pressing stoichiometric mixtures of Ca and C_{60} into pellets within tantalum cells, sealing the cells in quartz tubes under high vacuum, and then heating the reaction tubes at temperatures between 500 and 600°C. The reaction temperatures required for doping C_{60} with Ca can be compared with the much lower temperatures, 200–350°C, used for the vapor phase doping of C_{60} with K or Rb. The success of the Ca-doping reaction is due to the enhanced solid state diffusion resulting from the use of pressed pellet samples (vs. loose powders) and relatively high reaction temperatures. In principle, it would be possible to increase the solid state diffusion rates further by carrying out these intercalation reactions at even higher temperatures; however, increased reaction temperatures may also lead to the decomposition of C_{60} itself.

Similar procedures have also been used by Kortan *et al.* to prepare a barium-doped C_{60} superconductor that has a stoichiometry of Ba_6C_{60}.[90] An interesting structural feature of this material is that it forms a bcc lattice in contrast to all of the known alkali metal–doped superconductors. Unfortunately, the preparation of this new Ba-doped material appears to be even more difficult than Ca_5C_{60}. Briefly, extensively mixed powders of Ba and C_{60} were pressed into pellets in tantalum cells within an inert atmosphere glove box, the tantalum cells were sealed within quartz tubes at 10^{-6} torr, and then these reaction vessels were heated at temperatures up to 800°C. Interestingly, reactions carried out below 600°C were found to be irreproducible, and even at higher temperatures the yields of superconducting phase are generally low. We believe that these results underscore the need for alternative approaches to prepare intercalated C_{60} solids with high melting point metals. Additionally, it is also possible that other C_{60} superconductors may be discovered as new doping procedures are developed.

9. NEW DIRECTIONS

The alkali and alkaline earth metal–doped C_{60} superconductors are fascinating families of superconductors. Diversity in these materials is, however, limited since their structures and properties are determined by the sizes and charges of the metal dopants. There are several possible approaches that have been (or might be) used to expand the range of fullerene superconductors. Zhou and coworkers at Bell Laboratories have recently reported that ammonia (NH_3) can be cointercalated into alkali metal–doped C_{60} to produce new superconducting materials.[92] In these studies it was found that NH_3 may be absorbed by fcc M_3C_{60} compounds. Structural analysis of these materials suggests that the ammonia molecules tetrahedrally coordinate the alkali metal in the octahedral site of the fcc lattice, and thus cointercalation of NH_3 represents a method by which the effective size of the alkali metal dopants can be tuned. This new method may thus help to further understanding of superconductivity in the M_3C_{60} superconductors. More generally, cointercalation represents an approach that could lead to other new fullerene superconductors. For example, ammonia should also cointercalate into the alkaline earth–doped materials, and furthermore, it may also be possible to intercalate substituted amines (e.g., CH_3NH_2) that have different sizes and electron affinities.

Finally, it will also be worthwhile to explore C_{60} solids doped with metals such as lathanum and with two different metals, since these intercalation compounds could lead to materials with new structures and band fillings. Such changes may be needed to discover new fullerene solids with significantly different properties. Since the melting points of metals such as lanthanum are quite high, the development of new doping procedures to facilitate the exploration of these new systems will be necessary.

IV. New Fullerene Building Blocks

In the preceding sections we have reviewed the preparation of well-known fullerene clusters. Now we will introduce other carbon-based building blocks that have not been purified in macroscopic quantities. Although these materials cannot yet be used to prepare new solids, they do warrant consideration as potentially important building blocks for the future.

10. Endohedral Metal Fullerenes

Another general approach for that could be used to prepare doped fullerene solids involves encapsulating the metal dopant(s) within the

[92]O. Zhou, R. M. Fleming, D. W. Murphy, M. J. Rosseinsky, A. P. Ramirez, R. B. van Dover, and R. C. Haddon, *Nature* **362,** 433 (1993).

interior of the fullerene cluster. There are several features of metal encapsulated fullerenes that make these clusters important to consider for new materials research. First, encapsulated metal dopants are protected from reaction with environmental species such as oxygen and water and thus should be more stable than solids prepared by interstitial doping. Second, solids prepared from metal encapsulated fullerenes have a greater potential for diversity than is possible when doping is restricted to specific interstitial holes within the lattice. To explore the properties of solids based on metal encapsulated fullerenes requires the availability of macroscopic quantities of purified material. There are several approaches that can now be used to prepare metal encapsulated fullerenes, although methods to purify macroscopic quantities of these clusters have not been reported. Here we review the status of this work.

Metal encapsulated fullerenes were first detected in the mass spectra obtained from laser vaporization of lanthanum-, potassium-, and cesium-impregnated carbon rods.[7,10] More recently, the arc vaporization method of Krätschmer and Huffman has been applied to the preparation of macroscopic quantities of metal encapsulated fullerenes.[43,93–100] Application of this latter method simply requires that metal be incorporated into the graphite rods prior to the vaporization process; several methods have been used for this purpose. First, mixtures of metal oxide (e.g., La_2O_3), graphite powder, and graphite cement have been packed into holes drilled in graphite rods, and then heated from 200 to 1000°C.[94–96,99,100] This procedure yields mechanically robust and conducting rods that can be readily vaporized in an arc reactor. We believe, however, that it is difficult to investigate systematically the preparation of metal encapsulated fullerenes using this method, since the carbon–metal distribution is

[93]Y. Chai, T. Guo, C. Jin, R. E. Haufler, L. P. F. Chibante, J. Fure, L. Wang, J. M. Alford, and R. E. Smalley, *J. Phys. Chem.* **95**, 7564 (1991).

[94]M. M. Alvarez, E. G. Gillan, K. Holczer, R. B. Kaner, K. S. Min, and R. L. Whetten, *J. Phys. Chem.* **95**, 10561 (1991).

[95]M. M. Ross, H. H. Nelson, J. H. Callahan, and S. W. McElvany, *J. Phys. Chem.* **96**, 5231 (1992).

[96]E. G. Gillan, C. Yeretzian, K. S. Min, M. M. Alvarez, R. L. Whetten, and R. B. Kaner, *J. Phys. Chem.* **96**, 6869 (1992).

[97]H. Shinohara, H. Sato, M. Ohkohchi, Y. Ando, T. Kodama, T. Shida, T. Kato, and Y. Saito, *Nature* **357**, 52 (1992).

[98]R. D. Johnson, M. S. de Vries, J. Salem, D. S. Bethune, and C. S. Yannoni, *Nature* **355**, 239 (1992).

[99]L. Soderholm, P. Wurz, K. R. Lykke, D. H. Parker, and F. W. Lytle, *J. Phys. Chem.* **96**, 7153 (1992).

[100]H. Shinohara, H. Sato, Y. Saito, M. Ohkhchi, and Y. Ando, *J. Phys. Chem.* **96**, 3571 (1992).

very inhomogeneous in these rods. A second approach that yields a much more homogeneous mixture of metal and carbon involves forming carbon rods from intimate mixtures of metal oxide, graphite powder, and graphite cement in a metal die.[93,97,98] The rods obtained after initial curing of the graphite cement are poorly conducting and mechanically fragile, and thus they are unsuitable for arc vaporization. Vacuum annealing at high temperature (1000–1200°C), however, does produce conducting rods that are mechanically robust.

Metal–carbon rods obtained using the foregoing procedures have been used to prepare a variety of metal encapsulated fullerenes, including (1) metallofullerenes MC_{82} (M = Sc, Y, and the lanthanides) containing a single metal ion; (2) the dimetallofullerenes La_2C_{80} and Y_2C_{82}; and (3) the trimetallofullerene Sc_3C_{82}.[93–100] In general, the primary characterization method used to identify these metal encapsulated fullerene materials has been mass spectroscopy analysis of the soot obtained from arc vaporization of the metal–carbon rods. A typical example of a spectrum obtained following vaporization of 10% La_2O_3–carbon rods is shown in Fig. 16. In this spectrum a clear peak is observed at the mass corresponding to LaC_{82} as well as peaks corresponding to other fullerenes such as C_{60} and C_{70}; the LaC_{82} clusters represents approximtely 2–5% of the total fullerene products.

These results clearly demonstrate the presence of metallofullerene

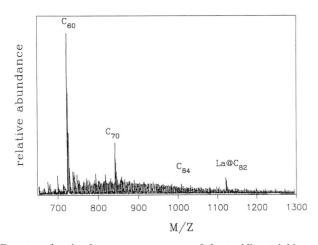

FIG. 16. Fast atom bombardment mass spectrum of the pyridine soluble extract of the soot produced by arc vaporization of lathanum-containing graphite electrodes. The presence of the lathanum encapsulated fullerene, $La@C_{82}$, is clearly evident in this spectrum.

clusters in the soot obtained from the vaporization of metal–carbon rods. The mere presence of metallofullerenes, however, does not necessarily make them useful for solid state studies. To prepare new materials based on metallofullerenes requires macroscopic quantities of the pure clusters. At present, however, there are no reported approaches that can be used to isolate pure metallofullerene clusters. Because solids based on these new clusters could yield exciting electrical, optical, or magnetic properties, it is important to consider possible ways in which large scale isolation can be achieved. First, separation would be greatly facilitated by increasing the overall yield of metallofullerene in the crude soot. Systematic studies of the relative metallofullerene to fullerene product yield as a function of the composition of the metal–carbon rods and the vaporization conditions will indicate how the yield of the desired metallofullerene clusters can be enhanced. Second, the stability of the metallofullerene clusters is not well understood at present. Because these clusters may decompose in an ambient environment (where most isolation procedures have been carried out), we believe that it would be worthwhile to investigate the purification of metallofullerenes using anaerobic conditions. Finally, the metallofullerenes are expected to be polar clusters (i.e., there is significant charge transfer from the metal to the fullerene),[101] and thus procedures used to isolate neutral fullerenes (e.g., C_{60} and C_{70}) may not be a good starting point for the isolation of metal-containing fullerenes. In this regard, chromatographic techniques that are suitable for the separation of large polar macromolecules should represent a promising avenue of investigation.

11. CARBON NANOTUBES

Carbon nanotubes are finite carbon structures that consist of concentric tubes of graphite sheets (Fig. 17).[102–106] Nanotubes differ significantly from the fullerene clusters already discussed since they possess a quasi-one-dimensional or needle like structure. These carbon tubes are, however, interesting building blocks to consider for the preparation of new solid state materials. For example, carbon nanotubes are predicted to be mechanically stiffer than any presently available carbon fiber

[101]K. Laasonen, W. Andreoni, and M. Parrinello, *Science* **258,** 1916 (1992).
[102]S. Iijima, *Nature* **354,** 56 (1991).
[103]T. W. Ebbesen and P. M. Ajayan, *Nature* **358,** 220 (1992).
[104]S. Iijima, T. Ichihashi, and Y. Ando, *Nature* **356,** 776 (1992).
[105]D. Ugarte, *Nature* **359,** 707 (1992).
[106]Z. Zhang and C. M. Lieber, *Appl. Phys. Lett.* **62,** 2792 (1993).

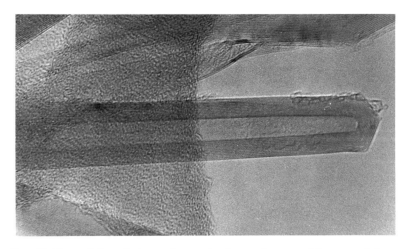

Fig. 17. Transmission electron micrograph of a carbon nanotube supported on an amorphous carbon grid; the micrograph shows a cross-sectional view of the tube. The nanotube has an outside diameter of 100 Å and consists of 12 concentric shells.

material.[107] Electronic structure calculations also predict that nanotubes will exhibit metallic or semiconducting properties depending on the diameter and helicity of the tubes.[108–111] These latter properties could be used to prepare solids with novel electronic and/or optical properties.

The preparation of carbon nanotubes was first reported by Iijima in 1991.[102] In this work it was shown that these needlelike structures were present in the deposit remaining after direct current evaporation of carbon rods in approximately 100 torr of argon. More recently, Ebbesen and Ajayan reported the development of a large scale synthesis of carbon nanotubes using a technique similar to that employed by Iijima.[103] Importantly, these latter investigations have demonstrated that high yields of nanotubes can be obtained when the inert gas pressure is 500 torr or greater during the vaporization process. The optimal nanotube preparation pressure is thus significantly higher than the 100–200 torr pressure that is best for fullerene formation.

To prepare homogeneous solids from these fascinating carbon tubes now requires the development of a method to separate macroscopic

[107]M. S. Dresselhaus, *Nature* **358,** 220 (1992).

[108]J. W. Mintmire, B. I. Dunlap, and C. T. White, *Phys. Rev. Lett.* **68,** 631 (1992).

[109]N. Hamada, S.-I. Sawada, and A. Oshiyama, *Phys. Rev. Lett.* **68,** 1579 (1992).

[110]R. Saito, M. Fujita, G. Dresselhaus, and M. S. Dresselhaus, *Appl. Phys. Lett.* **60,** 2204 (1992).

[111]R. Saito, M. Fujita, G. Dresselhaus, and M. S. Dresselhaus, *Phys. Rev. B* **46,** 1804 (1992).

quantities of nanotubes that have a uniform helicity and diameter. A significant hurdle in this regard is that the nanotubes are insoluble in organic solvents (in contrast to the fullerenes) and thus cannot be separated using standard chromatographic techniques. Several alternative strategies that might lead to advances in this separation problem include the following. First, variations in the current density and rate of carbon vaporization might lead to better control of the nanotube size distribution. Control of the size distribution could significantly reduce the difficulty of any purification process. Second, it may be possible to chemically functionalize or modify the nanotubes to facilitate separation and isolation. One interesting example of chemical modification involves the recently reported reactions of nanotubes with O_2 and CO_2 a elevated temperature.[112,113] In these studies it was found that either molecular species reacts with the tubes to remove the outer concentric carbon shells. Since the smallest inside shells of many nanotubes are the same, these reactions could be used to convert a crude mixture of material into relatively homogeneous nanotube samples. Finally, it is also interesting that Ajayan and Iijima have observed that heating nanotubes in the presence of liquid metals such as Pb results in the filling of the tubes with the metal.[114] Because Pb becomes superconducting at approximately 7 K, it would be possible to separate Pb-filled and virgin nanotubes magnetically at liquid helium temperatures. Since metal filling depends on the interior size of the tube, this approach could also lead to a technique for preparing homogeneous nanotube samples. Lastly, we expect that the metal nanotube species may constitute especially interesting building blocks for the preparation of new materials.

V. Concluding Remarks

In this chapter we have concentrated primarily on reviewing the preparation, isolation, and characterization of fullerene clusters, and the preparation of new solids based on these closed carbon clusters. This review has been done at a level such that physicists and chemists can now carry out experimental research in this field starting only from carbon rods. In the first part of the chapter, the relative merits and shortcomings of well-established techniques for preparing fullerenes from carbon rods,

[112]S. C. Tsang, P. J. F. Harris, and M. L. H. Green, *Nature* **362,** 520 (1993).
[113]P. M. Ajayan, T. W. Ebbesen, T. Ichihashi, S. Iijima, K. Tanigaki, and H. Hiura, *Nature* **362,** 522 (1993).
[114]P. M. Ajayan and S. Iijima, *Nature* **361,** 333 (1993).

isolating pure fullerenes, and characterizing the purity of these materials were discussed. These sections focused on developing efficient strategies to prepare and assess the purity of C_{60} since this cluster remains the most widely used building block for solids. Efficient and general approaches for preparing isotopically substituted fullerenes, which are essential materials for solid state research, were also discussed in this section. In the second major section of this chapter, we reviewed the properties of solid C_{60} and methods for preparing metal-doped C_{60} solids. This part of the review, which concentrated primarily on fullerene superconductors, provides readers with sufficient information to enter this fascinating area of condensed matter physics where a number of fundamental issues remain to be defined (e.g., see the chapter on fullerene superconductivity). Finally, in the last section of the chapter the prospects for materials based on two new carbon building blocks, metallofullerenes and nanotubes, were explored. While well-defined solids built up from metallofullerenes or nanotubes are not yet available, such materials could in the future provide a revolutionary advance in materials research as the discovery and preparation of copper oxide and C_{60} solids have done in the recent past.

SOLID STATE PHYSICS, VOLUME 48

Structure and Dynamics
of Crystalline C_{60}

J. D. AXE

Physics Department
Brookhaven National Laboratory
Upton, New York

S. C. MOSS

Physics Department
University of Houston
Houston, Texas

D. A. NEUMANN

Materials Science and Engineering Laboratory
National Institute of Standards and Technology
Gaithersburg, Maryland

149

ISBN 0-12-607748-7
ISBN 0-12-606048-7 (pbk.)

1. Introduction

1. OVERVIEW

This chapter concerns the structure and dynamics of crystalline C_{60} both in the orientationally disordered face-centered cubic (fcc) phase and below the first-order transition to the low temperature simple cubic (sc) ordered phase at $T_c \approx 260$ K. The issues that confront us are largely, though not exclusively, addressed through x-ray and neutron scattering on powder samples and single crystals. By now, large quantities of high purity C_{60} powders are available for neutron scattering, and excellent, nominally twin-free, single crystals have been grown from the vapor phase that are large enough for inelastic neutron scattering determinations of the phonon dispersion curves and for synchrotron x-ray studies of diffuse scattering. Many of the studies that require larger crystals, however, are either in progress or being planned, and we thus encounter the quandary faced by any review of a field in which definitive experiments often require some effort, considerable time, and very good samples; namely, that it is out-of-date by the time it appears. [Our discussion is confined to results available before June, 1993.] To minimize this dilemma we have adopted a point of view that is, in any case, consonant with our interests, and we concentrate on the underlying formalism and its consequences for orientational order and dynamics in molecular crystals, of which C_{60} is an elegant example. We also wish to show how the scattering data, often in combination with other measurements, may be used to elucidate the energetics of both the orientationally disordered fcc phase and the ordered sc phase below T_c.

The topics to be covered are listed in the Contents. Absent from this list is any discussion of intramolecular dynamics (the intramolecular structure is well known and briefly summarized). It is with the translational and, more exclusively, the orientational structure and dynamics that we are concerned. Salient among the issues are

- A formal description of orientational order in C_{60} and the attendant symmetry considerations;

• The use of scattering data to extract single particle (single C_{60} molecule) orientational order above and below T_c as well as orientational pair correlations above T_c;

• Neutron and related studies of rotational diffusion, librational dynamics, and phonons (excluding intramolecular modes) in which it is clear, both from formal considerations and available data, that at no temperature does the C_{60} molecule execute free (unhindered) rotation;

• A discussion of the single particle and pairwise orientational potentials as determined from static and dynamic scattering experiments and a comparison of these with current phenomenological calculations;

• The nature of the phase transitions, including the orientational order–disorder transition at ~260 K and the low temperature structure, as well as the putative orientational glass transition at lower temperatures (~90 K).

We may see from this brief list that scattering studies of C_{60} are interesting both on their own for what they tell us about this member of a remarkable new family of carbon phases and, more generally, for the insights they provide into an extremely interesting class of molecular solids. The formalism for dealing with these materials, while not new, is nonetheless somewhat complicated, and it is our intention here to present a unified version of this formalism at the outset, which is then illustrated, or "fleshed out," by the available data. In this way we hope to provide a context for the understanding not only of current results but also of future experiments on C_{60} and related molecular solids (for which there is already an extensive literature).

There are other reviews of the structure and dynamics of solid C_{60}, two of which have appeared in a special volume of the *J. Phys. Chem. Solids*: the first by P. A. Heiney[1] on structure, dynamics and ordering, and the second by J. R. D. Copley et al.[2] on neutron scattering from C_{60} and its compounds. In addition, Copley et al.[3] have a forthcoming review of structure and dynamics in *Neutron News*. The most comprehensive overview of the x-ray and neutron scattering formalism for orientationally disordered (fcc) C_{60} is in the paper of Copley and Michel.[4]

[1]P. A. Heiney, *J. Phys. Chem. Sol.* **53**, 1333 (1992).
[2]J. R. D. Copley, D. A. Neumann, R. L. Cappelletti, and W. A. Kamitakahara, *J. Phys. Chem. Sol.* **53**, 1353 (1992).
[3]J. R. D. Copley, W. I. F. David, and D. A. Neumann, *Neutron News* **4**(4), 20 (1993).
[4]J. R. D. Copley and K. H. Michel, *J. Phys.: Cond. Mat.* **5**, 4353 (1993).

II. Theoretical Considerations and Some Consequences

2. STATISTICAL MECHANICAL PRELIMINARIES

In this section we review briefly the ideas underlying the description of molecular rotations in solids and the cooperative phenomena associated with phase transformations involving molecular rotations. In the process, much of the basic notation used in subsequent sections will be introduced. The present understanding of this subject has developed through contributions from many workers. The seminal work in the field is that of James and Keenan on solid methane, which provided a rigorous foundation for a mean field description of orientational phase transformations for classical coupled rigid rotators, along with the symmetry considerations for simplifying the necessary calculations.[5] Subsequently, Michel and co-workers[6] have provided formal generalizations of this work, and they are foremost among the active groups in currently applying these treatments to C_{60}. The present discussion borrows heavily from these sources, as well as from the extensive work of Press and Hüller[7] on related materials.

The basic model of C_{60} developed here consists of a lattice of coupled rigid truncated icosahedal molecules, treated by methods of classical statistical mechanics. The orientation of the ith molecule will be described by Eulerian angles α_i, β_i, γ_i, illustrated in Fig. 1 and denoted collectively as ω_i. The Hamiltonian has kinetic and potential energy contributions,

$$H = \sum_i T_i + \sum_{i,j} V_{ij}(\omega_i, \omega_j), \tag{2.1}$$

where $V_{ij}(\omega_i, \omega_j)$ denotes the energy of a pairwise coupling of molecules i and j as a function of their orientation. The probability distribution for orientations of the ith molecule is given by an *orientational distribution function* $f_i(\alpha_i, \beta_i, \gamma_i) = f_i(\omega_i)$ such that $dP = f_i(\omega_i)\, d\omega_i$ is the probability of finding the orientation specified by ω_i within the range $d\omega_i$, and normalized such that

$$\int d\omega_i\, f_i(\omega_i) = \int d\alpha_i \int d\gamma_i \int d\beta_i \sin\beta_i f_i(\omega_i) = 1. \tag{2.2}$$

[5]H. M. James and T. A. Keenan, *J. Chem. Phys.* **31**, 12 (1959).
[6]See, for example, K. H. Michel and K. Parlinski, *Phys. Rev.* B **31**, 1823 (1985).
[7]W. Press and A. Hüller, *Acta Cryst.* A **29**, 252 (1973); A. Hüller and W. Press, *ibid.* **35**, 876 (1979); W. Press, H. Grimm, and A. Hüller, *ibid.* **35**, 881 (1979).

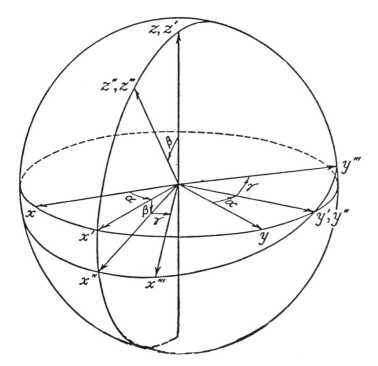

FIG. 1. A description of the three Euler angles by which a general rotation of a rigid body about a fixed point can be defined: a rotation α $(0 \leq \alpha \leq 2\pi)$ about z bringing y to y'; a rotation β $(0 \leq \beta \leq \pi)$ about y' bringing $z = z'$ to z''; a rotation γ $(0 \leq \gamma \leq 2\pi)$ about z'' bringing y'' to y'''.

The mean field approximation follows from the assumption that the orientational distributions of the molecules are independent and canonical to the distribution of the angular momentum variables. A straightforward calculation[5] shows that the entropy is given by

$$S(\omega_i) = -k \sum_i \int d\omega_i \, f_i(\omega_i) \ln \left[8\pi^2 f_i(\omega_i) \right] + S', \qquad (2.3)$$

and the internal energy by

$$E = \sum_{i,j} \int d\omega_i \int d\omega_j \, V_{ij}(\omega_i, \omega_j) f_i(\omega_i) f_j(\omega_j) + E', \qquad (2.4)$$

where the terms S' and E' come from the kinetic energy and are independent of $f_i(\omega_i)$. They consequently play no role in the present discussion and will subsequently be omitted.

3. DESCRIPTION OF MOLECULAR ORIENTATION

The mathematical description of the orientational state of rigid C_{60} molecules in an otherwise ordered solid naturally involves a discussion of the transformation properties of functions with respect to coordinate rotations. Two coordinate frames are specifically of interest. The crystal (global) frame F is fixed with respect to the fcc crystal, with Cartesian axes conventionally oriented along the three fourfold cubic symmetry directions. The molecular (local) frame F', is fixed rigidly to an individual molecule. By convention, the Cartesian axes of the molecule are usually chosen to be coincident with three orthogonal directions defined by two fold-symmetry axes in the C_{60} molecule, as in Fig. 2, which, when aligned with the Cartesian crystal axes, places the molecule in the *standard orientation*.

The rotational orientation of the two frames with respect to one another can be characterized by three real numbers. For example, as we shall see, one number can specify the magnitude of the rotation about an axis specified by two other polar coordinates. The more conventional and generally useful choices are the Eulerian angles $\omega = (\alpha, \beta, \gamma)$ of Fig. 1.[8]

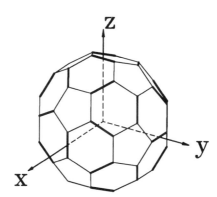

FIG. 2. View of a C_{60} molecule in one of the two standard orientations. In each of these, the twofold axes are aligned with cubic $\langle 100 \rangle$, $\langle 010 \rangle$, and $\langle 001 \rangle$ directions. The other standard orientation is obtained by rotating the molecule by 90° about any of these [100] axes. The double (6:6) bonds are represented by heavier lines.

[8]See, for example, A. R. Edmonds, "Angular-Momentum in Quantum Mechanics," Princeton Univ. Press, 1957.

Consider some field variable $f(r, \theta', \phi')$ expressed in the molecular frame. The point with coordinates $(r, \theta', \phi') \equiv (r, \Omega')$ in F' has different coordinates, say $(r, \theta, \phi) \equiv (r, \Omega)$, in F, and the effect of the rotation of the frame of reference from F' to F can be computed by expressing Ω' as a function of Ω, so that

$$f(r, \Omega') = f(r, \Omega'(\Omega)) = f'(r, \Omega), \qquad (3.1)$$

where Ω is in turn a function of ω.

It is useful to be able to expand a general function $f(r, \Omega')$ in terms of a set of orthogonal polynomials that have the simplest possible rotational transformation properties. These are the familiar spherical harmonic functions, $Y_{lm}(\Omega')$, which have the property of transforming under rotation only within the manifold of functions with the same l and with $-l \le m' \le l'$.

$$Y_{lm}(\Omega') = \sum_{m'} D^l_{m,m'}(\omega) Y_{lm'}(\Omega). \qquad (3.2)$$

The unitary (Wigner) matrices $D^l_{m,m'}(\omega)$ are orthonormal functions with properties well known from the theory of continuous groups.[9]

An additional simplifying option is to use as basis functions not the spherical harmonics themselves, but rather certain linear combinations of spherical harmonics that transform irreducibly under the icosahedral or cubic symmetries of the molecule or crystal, respectively. Let $T^A_l(\Omega')$ represent that combination of spherical harmonics of order l which transforms into itself under all the symmetry operations of an icosahedron and is normalized in such a way that

$$\int T^A_l(\Omega') T^{A*}_l(\Omega') \, d\Omega' = 1. \qquad (3.3)$$

The superscript A is shorthand for A_{1g}, the totally symmetric representation. The high symmetry of the C_{60} molecule severely restricts the allowed terms. For example, inversion symmetry requires that all odd l terms vanish. Similarly, no terms with $l = 2$, 4, or 8 exist. This is equivalent to the statement that an icosahedral molecule possesses vanishing quadrupole-, hexadecapole-, and 2^8-pole moments. In principle, for the remaining nonvanishing l values more than one such combination exists for a given l, but this does not occur in practice for $l \le 30$, and the notation is streamlined accordingly. There are various techniques for

[9]E. P. Wigner, Z. *Physik* **43**, 624 (1927).

generating these icosahedral harmonics, and the functions take a different form depending on how the coordinate axes are chosen with respect to the molecule.

The combinations of spherical harmonics that transform irreducibly with respect to the symmetry operations of a cube, the so-called cubic harmonics, were first introduced by von der Lage and Bethe and have been extensively tabulated.[10] Let $T_{l,\gamma}(\Omega)$ represent that combination of spherical harmonics of order l which transforms according to the γth irreducible representation of the cubic point group, m3m, normalized as in Eq. (3.3). (To prevent a proliferation of indices, the notation has been simplified. γ stands for the three indices necessary to specify uniquely the function under discussion. The first denotes the representation (A_{1g}, T_{2u}, etc.), another the subrepresentation index if the representation is degenerate, and finally a third index, if necessary, to account for the possibility that more than one such representation can be formed within a given l-manifold. These notational complications will be dealt with as the occasion arises.)

Any property such as the atomic or electronic density associated with the undistorted C_{60} molecule must have icosahedral symmetry and can therefore be expanded using the icosahedral harmonics $T_l^A(\Omega')$ when expressed in F'. On the other hand, depending upon how the molecule is oriented with respect to the crystal frame this quantity will have a different representation in F, but must in any event be expressible in terms of the cubic harmonics $T_{l,\gamma}(\Omega)$, since they represent a complete orthonormal set. The transformation between the two basis sets is as follows:

$$T_l^A(\Omega') = \sum_\gamma U_{l,\gamma}(\omega) T_{l,\gamma}(\Omega). \tag{3.4}$$

The molecular rotator functions (rotational basis functions) $U_{l,\gamma}(\omega)$ are closely related to the $D_{m,m'}^l(\omega)$ already discussed. They were first introduced and their properties discussed by James and Keenan.[5] They also form a complete orthonormal set, i.e.,

$$\int U_{l,\gamma}(\omega) U_{l',\gamma'}(\omega)\, d\omega = \frac{8\pi^2}{2l+1}\, \delta_{l,l'}\, \delta_{\gamma,\gamma'}, \tag{3.5}$$

and can thus be used to advantage to expand any arbitrary function of ω.

[10]F. C. von der Lage and H. A. Bethe, *Phys. Rev.* **71**, 61 (1947).

For example, the rotational pair potential introduced in Eq. (2.1) can be expressed as

$$V_{ij}(\omega_i, \omega_j) = \sum_{l,\, l',\, \gamma,\, \gamma'} v^{i,j}_{l\gamma,\, l'\gamma'} U^i_{l,\, \gamma}(\omega_i) U^j_{l',\, \gamma'}(\omega_j). \qquad (3.6)$$

Thermodynamic averages can be performed on the rotational basis functions in the usual way,

$$\bar{U}^i_{l,\, \gamma} = \int U^i_{l,\, \gamma}(\omega_i) f_i(\omega_i)\, d\omega_i, \qquad (3.7)$$

and the mean field energy, Eq. (2.4), can be rewritten as

$$E = \sum_{i,j} \sum_{l,l',\gamma,\gamma} v^{i,j}_{l\gamma,\, l'\gamma'} \bar{U}^i_{l,\, \gamma} \bar{U}^i_{l',\, \gamma'}. \qquad (3.8)$$

4. Order Parameters

The rotational basis functions $U^i_{l,\, \gamma}(\omega_i)$ described in Section 3 are sufficiently general to describe completely the rotational state of a rigid C_{60} molecule, and their orthogonality properties are very helpful in formal mathematical developments, but quite high orders of l are necessary to describe, say, the atoms of a molecule with reasonable resolution. Furthermore, although straightforward but tedious techniques exist for evaluating the functions, they are complicated trigonometric functions of ω. Fortunately, a good qualitative—and quite possibly an adequate quantitative—description of the phase transformation and the low frequency dynamics associated with it can be developed using only the thermodynamic averages of the basis functions $\bar{U}^i_{l,\, \gamma}$ introduced in Eq. (3.7). In particular, the $\bar{U}^i_{l,\, \gamma}$, or more exactly their spatial Fourier transforms, serve as *order parameters* for a discussion via a phenomenological Landau-like theory of possible phase transformations from the high symmetry disordered fcc (Fm3m) phase into the lower symmetry sc (Pa3̄) phase.

The fact that there are many order parameters (i.e., $\bar{U}^i_{l,\, \gamma}$ with many l) is unlike, say, an Ising ferromagnet with a single scalar order parameter and is perhaps an unfamiliar complication, but not an uninteresting one. A similar situation occurs in a Landau theory description of solidification. In that case the disordered phase is a liquid, and the atomic *translational* symmetry is broken upon solidification into an ordered lattice. The order parameters necessary to describe the ordered state are $\rho(\mathbf{k})$, the spatial Fourier components of the average atomic (or molecular) number density. If the transformation is only weakly first order, then a single value, $\rho(\mathbf{k}_0)$, associated with a sinusoidal charge- or mass-density wave, may dominate the description of the ordering near the phase boundary. But as the atoms become more localized in space, higher order components, $\rho(2\mathbf{k}_0), \ldots, \rho(n\mathbf{k}_0)$, grow due to nonlinear couplings in the Hamiltonian. (In practice, localization associated with solidification is so strong that the density wave description, while formally correct, is not particularly useful, except in the case of liquid crystals.)

In *orientational* ordering, the amplitudes of the spherical harmonics take the place of the Fourier amplitudes in solidification, and the higher order terms are required to describe the orientational (librational) localization. Low frequency orientational modes become the important dynamical variables, analogous to low frequency spatial density fluctuations (soft phonon modes). The point symmetry of the orientational modes are well defined. Consequently, the symmetry classification γ is a good quantum number and is useful in classifying the symmetry of the ordered phase. To the extent that l is an approximate quantum number in describing these fluctuations, $\bar{U}^i_{l,\gamma}$ with certain l values may dominate the transformation near the phase boundary. But l is not a rigorously good quantum number in either the ordered or the disordered phase. Consequently, a large set of order parameters are potentially necessary in discussing the orientational state of *both* the disordered and ordered phases.

Consider some useful properties of the orientational order parameters $\bar{U}^i_{l,\gamma}$. From Eq. (3.4),

$$\bar{T}^A_l(\mathbf{\Omega}'_i) = \int d\omega_i \, f_i(\mathbf{\omega}_i) \sum_\gamma U_{l,\gamma}(\mathbf{\omega}_i) T_{l,\gamma}(\mathbf{\Omega}_i)$$

$$= \sum_\gamma \bar{U}^i_{l,\gamma} T_{l,\gamma}(\mathbf{\Omega}_i), \tag{4.1}$$

showing that $\bar{U}^i_{l,\gamma}$ is the average projection that the icosahedral harmonic tied to the molecular frame $T^A_l(\mathbf{\Omega}'_i)$ makes on the corresponding cubic harmonic $T_{l,\gamma}(\mathbf{\Omega}_i)$ in the crystal frame, when the molecular orientations are weighted by the factor $f'_i(\mathbf{\omega}_i)$. Since any function with icosahedral symmetry can be expanded in terms of $T_{l,\gamma}(\mathbf{\Omega}_i)$, the orientational order parameters $\bar{U}^i_{l,\gamma}$ provide a way of describing the average projection of any symmetric molecular property onto the crystal frame.

As an example, consider the atom number density on the C_{60} molecule. It can be developed in a multipole moment expansion in terms of icosahedral harmonics,

$$\rho_i(\Omega_i') = \sum_{v=\text{atoms}} \delta(\Omega' - \Omega_{iv}') = \sum_l g_l T_l^A(\Omega_i'), \qquad (4.2)$$

where, using the orthogonality properties of the icosahedral harmonics,

$$g_l = \sum_v T_l^A(\Omega_{iv}'). \qquad (4.3)$$

The g_l are the 2^lth-pole moments of the atomic distribution. They are important in describing molecule diffraction, and in that context they have become known as the lth order *molecular form factors*.[4] To see how the atomic number density projects on the average onto the crystal frame, we use Eq. (3.4),

$$\bar{\rho}_i(\Omega_i) = \sum_l g_l \bar{T}_l^A(\Omega_i') = \sum_{l,\gamma} g_l \bar{U}_{l,\gamma}^i T_{l,\gamma}(\Omega_i) = \sum_{l,\gamma} C_{l,\gamma}^i T_{l,\gamma}(\Omega_i), \qquad (4.4)$$

showing that the coefficient of the cubic harmonic in the expansion, $C_{l,\gamma}^i = g_l \bar{U}_{l,\gamma}^i$, is the product of the molecular form factor, which is an intrinsic property of the molecule, with the appropriate orientational order parameter $\bar{U}_{l,\gamma}^i$. The same expansion can be used to describe any other property of the icosahedral C_{60} molecule. Suppose one wished to describe the average orientation of, say, the 12 fivefold axes of the truncated icosahedron. Defining, as in Eq. (4.2), a new density operator,

$$\rho_i(\Omega_i')_{5\text{-fold}} = \sum_{v=\text{axes}} \delta(\Omega_i' - \Omega_{iv}') = \sum_l g_1^{5\text{-fold}} T_l^A(\Omega_i'), \qquad (4.5)$$

by analogy with Eq. (4.4),

$$\bar{\rho}_i(\Omega_i)_{5\text{-fold}} = \sum_l g_l^{5\text{-fold}} \bar{T}_l^A(\Omega_i') = \sum_{l,\gamma} g_l^{5\text{-fold}} \bar{U}_{l,\gamma}^i T_{l,\gamma}(\Omega_i). \qquad (4.6)$$

The similarity of Eqs. (4.4) and (4.6) has important practical implications, which will be pursued in Section 8.

5. SYMMETRY CONSIDERATIONS AND LANDAU THEORY

The conventional fcc unit cell contains four molecules of C_{60}, centered at the lattice sites $\boldsymbol{R}_i = (0, 0, 0)$, $(1/2, 1/2, 0)$, $(0, 1/2, 1/2)$, $(1/2, 0, 1/2)$. In the high temperature disordered Fm3m structure, each molecule has an identical atomic density distribution, $\bar{\rho}_i(\boldsymbol{\Omega}_i) = \bar{\rho}_0(\boldsymbol{\Omega}_i)$, where $\bar{\rho}_0(\boldsymbol{\Omega}_i)$ has full cubic m3m symmetry. The interpretation of x-ray and neutron diffraction experiments[11-14] reveals the space group of the low temperature orientationally ordered phase to be Pa$\bar{3}$. In this phase each of the four molecules in the cubic unit cell is unique and can be described in terms of four Fourier components with wavevectors $\mathbf{k}_0 = (0, 0, 0)$, $\mathbf{k}_X = (1, 0, 0)$, $\mathbf{k}_Y = (0, 1, 0)$, and $\mathbf{k}_Z = (0, 0, 1)$. (Here wavevectors are expressed in units of $2\pi/a$). In this case,

$$\bar{\rho}_i(\boldsymbol{\Omega}_i) = \bar{\rho}_0(\boldsymbol{\Omega}_i) + \bar{\rho}_X(\boldsymbol{\Omega}_i)e^{i\mathbf{k}_X\cdot\boldsymbol{R}_i} + \bar{\rho}_Y(\boldsymbol{\Omega}_i)e^{i\mathbf{k}_Y\cdot\boldsymbol{R}_i} + \bar{\rho}_Z(\boldsymbol{\Omega}_i)e^{i\mathbf{k}_Z\cdot\boldsymbol{R}_i}. \quad (5.1)$$

The functions $(\bar{\rho}_X, \bar{\rho}_Y, \bar{\rho}_Z)$ are not independent but are related by symmetry, forming a T_{2g} irreducible representation of the point group m3m. [In fact, the three components transform like (xz, xy, yz), respectively.]

Evaluating Eq. (5.1) explicitly for the four molecules in the unit cell gives

$$
\begin{aligned}
\bar{\rho}(0, 0, 0) &= \bar{\rho}_0 + \bar{\rho}_X + \bar{\rho}_Y + \bar{\rho}_Z, \\
\bar{\rho}(\tfrac{1}{2}, \tfrac{1}{2}, 0) &= \bar{\rho}_0 - \bar{\rho}_X - \bar{\rho}_Y + \bar{\rho}_Z, \\
\bar{\rho}(0, \tfrac{1}{2}, \tfrac{1}{2}) &= \bar{\rho}_0 + \bar{\rho}_X - \bar{\rho}_Y - \bar{\rho}_Z, \\
\bar{\rho}(\tfrac{1}{2}, 0, \tfrac{1}{2}) &= \bar{\rho}_0 - \bar{\rho}_X + \bar{\rho}_Y - \bar{\rho}_Z,
\end{aligned}
\qquad (5.2)
$$

Notice that $\bar{\rho}_0(\boldsymbol{\Omega}_i)$, which describes the orientation of all of the molecules in the disordered phase, continues to have meaning even in the Pa$\bar{3}$

[11]P. A. Heiney, J. E. Fischer, A. R. McGhie, W. J. Romanow, A. M. Denenstein, J. P. McCauley, Jr., A. B. Smith, and D. E. Cox, *Phys. Rev. Lett.* **66,** 1922 (1991).

[12]R. Sachidanandam and A. B. Harris, *Phys. Rev. Lett.* **67,** 1467 (1991).

[13]W. I. F. David, R. M. Ibberson, J. C. Matthewman, K. Prassides, T. J. S. Dennis, J. P. Hare, H. W. Kroto, R. Taylor, and D. R. M. Walton, *Nature* **353,** 147 (1991).

[14]S. Liu, Y. J. Lu, M. M. Kappes, and J. A. Ibers, *Science* **245,** 410 (1991).

phase. It is the average atomic density when the definition of average is extended to include an average over the four molecules in the unit cell.

By analogy with Eq. (4.4), $\bar{\rho}_0(\mathbf{\Omega}_i)$ and $\bar{\rho}_X(\mathbf{\Omega}_i)$ can be written in terms of symmetry-adapted spherical harmonics as follows:

$$\bar{\rho}_0(\mathbf{\Omega}_i) = \sum_{l, A} g_l \bar{U}_{l, A} T_{l, A}(\mathbf{\Omega}_i) \tag{5.3}$$

and

$$\bar{\rho}_X(\mathbf{\Omega}_i) = \sum_{l, X} g_l \bar{U}_{l, X} T_{l, X}(\mathbf{\Omega}_i), \tag{5.4}$$

with similar expressions for $\bar{\rho}_Y$ and $\bar{\rho}_Z$. In Eq. (5.3) the dummy index A stands for all basis functions that transform like A_{1g}, while in Eq. (5.4), X (or Y or Z) stands for all T_{2g} basis functions transforming like xz (or xy or yz).

The mean field equations that determine the values of $\bar{U}_{l, \gamma}$ and that define the atomic density distributions given in Eqs. (5.3) and (5.4) will be discussed in Section 6. They require a knowledge of the intermolecular rotational potentials, and they are complex in principle because several values of l may be important. It is possible to postpone these complications and still make some progress in understanding the phase transformation by using general crystal symmetry considerations to construct a purely phenomenological Landau theory.[12,15]

We define three Landau order parameters, (η_X, η_Y, η_Z), such that

$$\int \bar{\rho}_X^2(\mathbf{\Omega}) \, d\mathbf{\Omega} = \sum_{l, X} g_l^2 \bar{U}_{l, X}^2 = \eta_X^2. \tag{5.5}$$

According to Landau theory, a properly constructed free energy density must be of the form

$$F_L(\eta_X, \eta_Y, \eta_Z) = \tfrac{1}{2} a_2 (\eta_X^2 + \eta_Y^2 + \eta_Z^2) + \tfrac{1}{3} a_3 \eta_X \eta_Y \eta_Z$$
$$+ \tfrac{1}{4} a_4 (\eta_X^2 + \eta_Y^2 + \eta_Z^2)^2 + \tfrac{1}{4} b_4 (\eta_X^4 + \eta_Y^4 + \eta_Z^4) + \cdots. \tag{5.6}$$

The cubic, quartic, and higher order terms are entropic in origin; i.e.,

[15] A. B. Harris and R. Sachidanandam, *Phys. Rev. B* **46**, 4944 (1992).

they arise from the expansion of the $f_i \ln f_i$ terms in Eq. (2.3), and all of the coefficients are temperature dependent. The allowed values of η_γ are found by minimizing Eq. (5.6): $\partial F_L / \partial \eta_\gamma = 0$, $\partial F_L^2 / \partial \eta_\gamma \, \partial \eta_{\gamma'} > 0$. If we set $a_3 = 0$ and assume positive values for the remaining coefficients, the minimum occurs for $\eta_X^2 = \eta_Y^2 = \eta_Z^2 = 0$, and it represents the high temperature disordered phase. Stable low temperature phase solutions occur only if $a_2 < 0$, and a continuous (second-order) phase transformation occurs at a temperature T_0 for which $a_2(T_0) = 0$. The low temperature phase has $\eta_X^2 = \eta_Y^2 = \eta_Z^2 \equiv \eta^2$ and represents the observed Pa$\bar{3}$ structure. As is well known, the cubic anisotropy ($a_3 \neq 0$) changes the transformation both qualitatively and quantitatively, causing the onset of order to be discontinuous (first order), as is confirmed experimentally, and increasing the actual transformation temperature, $T_c > T_0$.

As seen from Eq. (5.5), the order parameters η_Y are composites of linear combinations of the order parameters $\bar{U}_{l,\gamma}$ belonging to different l-manifolds. As the temperature is lowered, the composition changes in general toward more and higher l-value terms representing greater localization of the molecular orientations. Some of these changes will occur gradually with temperature, but another possibility, discussed by Michel,[16] is that the composition of the l-manifolds changes *discontinuously* with temperature. This would represent a first-order phase transformation without change in symmetry and is an interesting possible explanation for anomalous properties seen in C_{60} at temperatures ($T \approx 90$ K) well below that of the Fm3m $\rightarrow$ Pa$\bar{3}$ transformation temperature. [These anomalies have more often been discussed in terms of a freezing of the distribution of orientations between major and minor constituents, or rotation angles, leading to an orientational glassy state. We will cover this subject in some detail in Section 14.]

On the subject of other phase transformations, it is of passing interest to note that $b_4 < 0$ favors an entirely different tetragonal solution of Eq. (5.6) with, for example, $\eta_X^2 \equiv \eta^2$, $\eta_Y^2 = \eta_Z^2 = 0$. For such a solution the cubic term vanishes, allowing a continuous transformation from the m3m phase. There is no evidence for such a phase at present, but it is not impossible that it is stable at other temperatures and/or pressures.

6. MEAN FIELD SOLUTIONS

The discussion of the Landau theory introduced in Section 5 was phenomenological, motivated by symmetry considerations alone. Such

[16]K. H. Michel, *J. Chem. Phys.* **97**, 5155 (1992).

expressions can be derived microscopically from a mean field theory. Mean field theory can also be used to discuss other equilibrium properties such as the rotational susceptibility and orientational pair correlation functions, etc. In Section 2 the mean field expressions for the rotational energy and entropy and for the molecular orientational probability were introduced. Requiring that the variation of orientational free energy be stationary with respect to variations in the orientational distribution function, $\delta(E - TS)/\delta f_i(\omega_i) = 0$, provides a self-consistency relation between $f_i(\omega_i)$ and the orientational potential,

$$f_i(\omega_i) = N_i \exp\left[-\beta \sum_j \int d\omega_j \, V_{ij}(\omega_i, \omega_j) f_j(\omega_j) \right],   (6.1)$$

where $\beta = (k_B T)^{-1}$, k_B is the Boltzmann constant, and N_i is chosen to satisfy Eq. (1.2). Equation (6.1) can be rewritten as

$$f_i(\omega_i) = \frac{\exp[-\beta \bar{V}_i(\omega_i)]}{\int d\omega_i \exp[-\beta \bar{V}_i(\omega_i)]},   (6.2)$$

where, by definition, $\bar{V}_i(\omega_i)$ is the *effective single particle potential,* given by the crystal field (CF) acting on a single molecule, and called V_{CF} by Copley and Michel.[4] Taken together, $f_i(\omega_i)$ and $\bar{V}_i(\omega_i)$ provide a self-consistent mean-field description of the orientational state of the system. Equation (6.2) can be used to estimate $f_i(\omega_i)$ from an assumed potential or can be inverted to derive a mean field potential from experiment (see Section 9).

Assuming the rotational potential can be written as a pair potential as in Eq. (3.6), then

$$\bar{V}_i(\omega_i) = \sum_{l, \gamma} \bar{v}^i_{l, \gamma} U^i_{l, \gamma}(\omega_i),   (6.3)$$

where, from Eq. (3.7),

$$\bar{v}^i_{l, \gamma} = \sum_{j, l', \gamma'} v^{i, j}_{l, \gamma, l', \gamma'} \bar{U}^j_{l', \gamma'},   (6.4)$$

which shows how $\bar{V}_i(\omega_i)$, and thus $f_i(\omega_i)$, can be calculated if the orientational pair potentials are known.

Expanding the orientational distribution functions in terms of the rotational basis functions,

$$f_i(\boldsymbol{\omega}_i) = \sum_{l, \gamma} a^i_{l, \gamma} U^i_{l, \gamma}(\boldsymbol{\omega}_i). \tag{6.5}$$

Using the orthogonality of $U^i_{l, \gamma}(\boldsymbol{\omega}_i)$, Eq. (3.5), shows that

$$a^i_{l, \gamma} = \frac{2l + 1}{8\pi^2} \int U^i_{l, \gamma}(\boldsymbol{\omega}_i) f_i(\boldsymbol{\omega}_i) \, d\boldsymbol{\omega}_i, \tag{6.6}$$

and Eq. (3.7) allows Eq. (6.5) to be rewritten as

$$f_i(\omega_i) = \sum_{l, \gamma} \left(\frac{2l + 1}{8\pi^2} \right) \bar{U}^i_{l, \gamma} U^i_{l, \gamma}(\boldsymbol{\omega}_i). \tag{6.7}$$

Equation (6.7) is an important relation. Equation (4.4) showed that rotationally averaged crystal-frame quantities such as $\bar{\rho}_i(\boldsymbol{\Omega}_i)$ can be expressed in terms of the orientational order parameters. Equation (6.7) shows that the orientational order parameters can also be used to describe the orientational distribution function in *molecular* rotation space, ω_i. As will be explained in more detail in Sections 8 and 9, careful x-ray or neutron diffraction measurements can be used to find $\bar{\rho}_i(\boldsymbol{\Omega}_i)$, and thus $\bar{U}^i_{l, \gamma}$, experimentally. Then Eq. (6.7) can be used to deduce $f_i(\boldsymbol{\omega}_i)$, and Eq. (6.2) used to deduce $\bar{V}_i(\boldsymbol{\omega}_i)$ directly from experimental data.[17,18]

By inspection of Eqs. (3.7) and (6.2), the orientational order parameters are given by

$$\bar{U}^i_{l, \gamma} = \frac{\int d\boldsymbol{\omega}_i \, U^i_{l, \gamma}(\boldsymbol{\omega}_i) \exp[-\beta \bar{V}_i(\boldsymbol{\omega}_i)]}{\int d\boldsymbol{\omega}_i \, \exp[-\beta \bar{V}_i(\boldsymbol{\omega}_i)]}. \tag{6.8}$$

Using Eq. (6.3) to express $\bar{V}_i(\boldsymbol{\omega}_i)$ in terms of the $\bar{U}^i_{l, \gamma}$ leads to a set of

[17]P. C. Chow, X. Jiang, G. Reiter, P. Wochner, S. C. Moss, J. D. Axe, J. C. Hanson, R. K. McMullan, R. L. Meng, and C. W. Chu, *Phys. Rev. Lett.* **69**, 2943 (1992).
[18]P. C. Chow, P. Wochner, G. Reiter, S. C. Moss, and J. D. Axe (to appear).

coupled nonlinear self-consistency relations for the orientational order parameters with different order, l. While easily written down, these equations are not easily solved. There exists, however, a trivial exact temperature-independent solution of Eq. (6.8) for the spherically symmetric order parameter $\bar{U}^i_{l=0}$, which can be seen as follows. Clearly, the $l = 0$ spherical harmonic is invariant to coordinate rotations, so that $T^A_{l=0}(\mathbf{\Omega}') = T_{l=0}(\mathbf{\Omega}_i) = (4\pi)^{-1/2}$, and therefore from Eq. (3.4), $U^i_{l=0}(\mathbf{\omega}_i) = 1$. (The subscript $\gamma = A_{1g}$ has been omitted for simplicity.) Inserting this result into Eq. (6.8) yields the trivial solution, $\bar{U}^i_{l=0} = 1$, independent of i and temperature. This is the limiting high temperature solution for Eq. (6.8), and represents a state of true *random* disorder.

At any finite temperature, there will be corrections to the randomly disordered state. A high temperature expansion of Eq. (6.8), which avoids the complication of the exponential nonlinearities, has a useful though limited range of validity and will be discussed qualitatively here. Expanding to leading order in β produces linear coupled equations of the form

$$\sum_{j, l' \neq 0} (\beta^{-1} \delta_{l, l'} \delta_{i, j} + v^{ij}(\gamma)_{l, l'}) \bar{U}^j_{l', \gamma} = -\sum_j v^{ij}(A_{1g})_{l, 0} \delta_{\gamma, A_{1g}}. \quad (6.9)$$

[The fact that the pair potential must be totally symmetric and thus diagonal in γ has been used to simplify the notation, i.e., $v^{i,j}_{l\gamma, l'\gamma'} = v^{ij}(\gamma)_{l, l'} \delta_{\gamma, \gamma'}$]. Equation (6.9) applies to both the disordered $Fm3m$ and the ordered $Pa\bar{3}$ forms, but the structure of the equations is different in the two cases. For A_{1g} symmetry the equations are inhomogeneous, the right-hand side of Eq. (6.9) behaving like an external field, inducing nonvanishing $l \neq 0$ order at all finite temperatures. The left-hand side of Eq. (6.9) is like an inverse susceptibility for this field and has a typical Curie–Weiss form expected from a mean field theory. [For example, at temperatures such that $\beta v^{ij}(A_{1g})_{l, l'} \ll 1$, $\bar{U}^i_{l, A_{1g}} \approx -\beta \sum_j v^{ij}(A_{1g})_{l, 0}$.] Physically, this expresses the fact that *even spherically averaged neighbors on an fcc lattice produce a cubic rather than a completely spherically symmetric perturbation.*

By contrast, the self-consistency relation as applied to the T_{2g} symmetry order parameters that describe the $Pa\bar{3}$ phase is homogeneous. There is no external field to break the $m3m$ symmetry, and it must be generated spontaneously. An estimate of the temperature at which this occurs is obtained by Fourier transforming N coupled equations, Eq. (6.9) for each molecule, into a $\mathbf{k}$-space representation. The resulting eigenvalue problem has eigenvalues of the form $\beta = \beta_0(\gamma, \mathbf{k})$, which provide a lower bound on the temperature $T_0(\gamma, \mathbf{k})$ at which an instability of γ symmetry type and wavevector **k** can grow spontaneously. The corresponding eigenvector, which has contributions for all values of l, represents the unstable orientational mode itself.

Presumably, the single particle rotational potential is such that $T_0(T_{2g}, \mathbf{k} = \mathbf{k}_X)$ is higher than that for any competing instabilities of another symmetry or wavevector. Efforts to provide a global stability analysis along the lines that have been discussed are complicated by the lack of a simple, adequate pair potential (first principles or otherwise), by the number of potentially important order parameters $\bar{U}^i_{l,\gamma}$, and by the complexity of the equations that couple them. Michel[19] and his coworkers and Harris and Sachidanandam[15] have discussed approximate solutions of the mean field equations for assumed model potentials. Further discussion of model potentials will be deferred to later sections.

7. Scattering and Orientational Disorder

X-ray and neutron diffraction are the most direct experimental tools for studying the structure of C_{60} and its compounds. This section describes briefly the formalism necessary to interpret diffraction experiments in terms of the orientational order parameters introduced in the preceding sections.[4,7]

According to the first Born approximation, the scattering cross section from weakly scattering particles such as x-rays or neutrons is proportional to the ensemble average of the square of the Fourier transform of the atomic density distribution,

$$\frac{d\sigma}{d\Omega_Q} = \sum_{i,j} e^{i\mathbf{Q}\cdot(\mathbf{R}_i - \mathbf{R}_j)}\langle F_i(\mathbf{Q})F_j(-\mathbf{Q})\rangle, \tag{7.1}$$

where the atomic density of the crystal has been written as a sum over rigid molecules at lattice sites $\mathbf{R}_i$, $\rho(\mathbf{r}) = \sum_i \rho_i(\mathbf{r} - \mathbf{R}_i)$, and $F_i(\mathbf{Q})$ is the *molecular structure factor* (not to be confused with g_l), given by

$$F_i(\mathbf{Q}) = f_c(Q) \int \rho_i(\mathbf{r})e^{i\mathbf{Q}\cdot\mathbf{r}}\, d\mathbf{r}. \tag{7.2}$$

$f_c(Q)$ is either the carbon atomic x-ray form factor or the carbon neutron

[19]K. H. Michel, *Z. Phys. B—Cond. Mat.* **88**, 71 (1992).

scattering length (b_c), as appropriate, and $\rho_i(\mathbf{r})$ takes on the meaning of a density distribution of atoms over a single molecule. $\mathbf{Q}$ is the momentum transfer in the scattering event and has a magnitude, Q, of $4\pi \sin \theta / \lambda$, where 2θ is the angle between scattered and transmitted waves and λ is the x-ray or neutron wavelength.

The scattering described by Eq. (7.1) contains both sharp Bragg peaks due to the average long range order and diffuse scattering reflecting the fluctuations of individual molecular orientations. The two types of scattering are easily separated by adding and subtracting $|\langle F_i(\mathbf{Q})\rangle|^2$ in Eq. (7.1). In the Fm3m phase, the term containing $|\langle F_i(\mathbf{Q})\rangle|^2$ can be factored out of the sum since all molecules are identical on average. Consequently,

$$\frac{d\sigma}{d\Omega_Q} = \frac{d\sigma}{d\Omega_Q}\bigg]_{\text{Bragg}} + \frac{d\sigma}{d\Omega_Q}\bigg]_{\text{Diffuse}}, \qquad (7.3)$$

with

$$\frac{d\sigma}{d\Omega_Q}\bigg]_{\text{Bragg}} \sim N \sum_{G} \delta(\mathbf{Q} - \mathbf{G}) |\langle F_i(\mathbf{Q})\rangle|^2, \qquad (7.4)$$

where $\mathbf{G}$ is a vector of the reciprocal lattice, and the diffuse scattering cross section is broken into a "self" term, which arises from interference between atoms of the same molecule and which is independent of i,

$$\frac{d\sigma}{d\Omega_Q}\bigg]_{\text{Self}} \sim N\{\langle F_i(\mathbf{Q})^2\rangle - \langle F_i(\mathbf{Q})\rangle^2\}, \qquad (7.5)$$

and a "distinct" term, which contains interference between the atoms belonging to molecules on different sites,

$$\frac{d\sigma}{d\Omega_Q}\bigg]_{\text{Distinct}} \sim \sum_{i,j \neq i} e^{i\mathbf{Q} \cdot (\mathbf{R}_i - \mathbf{R}_j)}\{\langle F_i(\mathbf{Q})F_j(-\mathbf{Q})\rangle - |\langle F_i(\mathbf{Q})\rangle|^2\}. \qquad (7.6)$$

The molecular structure factor depends upon the orientation of the molecule in the crystal frame, a dependence that can be shown explicitly by introducing an expansion of the exponential in terms of cubic harmonics,

$$e^{i\boldsymbol{Q}\cdot\mathbf{r}} = 4\pi \sum_{l,\,\gamma} i^l j_l(QR) T_{l,\,\gamma}(\Omega_Q) T_{l,\,\gamma}(\Omega). \tag{7.7}$$

Equation (7.7) is the analog of the more familiar partial wave expansion in terms of spherical harmonics, from which it can be easily derived. $j_l(QR)$ is the spherical Bessel function. Writing $\rho_i(\mathbf{r}) = \delta(r - R)\,\rho_i(\Omega_i)$, where R is the C_{60} molecule radius, and using Eqs. (4.1), (4.2), and (7.7), it is straightforward to show that Eq. (7.2) can be rewritten as

$$F_i(\boldsymbol{Q}) = 4\pi f_c(Q) \sum_{l,\,\gamma} i^l g_l j_l(QR) T_{l,\,\gamma}(\Omega_Q) U^i_{l,\,\gamma}(\boldsymbol{\omega}_i). \tag{7.8}$$

The dependence on molecular orientation is expressed in terms of the orientational basis functions $U^i_{l,\,\gamma}(\boldsymbol{\omega}_i)$. Similarly, the various ensemble averages of $F_i(\boldsymbol{Q})$ that occur in Eqs. (7.1)–(7.6) can be rewritten to involve the orientational order parameters $\bar{U}^i_{l,\,\gamma}$ and pair correlations $\langle U^i_{l,\,\gamma} U^j_{l',\,\gamma'} \rangle$. In this way the scattering experiments make direct contact with the fundamental variables describing the average orientational order and the fluctuations about that average.

Figure 3 shows the behavior of $j_l(QR)$ for the first few allowed values

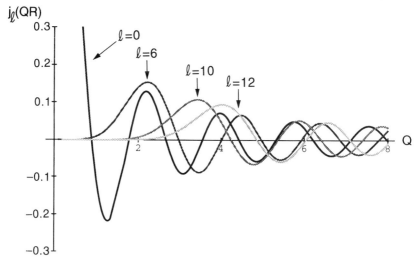

FIG. 3. The momentum (Q) dependence of the spherical Bessel functions, $j_l(QR)$. For successively higher l the first peak shifts to larger Q.

of l and for a range of Q accessible to 1-Å radiation in a typical scattering experiment. The patterns, and in particular the positions of the first maximum for each order, are quite distinctive. This is most easily observed in the diffuse scattering data, where at high temperatures the influence of the intermolecular correlations, while important, is small. The diffuse scattering is then dominated by the self-term, the Q dependence comes mainly from the Bessel functions, and the contributions from various orders can be read off from the scattering data nearly by inspection, as can be seen from Fig. 4, in which a single crystal scan has been selected to avoid all of the fcc Bragg peaks.[20] This greatly simplifies the identification of the contributions of the various orders l to the molecular structure factor and thus ultimately to the atomic probability density and to the rotational potential.

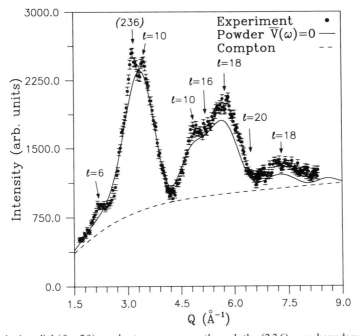

FIG. 4. A radial $(\theta - 2\theta)$ synchrotron x-ray scan through the $(2\,3\,6)$ zone boundary point in a single crystal at room temperature; the independently measured Compton scattering is included and the solid line is a calculation [see Eq. (11.2)] of the powder-averaged intensity in the absence of a crystal field potential. The arrows indicate the calculated peak positions in $j_l(QR)$. From Wochner et al.[20]

[20]P. Wochner, P. C. Chow, M. Meents, A. Vigliante, G. Reiter, S. C. Moss, J. D. Axe, P. Zschack, M. Nelson, J. Z. Liu, J. W. Dykes, and R. N. Shelton (to appear).

The first maximum in $j_l(QR)$ occurs for $Q_{max} \approx l/R$, and the sensitivity to such terms drops off significantly for $Q < Q_{max}$. In practice this means that the contribution of spherical harmonics with l greater than about 30 cannot effectively be probed with $\lambda \gtrsim 1$ Å. Experiments using much shorter wavelength radiation (epithermal neutrons) have been used to increase the **r**-space resolution as discussed in Section 8. Under those conditions, both the Bragg and distinct scattering terms are negligible at high Q since the intermolecular correlations are lost. The scattering resembles that from a high density "gas" of independent molecules and is dominated by the diffuse "self" term. This will be illustrated in more detail in Section 8.

III. Experimental Aspects of the Scattering from the fcc Phase

8. ORIENTATIONAL DISORDER IN FCC C_{60}

The starting point for our discussion is the structure of the C_{60} molecule. The lengths of the longer (6:5) single bond and the shorter (6:6) double bond in the truncated icosahedron have been determined via a variety of structural probes including high-Q neutron diffraction at a pulsed neutron source,[21-24] single crystal x-ray diffraction,[14,25] gas phase electron diffraction,[26] and NMR,[27] along with theoretical calculations.[28] The gas phase electron diffraction is noteworthy as it probes the structure of a single isolated molecule. As we can see from Table I, it appears that, within error, these bond lengths do not change on going from the gas to the solid state or on transforming from the orientationally disordered

[21]R. Hu, T. Egami, F. Li, and J. S. Lannin, *Phys. Rev. B* **45**, 9517 (1992).

[22]A. K. Soper, W. I. F. David, D. S. Sivia, T. J. S. Dennis, J. P. Hare, and K. Prassides, *J. Phys.: Cond. Mat.* **4**, 6087 (1992).

[23]F. Li and J. S. Lannin, Int. Symp. on Phys. and Chem. of Finite Systems, (P. Jena, S. N. Khanna, and B. K. Rao, eds.) Kluwer, Dordrecht (1992), p. 1341.

[24]F. Li, D. Ramage, J. S. Lannin, and J. Conceicao, *Phys. Rev. B* **44**, 13167 (1991).

[25]H.-B. Bürgi, E. Blanc, D. Schwarzenbach, S. Liu, Y. Lu, M. M. Kappes, and J. A. Ibers, *Angew. Chem. Int. Ed. Engl.* **31**, 640 (1992).

[26]K. Hedburg, L. Hedburg, D. S. Bethune, C. A. Brown, H. C. Dorn, R. D. Johnson, and M. de Vries, *Science* **254**, 410 (1991).

[27]R. D. Johnson, D. S. Bethune, and C. S. Yannoni, *Accts. Chem. Res.* **25**, 169 (1992).

[28]M. Haeser, J. Almhof, and G. E. Scuseria, *Chem. Phys. Lett.* **181**, 497 (1991).

TABLE I.   CARBON–CARBON NEAREST–NEIGHBOR DISTANCES IN
SOLID C$_{60}$ AS DETERMINED BY VARIOUS EXPERIMENTAL TECHNIQUES; A REPRESENTATIVE
THEORETICAL RESULT IS ALSO INCLUDED

TECHNIQUE	6:6 SHORT BOND LENGTH (Å)	6:5 LONG BOND LENGTH (Å)	REF.
High-Q neutron diffraction (295 K)	1.401(5)	1.440(5)	22
High-Q neutron diffraction (20 K)	1.406(5)	1.440(5)	22
High-Q neutron diffraction (IPNS)	1.40(1)	1.45(1)	21
High resolution neutron diffraction	1.391(20)	1.455(12)	13
Single crystal x-ray diffraction	1.399(7)	1.455(5)	14
Gas phase electron diffraction	1.401(10)	1.458(6)	26
Nuclear magnetic resonance	1.400(15)	1.450(15)	27
Ab initio calculations	1.405	1.455	28
Experimental average	1.402(3)	1.446(3)	

From Copley et al.[3]

state above ~260 K to the orientationally ordered state below. This constancy of bond length is perhaps not surprising inasmuch as the intermolecular interactions are mainly of the van der Waals variety and do not involve charge transfer. Nonetheless, it is precisely in the details of this interaction potential that we are interested.

Although there was some initial confusion as to the room temperature structure of C$_{60}$, in part due to sample impurity effects, measurements by Fleming et al.[29] on single crystals established the diffraction symmetry to be fcc, Fm3m. Since icosahedral molecular symmetry is inconsistent with the local cubic m3m symmetry, some form of statistical disorder must be assumed by the molecules to achieve average cubic symmetry. In the so-called standard orientation (see Fig. 2), which is in fact the most symmetrical orientation possible with respect to a cube, the C$_{60}$ molecule has the required threefold axes along the $\langle 111 \rangle$ directions but has twofold rather than fourfold axes along the $\langle 100 \rangle$ directions. Rotation of the molecule by 90° about any [100] axis produces its *merohedral* image, and an equal superposition of these two images restores the needed fourfold symmetry. Thus, a minimal model producing fcc symmetry involves statistical disorder between these two merohedrally related configurations. At the other extreme, a model with the maximal disorder invokes

[29]R. M. Fleming, T. Siegrist, P. M. Marsh, B. Hessen, A. R. Kortan, P. W. Murphy, R. C. Haddon, R. Tycko, G. Dabbagh, A. M. Mujsce, M. L. Kaplan, and S. M. Zahurak, *Mat. Res. Soc. Symp. Proc.* **206**, 691 (1991).

equal probability for any random orientation of the C_{60} molecule, since the resulting average molecular symmetry is spherical and obviously compatible with that of a cube.

As different as these two models appear, it is not a completely trivial matter to distinguish between them crystallographically. Powder diffraction measurements can be fit with either model with comparable crystallographic goodness-of-fit parameters, $R \sim 8-10\%$.[1] The Bragg peaks are not the only information in the diffraction pattern, however. As noted in Section 7, the scattering intensity in the disordered fcc phase of C_{60} can be partitioned into Bragg and diffuse scattering contributions. (This familiar statement applies equally to site disorder, structural or displacive disorder, and orientational disorder, and a discussion of these topics can be found in a number of reference texts and reviews.[30-34]) Diffuse neutron scattering measurements are consistent with random diffusive disorder rather than twofold merohedral disorder.[2,35] These, as well as NMR measurements,[36,37] show orientational correlation times, $\tau \approx 10$ psecs, only three times longer than that calculated for free rotation and shorter than the value measured for C_{60} in solution.

Additional powder measurements have been done at a pulsed neutron source where the accessible range of Q is inherently greater than at a reactor. Selected results are given in Fig. 5, in which the data of Li and Lannin,[23] collected at the Argonne National Laboratory's Pulsed Neutron Source (IPNS), are compared with reactor-based data[38] from the National Institute of Standards and Technology (NIST). The NIST data of

[30]B. E. Warren, X-ray Diffraction (Addison-Wesley, Reading, Mass. 1969; reprinted by Dover Publications, New York, 1990).

[31]A. Guinier, "X-ray Diffraction in Crystals, Imperfect Crystals and Amorphous Bodies," W. H. Freeman, San Francisco, 1963.

[32]W. H. Zachariasen, "Theory of X-ray Diffraction in Crystals" (General Publishing Co. Ltd., Toronto, 1945; reprinted by Dover Publications, New York, 1967.

[33]R. W. James, The Optical Principles of the Diffraction of X-rays. Bell and Hyman, Ltd., England, 1948; reprinted by Ox Bow Press, Woodbridge, 1982.

[34]M. A. Krivoglaz, "Theory of X-ray and Thermal Neutron Scattering by Real Crystals." (Plenum Press, New York, 1969).

[35]D. A. Neumann, J. R. D, Copley, R. L. Cappelletti, W. A. Kamitakahara, R. M. Lindstrom, K. M. Creegan, D. M. Cox, W. J. Romanov, N. Coustel, J. P. McCauley, N. C. Maliszewskyj, J. E. Fischer, and A. B. Smith, Phys. Rev. Lett. 67, 3808 (1992).

[36]R. Tycko, G. Dabbagh, R. M. Fleming, R. C. Haddon, A. V. Makija, and S. M. Zahurak, Phys. Rev. Lett. 67, 1886 (1991).

[37]R. D. Johnson, C. S. Yannoni, H. C. Dorn, J. R. Salem, and D. S. Bethune, Science 255, 1235 (1992).

[38]J. R. D. Copley, D. A. Neumann, R. L. Cappelletti, W. A. Kamitakahara, E. Prince, N. Coustel, J. P. McCauley, Jr., N. C. Maliszewskyj, J. E. Fischer, A. B. Smith, K. M. Creegan, and D. M. Cox, Physica B 180/181, 706 (1992).

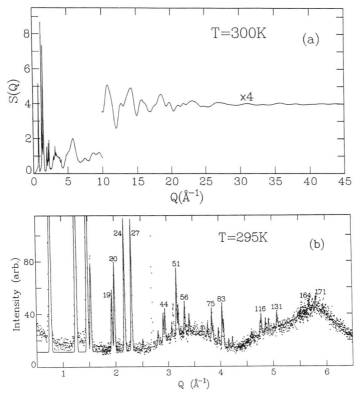

FIG. 5. (a) The total scattering function $S(Q)$ for a powder sample of C_{60} in the fcc phase. The liquidlike structure factor at high Q is the diffuse scattering from orientationally disordered C_{60} molecules powder-averaged over angle. From Li and Lannin.[23] (b) The room temperature neutron diffraction pattern of C_{60} powder. The solid line is a calculation of the scattering expected for orientationally disordered C_{60} molecules located on an fcc lattice as in Fig. 4 $[\bar{V}(\omega) = 0]$. Several of the Bragg peaks have been labeled with the quantity $(h^2 + k^2 + l^2)$. Peaks at ~2.7 Å^{-1} and ~3.1 Å are due to aluminum. From Copley et al.[38]

Copley et al.[38] show clearly a set of powder Bragg peaks, labeled by the sum of the squares of the cubic Miller indices $(h^2 + k^2 + l^2)$, over a diffuse background, which was discussed earlier (see Fig. 4) and to which we shall return. What is of interest here is the range covered by the NIST data, which essentially cuts off at $Q \approx 6.5$ Å^{-1}. By contrast the IPNS data extends to $Q \approx 45$ Å^{-1} and shows distinct liquidlike oscillations for $Q \gtrsim 6$ Å^{-1} In other words the Bragg-like portion of the data is rapidly damped in Q-space, due to rotational disorder, and we are left, at large Q, with only the diffuse scattering from a spinning C_{60} molecule, powder-averaged over angle. The Fourier transform of this entire scattering pattern including *both* Bragg and diffuse data shows discrete (sharp) spatial correlations only for the intramolecular distances. In addition, it is almost identical to the transform of the scattering pattern from the librating ordered structure taken at 10 K; i.e., the intramolecular distances and diameter are essentially unaltered on cooling. A more detailed analysis of similar data was also done by Li and coworkers[24] and more recently at the Rutherford Laboratory (ISIS) by Soper *et al.*[22]; these confirm the constancy in intramolecular distance over the whole range of temperature, although Soper *et al.*[22] suggest a change in intermolecular distance on passing through the phase transition—a fact confirmed by accurate lattice parameter measurements[39] and dilatometry.[40]

While these studies indicate that the orientational distribution of the molecules in fcc C_{60} is more nearly spherical than merohedrally distributed over distinct orientations, the most detailed information to date concerning the orientational distribution in fcc C_{60} has been obtained with single crystal x-ray diffraction. In these experiments of Chow *et al.*,[17] over 1100 Bragg peaks were measured at the National Synchrotron Light Source (NSLS) and the resulting 300 unique peak intensities were fit to elastic structure factors derived from Eq. (7.8):

$$\langle F_i(\mathbf{Q}) \rangle = 4\pi f_c(Q) \sum_{l,\gamma} i^l g_l j_l(QR) T_{l,\gamma}(\mathbf{\Omega}_Q) \bar{U}^i_{l,\gamma}. \tag{8.1}$$

The coefficients $C_{l,\gamma}$ in Eq. (4.4) for the even order cubic harmonics with $l \le 30$ were permitted in the fit and are given in Table II. We see that, compared with $C_0 = 1000$ they are all small, but are certainly statistically

TABLE II. THE VALUES OF THE CUBIC HARMONIC COEFFICIENTS $C_{l,\gamma}$ DETERMINED FROM SINGLE CRYSTAL BRAGG DATA[a]

$C_{0,1}$	$C_{6,1}$	$C_{10,1}$	$C_{12,1}$	$C_{12,2}$	$C_{16,1}$
1000	−23.0(17)	12.8(14)	9.4(17)	26.0(13)	−7.7(23)
$C_{16,2}$	$C_{18,1}$	$C_{18,2}$	$C_{4,1}$	$C_{8,1}$	$C_{14,1}$
−0.4(20)	−3.0(24)	23.7(20)	−2.7(17)	−3.4(22)	1.4(16)

[a] The values allowed by the icosahedral molecular symmetry are given on the top line. The additional three cubic harmonic coefficients, whose values mush vanish for icosahedral symmetry, are on the second line and are zero within experimental error. From Chow *et al.*[17]

[39]W. I. F. David, R. M. Ibberson, and T. Matsuo, *Proc. Roy. Soc. A* **442**, 129 (1993).
[40]F. Gugenberger, R. Heid, C. Meingast, P. Adelmann, M. Braun, H. Wühl, M. Haluska, and H. Kuzmany, *Phys. Rev. Lett.* **69**, 3774 (1992).

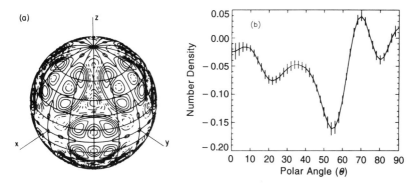

FIG. 6. (a) Atomic number density on a spherical molecular surface evaluated from coefficients $C_{l,\gamma}$, obtained by refinement of single crystal Bragg data, with contour intervals of $\pm 3\sigma = 0.024$. The C_0 term was omitted to emphasize differences from spherical symmetry, and dashed contours indicate a density deficit ($\sim 16\%$ maximum) while solid contours indicate an excess ($\sim 10\%$ maximum) over the uniform density. (b) A θ-plot of $(\rho(\Omega) - \rho_0)/\rho_0$, where ρ_0 is the uniform density, along constant $\phi = 45°$ ($x = y$) through the $\langle 001 \rangle$, $\langle 111 \rangle$, and $\langle 110 \rangle$ directions including estimated standard deviations. Number density thus refers to a fractional excess or deficit over the uniform density. From Chow *et al.*[17]

significant for all l values except 4, 8, and 14, namely only for those conforming to icosahedral symmetry. The resulting normalized atomic density distribution (with the isotropic component removed) is shown in Fig. 6 for a molecular sphere of radius 3.5429 Å. While the distribution is to first approximation spherical, there is a density deficiency of about 16% in the [111] directions and four lobes of about 10% excess density in the [110] directions, which actually comprise a hexagon about $\langle 111 \rangle$.

While Fig. 6 demonstrates unambiguously that some orientations are more probable than others, it is not easily seen which orientations they are. A clearer understanding of the significance of these results can be had by using the orientational order parameters so obtained to derive the probability density for the orientation of the molecular symmetry axes, as explained in detail in Section 4, Eq. (4.6). Figure 7(a) shows, for example, the distribution of fivefold axes that are consistent with the atomic density distribution of Fig. 6. There is a pronounced tendency for fivefold axes to point toward nearest neighbors, i.e., along [110] directions, which is 8.8 times more probable than along the least probable [111] directions. Similar distributions can be derived for the three- and twofold axes. The threefold axes in Fig. 7(b) prefer to point along [111] and to avoid [110] (the anisotropy ratio is about 3.0). As with the fivefold distribution, the

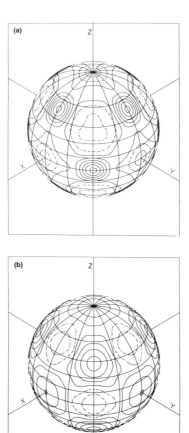

FIG. 7. (a) $\rho(\Omega)_{5\text{-fold}}$ is the probability density for the orientation of the 12 fivefold axes (pentagonal faces) of the C_{60} molecule. It can be derived from the atomic number density, $\rho(\Omega)$, and is more easily interpreted. The values shown here are for the fcc phase and are consistent with $\rho(\Omega)$ shown in Fig. 6.; (b) $\rho(\Omega)_{3\text{-fold}}$, the probability density for the orientation of the 20 threefold axes (hexagonal faces of the C_{60} molecule, consistent with $\rho(\Omega)$ shown in Fig. 6.

threefold distribution is very clean, with single maxima and minima directly along the crystal symmetry directions.

It is interesting to ask why the five- and threefold axes show more distinctive "orientation" than the atoms. The first reason is obvious. The symmetry axes are fewer in number (12 fivefold axes, 20 threefold axes) than the 60 atoms, and any nonuniform distribution of them will thus tend to be more distinct. The second reason is less obvious but more interesting. The anisotropy of the distributions of both the five- and threefold axes are dominated by $l = 6$ cubic harmonics, whereas $l = 6$ terms are nearly negligible in the atomic distribution. The root cause is geometry—the truncated icosahedral atomic array has an anomalously small 2^6-pole moment, $g_{l=6}$. Thus, the strong contribution of $l = 6$ terms, evident in the axial distributions, is obscured in the atomic density distribution. The dominance of $l = 6$ terms in the cubic single particle potential has been predicted by Michel et al.[41]

That the most pronounced orientational preference is for fivefold axes to point along [110] directions can be understood qualitatively by a simple steric argument. Compact packing can be achieved either by orienting flat hexagonal or pentagonal faces (three- or fivefold axes) toward the nearest neighbors. The former are larger and more numerous, but line up less well with the neighbors. There are 12 fivefold axes, just as there are 12 nearest neighbors, and both are distributed approximately uniformly over a sphere. [The angle between adjacent fivefold axes in the icosahedron is 63.44°, compared with 60° for [110] cubic axes.] Of course, the two geometries are fundamentally different, but on average the icosahedron can accommodate itself to the cube in this way by reorientations of rather small amplitude. This distribution of fivefold axes determines completely the entire molecular orientation and, in particular, the orientation of the threefold axes along [111] directions.

Although the symmetry axes have a high degree of orientation, even small amplitude orientational fluctuations are capable of smearing out the distribution of atoms because they are so numerous and so nearly equally spaced. The angle subtended by near-neighbor C_{60} atoms is approximately 22°, and fluctuations of half of this amplitude would be very effective in homogenizing their distribution. The full width of the distribution of fivefold aces at half maximum is about 24°. Thus, there is no contradiction between the rather striking anisotropies in the orientation shown in Fig. 7 and the finding that the atoms have a nearly spherical distribution. [In Part IV we will discuss in more detail the orientational configurations in the low temperature $Pa\bar{3}$ sc phase. There it will be shown that the coordinated orientations of the molecules favor the placing of a majority of the shorter hexagon–hexagon (6:6) double bonds opposite neighboring pentagons (composed entirely of the longer (6:5) single bonds) with a minority opposite hexagons. In other words, above the transition the pentagons maintain their low temperature orientational preference.

[41]K. H. Michel, J. R. D. Copley, and D. A. Neumann, *Phys. Rev. Lett.* **68**, 2929 (1992).

Subsequent to the foregoing single crystal study,[17] the powder analysis has been pushed harder to retrieve the coefficients $C_{l,\gamma}$ of the symmetry-adapted spherical harmonics both above and below the orientational ordering transition (below naturally requires a larger set, taken only with the fcc-allowed reflections and not with the additional simple cubic, mixed index, reflections). This has been accomplished using high resolution neutron powder diffraction.[39] David et al.[39] find reliable, nonzero, coefficients for $l = 6$ and $l = 10$ at room temperature and construct a nuclear map density in good agreement with Fig. 6 but with understandably poorer statistics.

So far we have emphasized the use of spherical harmonic expansions to obtain the orientational distribution of the molecules. Two other approaches to obtaining this information have been applied to date. In the first of these, Bürgi et al.[42] have parameterized the density distribution in terms of a combination of 61% of eight symmetry-related images of the C_{60} molecule taken together with 39% freely spinning molecules. The first component is obtained by imposing m3m symmetry on the major constituent found for the low temperature structure, namely, a (6:6) bond facing a pentagon,[43] and librating about this position by an rms of 7.8(1)°, representing a strongly anharmonic libration. The minor orientation of a (6:6) bond facing a hexagon is not found. This fit yields an R-factor quite comparable with that of Chow et al.[17] and represents an interesting alternative way of modeling the data, especially with regard both to the low temperature structure and to the polar plots in Fig. 7. While this approach is certainly capable of describing the diffraction data, it implies two components in the rotational dynamics. This is probably misleading, since at room temperature the dynamical measurements are better described as a single (hindered) rotational diffusion process rather than a jump reorientation of librating molecules among preferred sites.[35–37]

Finally, Papoular et al.[44] have done a combined x-ray and neutron study of C_{60} single crystals. They first phased the structure using a uniform spherical shell and then employed maximum entropy methods to reconstruct the three-dimensional number density over a C_{60} molecular sphere at room temperature. Without resorting to a spherical harmonic

[42]H.-B. Bürgi, R. Restori, and D. Schwarzenbach, *Acta Cryst.* B **49,** 832 (1993).

[43]W. I. F. David, R. M. Ibberson, T. J. S. Dennis, J. P. Hare, and K. Prassides, *Europhys. Lett.* **18,** 219 (1992); an addendum appears in *Europhys. Lett.* **18,** 735 (1992).

[44]R. J. Papoular, G. Roth, G. Hager, M. Haluska, and H. Kuzmany, Proc. Int. Winterschool on Electronic Properties of Novel Materials (H. Kuzmany, ed.) Springer Series on Solid State Science (in press).

analysis, these authors have found clear evidence for a modulation of the number density that is in good qualitative agreement with the map of Fig. 6.

9. THE ORIENTATIONAL POTENTIAL $\bar{V}(\omega)$ IN THE FCC PHASE

The clear inference from all these studies, including the dynamical results to be discussed in Section 10, is that the C_{60} molecule, above its orientational ordering transition, is undergoing hindered rotation in the cubic field of its neighboring molecules. As discussed in Section 6 and in Chow et al.,[17,18] it is then a straightforward, albeit rather cumbersome, task to extract the mean field potential $\bar{V}(\omega)$ using the measured coefficients, $C_{l,\gamma}$. Via Eq. (6.3),

$$\bar{V}(\omega) = \sum_{l,\gamma} \bar{v}_{l,\gamma} U_{l,\gamma}(\omega), \tag{9.1}$$

and the expression in Eq. (4.4),

$$C_{l,\gamma} = g_l \bar{U}_{l,\gamma}$$

we can then write

$$C_{l,\gamma} = \frac{g_l \int U_{l,\gamma}(\omega) e^{-\beta \sum_{l,\gamma} \bar{v}_{l,\gamma} U_{l,\gamma}(\omega)} \, d\omega}{\int d\omega \, e^{-\beta \bar{V}(\omega)}}. \tag{9.2}$$

Equation (9.2) can be solved iteratively for the coefficients $\bar{v}_{l,\gamma}$, which reproduce the measured values of $C_{l,\gamma}$ in Table II. The calculation takes advantage of the high icosahedral symmetry and tabulated molecular rotator functions[45] but is nonetheless rather computer intensive when carried out to $l = 18$ on a fine (1°) grid in ω space. The results can be plotted either in this Euler angle space as contours in, say, α, β

---

[45]C. J. Bradley and A. P. Cracknell, "The Mathematical Theory of Symmetry in Solids," (Clarendon, Oxford, 1972).

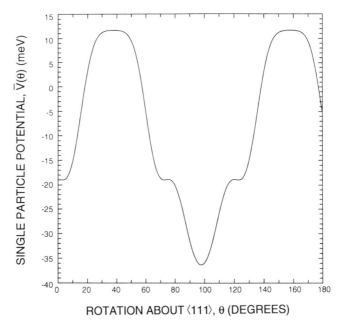

ROTATION ABOUT ⟨111⟩, θ (DEGREES)

FIG. 8. Single particle potential at room temperature derived from single crystal Bragg intensities. θ = 0 represents the "standard" reference configuration (Fig. 2) with hexagonal faces along [111] directions. The deep minimum near 98° aligns pentagonal faces approximately along nearest-neighbor [110] directions. This feature persists in the low temperature sc phase. From Chow et al.[18]

(γ = fixed), or as $\bar{V}(\theta)$, where θ is a rotation angle about selected crystal axes, which, while it does not cover all of the ω space, provides a physically more transparent picture.[18] As discussed in Section 8 and anticipated by Bürgi et al.,[42] Fig. 8 shows a potential minimum on rotation about ⟨111⟩ of ~98° from the standard orientation of Fig. 2 in which pentagons are aligned in [110] directions much as in the low temperature phase[43] to be discussed in Section 12. There is also a barrier height of ~46 meV at room temperature, in very good agreement with the NMR and neutron quasi-elastic rotational diffusion constants to be discussed in Section 10. The shoulder at ~78° is not clearly related to the low temperature structure.

Using the measured coefficients $C_{l,\gamma}$,[17] Lamoen and Michel[46] have also explored the mean field (crystal field) potential in solid C_{60} at room temperature. The molecules were modeled as rigid bodies with atoms and single (6:5) bonds treated as single interaction centers, while double

[46]D. Lamoen and K. H. Michel, Z. Phys. B **92,** 323 (1993).

(6:6) bonds were described using a distribution of centers. By expanding the crystal field in cubic harmonics, values of $C_{l,\gamma}$ could be compared with experiment[17] in order to extract an approximate potential using only the linear terms in the expansion of the exponential of $\bar{V}(\omega)$ in Eq. (9.2) and a limited range of $l = 6, 10, 12$. A pronounced potential minimum at ~98° about $\langle 111 \rangle$ was thereby calculated by Lamoen and Michel,[46] which agrees well with the results of Chow et al.[18] and with the low temperature structure.[43] Lamoen and Michel[46] observed, as well, a splitting of the potential barrier in Fig. 8, which suggested the presence of a minor orientation in the fcc phase in apparent agreement with David et al.[43] for the low temperature structure. This finding, however, would seem to be an artifact of their (truncated) analysis, as it is clearly not present in the full calculation in Fig. 8. The minor orientation is also not found in the fit of Bürgi et al.[42] to the room temperature data of Ref. 17. The shoulder in Fig. 8, at an angle of ~78°, remains in the calculation of Lamoen and Michel.[46]

10. Disorder Dynamics: Reorientational Diffusion

Figure 9 shows a representative inelastic neutron scattering spectrum obtained at $Q = 5.65\,\text{Å}^{-1}$ and $260\,\text{K}$.[35,47] The single quasi-elastic line, which is much broader than the resolution, is characteristic of diffusive motion, directly demonstrating that orientational disorder above the transition is dynamic. Given that the molecuar centers lie on an fcc lattice, it is clear that the diffusive motion involves angular displacements of the molecules about their centers, especially as the rms translational displacement of the molecule is ~1% of the lattice parameter.[17]

The reorientational character of diffusive motion in C_{60} is reflected in the Q dependence of the quasielastic scattering. Sears[48] has developed a model in which the rotations of adjacent molecules are uncorrelated and individual molecules undergo rotational diffusion about their fixed centers. This, of course, ignores the cubic crystal field discussed previously, and the model therefore assumes that all of $C_{l,\gamma}$ are zero except for C_0. In a classical treatment, the angular motion of each molecule satisfies the differential equation

$$D_R \nabla^2_\omega P(\omega, \omega_0, t) = \frac{\partial}{\partial t} P(\omega, \omega_0, t), \tag{10.1}$$

[47]B. Renker, F. Gompf, R. Heid, P. Adelmann, A. Heiming, W. Reichardt, G. Roth, H. Schober, and H. Rietschle, Z. Phys. B—Cond. Mat. 90, 325 (1993).

[48]V. F. Sears, Can. J. Phys. 45, 237 (1967).

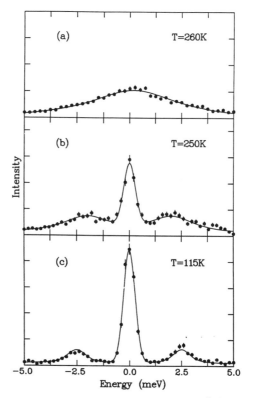

FIG. 9. Inelastic neutron scattering spectra taken at $Q = 5.65$ Å^{-1} at 260, 250, and 115 K. Above the orientational order–disorder transition (257 K, for this sample), the spectrum consists of a single Lorentzian centered at zero energy transfer, characteristic of diffusive motion. Below the transition, the spectra show peaks at nonzero energy transfers, corresponding to librations of C_{60} molecules about their equilibrium positions. These peaks shift to larger energy transfers and become narrower as the sample is cooled. From Neumann et al.[35,53]

where D_R is the rotational constant, ∇^2_ω is the Laplacian opeator in the space of θ and ϕ, and $P(\omega, \omega_0, t)$ is the probability that a molecule has orientation ω at time t, having had orientation ω_0 at time 0. Equation (10.1) implies that the molecule tumbles through a continuum of orientational angles and does not, for example, undergo rotational jumps between some set of discrete orientations. This equation has a solution of the form

$$P(\omega, \omega_0, t) = \sum_{l, m} \frac{(2l + 1)}{8\pi^2} e^{-l(l+1)D_R t} \sum_{m'} D^l_{m, m'}(\omega) D^l_{m, m'}(\omega_0). \quad (10.2)$$

The scattering is given by the Fourier transform of the intermediate scattering function $I_R(Q, t)$:

$$I_R(Q, t) = \frac{1}{8\pi^2} \int \int e^{-i\boldsymbol{Q}\cdot\mathbf{r}_n} e^{i\boldsymbol{Q}\cdot\mathbf{r}_{n'}} P(\boldsymbol{\omega}, \boldsymbol{\omega}_0, t) \, d\boldsymbol{\omega}_0 \, d\boldsymbol{\omega}. \quad (10.3)$$

Using the partial wave expansion Eq. (7.7) as well as the spherical harmonic addition theorem yields

$$I_R(Q, t) = \sum_{l=0}^{\infty} (2l + 1) j_l^2(QR) e^{-l(l+1)D_R t} \sum_{n, n'} P_l(\cos \theta_{nn'}), \quad (10.4)$$

where P_l is a Legendre polynomial and $\theta_{nn'}$ is the angle between the position vectors joining the molecular center to atoms n and n' within a single molecule. The Fourier transform of Eq. (10.4) is the scattering function $S_R(Q, \omega)$, which is given by

$$S_R(Q, \omega) = \frac{1}{\pi} \sum_{l=0}^{\infty} S_l \frac{\tau_l}{1 + \omega^2 \tau_l^2}, \quad (10.5)$$

where

$$\tau_l^{-1} = l(l + 1) D_R \quad (10.6)$$

and

$$S_l = (2l + 1) j_l^2(QR) \sum_{nn'=1}^{60} P_l(\cos \theta_{nn'}). \quad (10.7)$$

The sum over the Legendre polynomials accounts for the fact that motions among atoms within a single molecule are correlated, and it therefore reflects the molecular geometry. As we have seen, the only terms that are allowed by icosahedral symmetry are those with $l = 6$, 10, 12, 16, 18, ...

Using the conventional identity discussed earlier, the total rotational scattering $S_R(\boldsymbol{Q}) = \int S_R(\boldsymbol{Q}, \omega) \, d\omega$ is found to be $\langle |F(\boldsymbol{Q})|^2 \rangle - \langle |F(\boldsymbol{Q})| \rangle^2$,

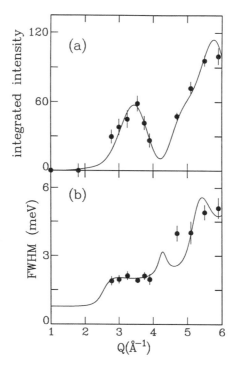

FIG. 10. Q dependence of (a) the integrated intensity and (b) the full width at half maximum of the quasi-elastic scattering at 260 K. The solid line was obtained using Eqs. (10.5)–(10.7) and a rotational diffusion constant $D_R = 1.4 \times 10^{10} \, \text{sec}^{-1}$. From Neumann *et al.*[35]

which accounts for the diffuse scattering observed in the fcc phase[38] (Fig. 5b). This same function describes the intensity of the energy-integrated quasi-elastic scattering (Fig. 10a).[35] Therefore, all the diffuse scattering is dynamic, and there is no truly elastic component. This model also predicts the Q dependence of the widths of the quasielastic peaks in Fig. 10(b), in which symbols are data taken at 260 K[35] while the solid line is obtained for $D_R = 1.4 \times 10^{10}$/sec at 260 K. NMR measurements have also been used to determine the "correlation time," τ_{NMR}, of the rotational dynamics. The values of $\tau_{\text{NMR}} = 12$ psec obtained at 300 K[36] and 9 psec, at 283 K[27,37] are consistent with the results reported here since for rotational diffusion $D_R = 1/(6\tau_{\text{NMR}})$.[27,37] The NMR measurements reported to date, however, contain little if any spatial information about the diffusion mechanism; for example, they do not distinguish between reorientational jumps and rotational diffusion. Nonetheless, the temperature dependence of D_R and τ_{NMR} is consistent with a thermally activated process rather than free rotation; i.e., D_R obeys the Arrhenius equation. There is, however, considerable spread in the values of the activation energies from 35(15) meV to 60 meV.

The foregoing model for the rotational dynamics in the orientationally disordered phase of C$_{60}$ cannot be strictly correct. Experimental proof of this come from the x-ray results of Chow et al.,[17] which demonstrate that the molecule is not spherical, as well as from the x-ray diffuse scattering studies of single crystals, to be discussed in Section 11, which show that this diffuse scattering is not isotropic.[20,49] An accurate calculation of the effect that the crystal field has on the rotational dynamics requires an accurate single particle potential $\bar{V}(\omega)$, as presented in Fig. 8. Moreover, we still need to account for correlations between the motions of adjacent molecules. A good deal of information concerning these correlations can be ascertained using the energy-integrated single crystal x-ray diffuse scattering. In principle, though, the most detailed results are obtainable by measuring the complete scattering function $S_R(Q, \omega)$ using quasielastic neutron scattering. Unfortunately, this awaits larger single crystals than are currently available.

11. X-ray Diffuse Scattering

In Section 10 we noted that the energy-integrated diffuse scattering is directly measured by x-rays. This is simply because the x-ray energy is ~10 keV, while the energy scale for reorientational (or translational) motion in C$_{60}$ is on the order of several meV. The formalism for this scattering is covered by Copley and Michel[4] and has been summarized briefly in Section 7. Here we discuss mainly single crystal results,[20,49] which continue to be collected and analyzed.[20] These results, aside from the powder-averaged diffuse neutron scattering data, which is fit very well assuming completely uncorrelated and unhindered rotation of the C$_{60}$ molecules,[38] provide the most detailed information on intermolecular interactions currently available.

In the first place, we can expect that the "self" part of the orientational diffuse scattering will show a pronounced structure and anisotropy in reciprocal space as long as there is a nonzero crystal field, or single particle, potential $\bar{V}(\omega)$. This is, of course, true even in the absence of any of the orientational correlations that appear in the "distinct" terms in Eqs. (7.3) and (7.6), because the two contributions are separable. The

[49]R. Moret, S. Ravy, and J.-M. Godard, J. Phys. I France 2, 1699 (1992).

foregoing is perhaps an unfamiliar statement to those who are used to thinking of the "self" part as Laue scattering, which for, say, a binary alloy is a monotonic function in reciprocal space with no anisotropy. In this case, however, the Laue contribution to $[\langle|F(Q)|\rangle^2 - \langle|F(Q)|^2\rangle]$ arises from uncorrelated C_{60} molecules for which there are prominent *intramolecular* interferences. The calculation of this "self" part proceeds directly from the definition of the terms in the scattering equation. In particular, returning to the *powder-averaged* Laue scattering, we have, from Copley and Michel,[4]

$$\left.\frac{d\sigma}{d\Omega_Q}\right]_{\text{self,powder}} = 4\pi N f_c^2 \sum_{l\neq 0} \left\{ \sum_\gamma \overline{(U_{l,\gamma})^2} - \sum_{\gamma\in A_{1g}} \bar{U}_{l,\gamma}^2 \right\} j_l^2(QR) g_l^2. \quad (11.1)$$

In the absence of a crystal field, $\bar{V}(\omega) = 0$, $\bar{U}_{l,\gamma} = 0$ for $l \neq 0$, and

$$\overline{(U_{l,\gamma})^2} = 1/(2l+1).$$

Since there are $2l + 1$ components for a given l, Eq. (11.1) reduces to

$$\left.\frac{d\sigma}{d\Omega_Q}\right]_{\text{self, powder}} = 4\pi N f_c^2 \sum_{l\neq 0} [g_l j_l(QR)]^2, \quad (11.2)$$

where f_c is the x-ray scattering factor, or neutron scattering length (b_c), for a carbon atom.

Equation (11.2) may be readily evaluated and was used by Copley and coworkers[35,38] to describe the diffuse background in their neutron powder pattern taken at room temperature and is shown in Fig. 5(b), where the broad features in the *l*th-order Bessel functions may be identified. [Note that this equation is essentially the energy-integrated version of Eqs. (10.5)–(10.7), which were found to describe the quasi–elastic scattering in the fcc phase.[35]] We also display this powder-averaged calculation in the single crystal radial scan of Wochner et al.[20] in Fig. 4, which passes through the (2 3 6) zone boundary point but otherwise misses all other reciprocal lattice points. In Fig. 4 the *l*-identification is particularly clear. [This scan also passes through the (4 6 12) fcc reflection at $Q(4\,6\,12) = 2Q(2\,3\,6)$, but a very slight crystal misorientation still captures the broad diffuse peak but misses the Bragg reflection, which is weak in any case because it comes near a minimum in the continuous function $\langle F(Q)\rangle^2$.]

Clearly the powder-averaged, isotropic "self" term for the diffuse scattering provides a qualitative fit to a radial scan in an arbitrary direction in Q space. It does not, however, completely describe the diffuse background due either to "self" or, in particular, to "distinct" correlations. This was first demonstrated by Moret *et al.*[49] using monochromatic Laue photography on a C$_{60}$ single crystal. Their x-ray photographs indicate a radially modulated diffuse scattering (as in Figs. 4 and 5b) that is also *azimuthally* anisotropic from room temperature to above the transition temperature at ~260 K. That this must be so follows directly from the presence of a nonzero $\bar{V}(\omega)$, which in turn means that $\bar{U}_{l,\gamma} \neq 0$ for $l \neq 0$ and that $\overline{(U_{l,\gamma})^2}$ is no longer simply evaluated. Copley and Michel[4] present a detailed procedure for evaluating both the "self" and "distinct" terms in the diffuse scattering, and it suffices to note here that considerable effort is required.

In Fig. 4 the diffuse peak at (2 3 6) is easily identified, and it is explored more carefully in Fig. 11 in rocking curves taken through the (2 3 6) at

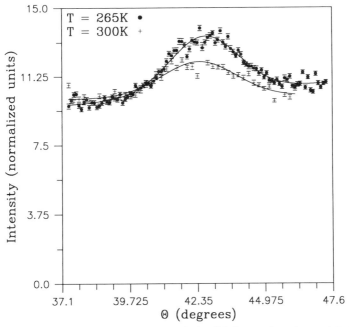

FIG. 11. Single crystal rocking curves through the (2 3 6) zone boundary point at 300 K and 265 K, which show the "distinct" part of the diffuse scattering. The Compton scattering has been experimentally removed, and the background is due, as in Fig. 4, to the "self" part of the diffuse scattering. These plots show that the scattering (a) is clearly peaking at the (2 3 6) point, (b) increases by ~30% on cooling to above T_c (at ~260 K), and (c) is asymmetric due to the (temperature dependent) "self" term. From Wochner *et al.*[20]

~300 K and at 265 K just above the phase transition.[20] In these plots the independently measured Compton scattering has been removed, and the background is due solely to the "self" scattering, which peaks at $l = 10$, which is why the $(2\,3\,6)$ point was chosen. The asymmetry in Fig. 11 is due to a background asymmetry in the "self" term. In a similar rocking scan in the vicinity of $l = 18$, taken through the $(2\,5\,12)$ point at 270 K,[20] there appears to be no measurable peak, which indicates, as suggested by Michel et al.,[41] that $\bar{U}_{10}^{\gamma}$ is perhaps the most prominent noncubic (T_{2g}) order parameter.

The "self" scattering has also been evaluated throughout a $\bar{1}10$ plane in reciprocal space covering the $(2\,3\,6)$, $(3\,4\,6)$, $(2\,3\,7)$, and $(3\,4\,7)$ zone boundary points.[20] A preliminary calculation of this "self" scattering, using the measured $\bar{V}(\omega)$, shows there indeed is a pronounced anisotropy. The "distinct" scattering at the zone boundary points is associated with the development of orientational pair correlations in analogy with diffuse short-range order scattering in other systems. The scattering at $(2\,3\,6)$, while it peaks at a forbidden fcc position, remains quite diffuse down to the phase transition point of ~260 K, increasing by only ~30% as the temperature is lowered (Fig. 11); i.e., it is not critical scattering because the orientational ordering transition is strongly first order. Nonetheless, from a careful evaluation of this "distinct" scattering at $l = 10$ we may extract an effective mean-field orientational interaction potential, which is expressible as an interaction matrix because of its tensor nature. For $\gamma = T_{2g}$, this relationship is given by:[4]

$$\overline{U_{10}^{\gamma}(\mathbf{q})U_{10}^{\gamma'}(-\mathbf{q})} = \frac{\overline{U_{10}^{\gamma^2}}}{1 + \beta \overline{U_{10}^{\gamma^2}} J_{\gamma,\,\gamma'}(\mathbf{q})} \tag{11.3}$$

where $\mathbf{q}$ is the fluctuation wavevector, $\beta = 1/k_B T$ and $J_{\gamma,\,\gamma'}(\mathbf{q})$ is the Fourier transform of the pair potential $v_{\gamma,\,\gamma'}^{i,j}(T_{2g})$ for $l = 10$. Here $\gamma \equiv (X, Y, Z)$, the triply degenerate subrepresentations of T_{2g} [Eq. (5.4)].

Equation (11.3) appears quite similar to conventional expressions for the mean-field susceptibility, or diffuse scattering intensity, of alloy or magnetic systems[34,50-52] above their phase transition points. In principle, it permits an evaluation of the orientational pair interaction potential from diffuse scattering data. In distinction to these other systems,

[50]R. H. Brout, "Phase Transitions." W. A. Benjamin, New York, 1965.

[51]D. de Fontaine, in "Solid State Physics, Vol. 34, eds. H. Ehrenreich, F. Seitz, and D. Turnbull. Academic, New York, 1979.

[52]P. C. Clapp and S. C. Moss, Phys. Rev. 142, 418 (1966).

however, the C_{60} crystal is, as in Section 5, characterized by a set of order parameters, each of which is associated with a separate spherical harmonic function. Fortunately, the lth contribution to the fluctuation spectrum may be separated from the l'th contribution in the diffuse scattering. [Indeed, $l = 10$ appears to be important, while $l = 18$ does not.]

IV. Experimental Aspects of the sc Phase

12. Low Temperature Structure

At low temperatures, solid C_{60} adopts an orientationally ordered simple cubic structure, whose space group was determined to be $Pa\bar{3}$,[12,14] in which the molecules remain on the fcc sites but assume distinct equilibrium orientations. Figure 12 shows the four C_{60} molecules aligned

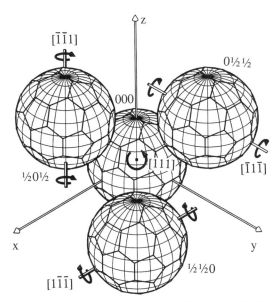

FIG. 12. The four molecules in the unit cell of C_{60} shown in the same standard orientation, with twofold axes aligned parallel to the cube edges. Starting from this orientation, molecules at 000, 1/2 1/2 0, 1/2 0 1/2, 0 1/2 1/2 are rotated by the same angle about local axes $\langle 111 \rangle$, $\langle 1\bar{1}1 \rangle$, $\langle \bar{1}1\bar{1} \rangle$, and $\langle \bar{1}1\bar{1} \rangle$, respectively. The sense of rotation is indicted. From Copley *et al.*[3]

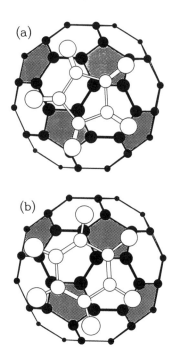

FIG. 13. Two views along an axis joining centers of nearest neighbor C_{60} molecules. In (a) the angle $\Gamma = \Gamma_1$ and a pentagon on the nearer molecule faces a 6:6 short bond on the more distant molecule; in (b) $\Gamma = \Gamma_1 + 60°$ and a hexagon faces the (6:6) bond on the more distant molecule. From Copley et al.[2]

in one of the two equivalent standard orientations, in both of the twofold molecular axes are aligned along the crystallographic axes as noted in Section 2. The Pa$\bar{3}$ space group requires that each of the four molecules in Fig. 12 be rotated about different [111] directions preserving the $\bar{3}$ axis along [111] required for cubic symmetry. Analyses of the x-ray[12] and neutron[13] powder diffraction patterns showed in addition that the principal rotation angle was about 98°. This angle is such that six of the pentagonal faces of a given molecule face the short (6:6) bonds on the neighboring molecules, as shown in Fig. 13(a). Following this work were more detailed single crystal[25] and powder[43] analyses, which indicated a secondary rotation of ~38° in which the short (6:6) bonds face neighboring hexagons (Fig. 13b). While the details of these temperature-dependent configurations and the attendant disorder in the sc phase are treated presently, we note here that there arises naturally a set of domains in the sc phase that, even in a single crystal, perforce limit our ability to observe all of the symmetry associated with Pa$\bar{3}$. These domains, often referred to as merohedral twins, are associated with the choice of rotation sequence in the low temperature structure. By this we mean that the choice of a particular set of [111] axes is arbitrary and thus varies among the macroscopic domains. The possibilities yield four distinct domains for the same orientational order, and a random array of these domains imposes a higher symmetry than Pa$\bar{3}$ on the diffraction pattern.

13. The Orientational Order–Disorder Transition

The high temperature, orientationally disordered fcc phase is separated from the low temperature, ordered sc phase by a first-order transition that occurs at ~260 K at ambient pressure. This transition manifests itself in diffraction measurements with the appearance of additional Bragg peaks (h, k, l of mixed parity), whose intensities are proportional to $\bar{U}_{l,\gamma}^2$ and which are particularly obvious at large Q.[11,13,38,43] Furthermore, the transition shows up both as a reduction in the diffuse scattering[38] and in inelastic scattering measurements. The broad Lorentzian in Fig. 9 at 260 K evolves into an elastic peak and two side peaks as the sample is cooled through its transition temperature T_c, and the side peaks become better defined and move to higher energy as T is further decreased. The Q dependence of the integrated intensity of the side peaks at 115 K and 240 K is similar to the Q dependence of the intensity of the quasi-elastic scattering at 260 K and 520 K (Fig. 14). This similarity indicates that these scattering data have a common origin, namely orientational disorder, and the peaks observed below T_c are due to librations of C_{60} molecules.[47,53] This is confirmed by calculations of the intensity of the thermal diffuse scattering associated with the librational motions. Monte Carlo results, which yield realistic root mean square librational amplitudes, show that the energy-integrated diffuse intensity increases with increasing amplitude, and therefore with increasing temperature, more slowly at large Q than at small Q, consistent with a librational Debye–Waller factor.[2,53] Amplitudes deduced from Fig. 14 suggest that the first-order orientational transition at ~260 K occurs when the rms amplitude of libration is about 10°.

This phase transition has been observed for both polycrystalline and

[53]D. A. Neumann, J. R. D. Copley, W. A. Kamitakahara, J. J. Rush, R. L. Cappelletti, N. Coustel, J. E. Fischer, J. P. McCauley, A. B. Smith, K. M. Creegan, and D. M. Cox, *J. Chem. Phys.* **96**, 8631 (1992).

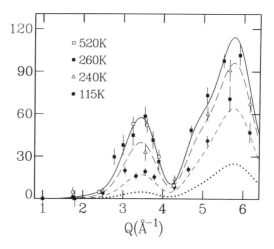

FIG. 14. Q dependence of the energy-integrated intensity of the low energy inelastic and quasi-elastic scattering measurements on C_{60} below and above the orientational order-disorder phase transition. The lines show the results of Monte Carlo calculations of the integrated intensity for rms librational amplitudes of 0.05 (dotted), 0.10 (short dash), and 0.20 radius (long dash). The solid line shows the integrated intensity for complete rotational disorder. After Copley *et al.*[3]

single crystal samples using a wide variety of other techniques including differential scanning calorimetry;[11,54] muon spin rotation and relaxation;[55] dielectric response;[56] sound velocity and attenuation;[57] and specific heat (Fig. 15).[58] These measurements all indicate that the transition is first order and, for sublimation grown single crystals, T_c is very nearly 260 K with a spread of about 2 K among the various measurements. In general, powder samples display a somewhat lower T_c and the spread is larger. It has been well established that there is a correlation between the transition temperature and the sample purity, with purer samples having

[54]A. Dworkin, H. Szwarc, S. Leach, J. P. Hare, T. J. S. Dennis, H. W. Kroto, R. Taylor, and D. R. M. Walton, *C. R. Acad. Sc. Paris Séris II* **312**, 979 (1991).

[55]R. F. Kiefl, J. W. Schneider, A. MacFarlane, K. Chow, T. L. Duty, T. L. Estle, B. H. Jitti, R. L. Lichti, E. J. Ansaldo, C. Schwab, P. W. Percival, G. Wei, S. Wlodek, K. Kojima, W. J. Romanow, J. P. McCauley, N. Coustel, J. E. Fischer, and A. B. Smith, *Phys. Rev. Lett.* **68**, 2708 (1992).

[56]G. B. Alers, B. Golding, A. R. Kortan, R. C. Haddon, and F. A. Theil, *Science* **257**, 5116 (1992).

[57]X. D. Shi, A. R. Kortan, J. M. Williams, A. M. Kini, B. M. Savall, and P. M. Chaikin, *Phys. Rev. Lett.* **68**, 827 (1992).

[58]T. Matsuo, H. Suga, W. I. F. David, R. M. Ibberson, P. Bernier, A. Zahab, C. Fabre, A. Rassat, and A. Dworkin, *Solid State Comm.* **83**, 711 (1992).

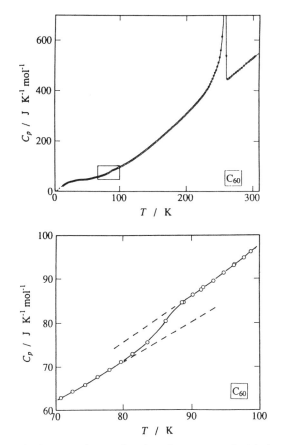

FIG. 15. The molar heat capacity as a function of temperature in (a); the most prominent feature at ~260 K arises from the orientational order–disorder transition. A small anomaly is also seen at the orientational glass transition at ~85 K. (b) displays an expanded view of the data around this temperature. From Matsuo et al.[58]

higher T_c's.[54,59] Also, it is becoming clear[60–64] that molecules of the ambient atmosphere to which the sample has been exposed may be

[59]T. Atake, T. Tanaka, H. Kawaji, K. Kikuchi, K. Saito, S. Suzuki, I. Ikemoto, and Y. Achiba, *Chem. Phys. Lett.* **196,** 321 (1992).

[60]T. Arai, Y. Murakami, H. Suematsu, K. Kikuchi, Y. Achiba, and I. Ikemoto, *Solid State Comm.* **84,** 827 (1992).

[61]R. A. Assink, J. E. Schirber, D. A. Loy, B. Morosin, and G. A. Carlson, *J. Mater. Res.* **7,** 2136 (1992).

[62]G. A. Samara, L. V. Hansen, R. A. Assink, B. Morosin, J. E. Schirber, and D. Loy, *Phys. Rev. B* **47,** 4756, (1993).

[63]S. A. Fitzgerald and A. J. Sievers, *Phys. Rev. Lett.* **70,** 3175 (1993).

[64]S. Huant, J. B. Robert, and G. Chouteau, *Phys. Rev. Lett.* **70,** 3177 (1993).

intercalated into the large interstitial sites of the structure (especially under pressure). Since this process would in principle be faster for the smaller grains present in a polycrystalline sample than for a single crystal, one can understand the lower transition temperatures typically seen in powders as being due to impurities introduced by handling the sample. Careful comparisons of the transition temperature for a polycrystalline sample that has been baked in a vacuum and never exposed to any atmosphere with that of the same sample after exposing it to various atmospheres may answer some important questions concerning sample quality.

The orientational order–disorder transition is more complicated than the preceding discussion might indicate. For instance, the transition actually occurs over a temperature range of 5–10 K, and there is a coexistence of the two phases in this region as shown in Fig. 16, from Wochner et al.[65] This effect has been observed in a variety of x-ray and neutron scattering experiments on both powders[66] and single crystals,[65,67] making it unlikely that it is an experimental artifact. Nonetheless, it is well known that in the absence of applied fields, simple thermodynamic arguments (the Gibbs rule) show that two-phase behavior is impossible for a single-component system over a range of temperature. Thus, either some impurities provide the second component, allowing two phases to coexist, or internal stresses provide the necessary additional field. The first seems rather unlikely, since the system (C_{60} + impurities) must separate into regions of higher and lower impurity densities, requiring the impurities to migrate over macroscopic distances in order to produce the two sharp Bragg peaks that are observed in Fig. 16(a). Furthermore, the coexistence is essentially the same for single crystals and powder samples, which probably have different densities of impurities. On the other hand, the transition is accompanied by an abrupt change in lattice constant of $0.342(8)\%$[39,40,65,66] (Figs. 16 and 17). Results on several single crystal samples also show significant thermal hysteresis of the transition temperature obtained with a variety of techniques,[40,56,57,67] while the original powder results did not.[66] The most recent synchrotron single crystal data[65] in Fig. 16 on both the intensity of a superstructure (mixed parity)

[65]P. Wochner, L. Mailänder, P. C. Chow, X. Jiang, S. C. Moss, and J. C. Hanson (in progress).

[66]P. A. Heiney, G. B. M. Vaughan, J. E. Fischer, N. Coustel, D. E. Cox, J. R. D. Copley, D. A. Neumann, W. A. Kamitakahara, K. M. Creegan, D. M. Cox, J. P. McCauley, Jr., and A. B. Smith III, Phys. Rev. B **45**, 4544 (1992).

[67]R. Moret, P. A. Albouy, V. Agafonov, R. Ceolin, D. André, A. Dworkin, H. Szwarc, C. Fabre, A. Rassat, A. Zahab, and P. Bernier, J. Phys. I France **2**, 511 (1992).

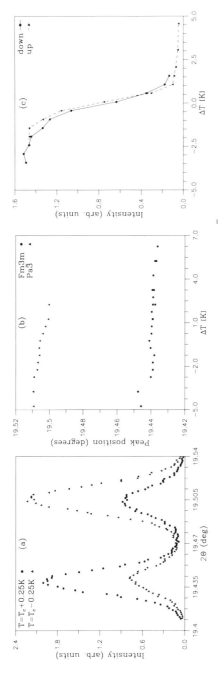

FIG. 16. Radial plots (a) through the (444) reflection (common to both Fm3m and Pa3̄) on passing through the phase transition at $T_c \sim 260$ K; (b) the peak resolutions in 2θ for the low temperature (Pa3̄) and high temperature (Fm3m) phases, obtained from plots such as in (a), which show not only coexistence but a larger thermal expansion for Pa3̄; (c) integrated intensity of the (7 2 1) (mixed parity) peak near $l = 10$ on passing through the phase transition in which there essentially is no hysteresis. From Wochner et al.[65]

195

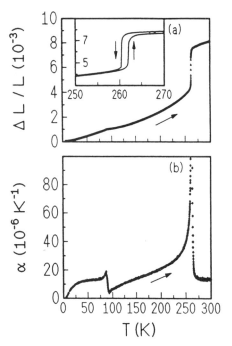

FIG. 17. Temperature dependence of the (a) relative linear thermal expansion of $\Delta L/L(T)$ and (b) thermal expansivity $\alpha(T)$ measured by dilatometry. Note that the peak in $\alpha(T)$ at $\sim$260 K has been cut off ($\alpha_{max} \approx 600 \times 10^{-6}$ K). This shows the change in lattice constant that accompanies the orientational order–disorder transition and the change in α at the orientational glass transition. The inset in (a) displays the observed hysteresis in the transition temperature. From Gugenberger et al.[40]

peak (Fig. 16c) and the lattice parameter change (Fig. 16b) indicate very little hysteresis through the coexistence region, which is nonetheless broad. There is therefore an inherent problem in using the intensity of a superlattice peak to determine an order parameter through the transition, because it may simply reflect the volume fraction of the low temperature phase within this coexistence region. Only when the intensity of the high temperature phase vanishes does the superlattice intensity reflect correctly the thermodynamic order parameter.

These results suggest that the macroscopic constraint suffered by a single crystal is large and can significantly affect the transition. The best crystals show a large increase in mosaic spread (from 0.02° to 0.15°) in their initial cooling through the transition temperature.[65] It therefore seems likely that internal coherency stress plays a crucial role in the orientational phase transition. Perhaps this behavior may be viewed within a context similar to martensitic transformations where large stresses induce mosaic fragmentation[68] and where phase coexistence has been observed over hundreds of Kelvin. In the present case in which the transformation is cubic to cubic, substantial stresses are set up, which may act as internal fields and provide the necessary relaxation of the Gibbs criterion. More work, using single crystal samples, is required to definitively answer these questions concerning the phase transition, and efforts in this regard continue.[65]

These observations, together with the large pressure dependence of T_c,[62,69,70] require significant coupling between the translational and orientational degrees of freedom in solid C_{60}. Since the condensation of the active orientational modes occurs at the zone boundary X-point rather than the zone center Γ-point of the Brillouin zone of the fcc phase,[41] however, there can be no bilinear coupling between the orientational fluctuations and the long wavelength displacements as is the case, say, for KCN.[71] Motivated by these considerations, Lamoen and Michel[72] have investigated the origin of the coupling between the lattice constant and the orientations and have demonstrated that in solid C_{60} such a coupling can occur between the long wavelength acoustic modes and the square of the order parameter.[72] Physically, the coupling is to the square of the order parameter because long wavelength deformations (i.e., at the zone center) can only couple to zone boundary orientational fluctuations (i.e, at the X-point) if two modes with opposite wave vectors are included. Then one can show[72] that the change in lattice constant Δa ($= a_- - a_+$), observed in Figs. 16 and 17, is given by

$$\Delta a = -8/a\,|\lambda'|\,K_0\eta^2 \qquad (13.1)$$

where a is the lattice constant, λ' is the appropriate matrix element (for a Lennard–Jones potential this has a value of about $-5\ \text{K}/\text{Å}$ for the $l = 10$ manifold of C_{60}), K_0 is the bare compressibility, and η is the Landau order parameter discussed in Section 5. Equation (13.1) also predicts that the thermal expansion just below the transition should be greater than just above, as observed in Figs. 16 and 17.

Despite the first-order nature of the transition, significant precursor

[68]A. L. Roitburd, *Phys. Stat. Sol.* **16A,** 329 (1973).

[69]G. A. Samara, J. E. Schirber, B. Morosin, L. V. Hansen, D. Loy, and A. P. Sylwester, *Phys. Rev. Lett.* **67,** 3136 (1991).

[70]G. Kriza, J.-C. Ameline, J. Jerome, A. Dworkin, H. Szwarc, H. Fabre, D. Schutz, A. Rassat, and P. Bernier, *J. Phys. I* **1,** 1361 (1991).

[71]K. H. Michel and J. M. Rowe, *Phys. Rev. B* **32,** 5827 (1985).

[72]D. Lamoen and K. H. Michel, *Phys. Rev. B* **48,** 807 (1993).

effects have been observed in a number of measurements including the elastic response, the heat capacity (Fig. 15), and the thermal expansion (Fig. 17) as the transition is approached from below.[40,57,58,73] It should be noted, however, that the time scale of these precursor effects differs considerably among the experiments. In particular, Schranz et al.[73] observe a precursive softening, as predicted by Lamoen and Michel[72] for C_{11}, followed by an abrupt rise in the compressive [111] elastic constant, measured at 1 Hz, while Shi et al.[57] observed only a discontinuous drop in their single crystal sound velocity measured at 10 kHz via a vibrating reed technique. Both of these frequencies would appear to be slower than the inverse reorientational time for a C_{60} molecule at 260 K, as discussed previously, and therefore within a bandwidth expected to be influenced by molecular reordering.

14. ORIENTATIONAL DISORDER IN THE SC PHASE AND THE
    GLASS TRANSITION

The square of the order parameter in Fig. 18 is measured below the transition using x-ray diffraction from a single crystal and is rather

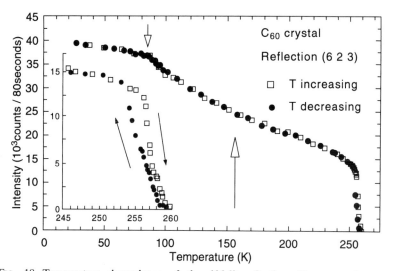

FIG. 18. Temperature dependence of the (6 2 3) reflection. The unusual concave temperature dependence between ~85 K and ~250 K is due to orientational disorder in the low temperature phase. An expanded view of the hysteresis near the orientation order–disorder is shown in the inset. From Moret et al.[49]

[73]W. Schranz, A. Fuith, P. Dolinar, H. Warhanek, M. Haluska, and H. Kuzmany, Phys. Rev. Lett. 71, 1561 (1993).

unusual, displaying a concave temperature dependence.[67] As discussed earlier in connection with the diffuse scattering, the (6 2 3) reflection studied by Moret et al.[67] comes at the $l = 10$ peak in the "self" part of the diffuse scattering in Fig. 4.[20] The order parameter explored at this reflection is therefore dominated by $\bar{U}_{10}^{\gamma}$, and the superlattice intensity is thus proportional to $\bar{U}_{10}^{\gamma 2}$. The importance of this particular harmonic in the ordering transition has been discussed theoretically,[16,41,74] although no theoretical treatment of its temperature dependence yet exists. In addition, there is more diffuse scattering below the transition than can be explained by thermal diffuse scattering from the librations,[2] and the NMR ^{13}C line shows no discontinuous increase in width as the sample is cooled through the orientational ordering transition.[75,76] The explanation for this series of observations is the existence of significant structural disorder between the major and minor orientations mentioned in the foregoing. The temperature dependence of their relative populations was investigated using high resolution neutron diffraction.[39,43] Data for a series of temperatures were analyzed on the basis of a model in which each molecule has one of the two possible orientations shown in Fig. 13, with occupancies P and $1 - P$, respectively. The temperature dependence of P, as deduced from this analysis, is shown in Fig. 19. It appears that the "hexagon-facing" (HF) configuration (Fig. 13b) corresponds to a local minimum in the orientational potential, whereas the "pentagon-facing" (PF) configuration (Fig. 13a) corresponds to a global minimum and is therefore the majority orientation that the molecules increasingly adopt as the system is cooled. It should be emphasized that these two configurations are not to be treated as two phases; rather they coexist within a single phase with a distinct lattice parameter, which does, nonetheless, depend upon P.[39,40] An unusual aspect of this dependence of lattice parameter on P is that the minority orientation brings nearest-neighbor molecules into closer contact along [110] than the ground state majority orientation!

Above the discontinuity at ~85 K, the C$_{60}$ molecules can reorient quickly, and thermodynamic equilibrium is achieved on the laboratory time scale. Therefore, the relative population of the HF and PF orientations is determined by Boltzmann statistics:

$$\frac{1-P}{P} = e^{-\Delta\bar{V}/k_{\mathrm{B}}T}, \tag{14.1}$$

[74]R. Heid, Phys. Rev. B **47**, 15912 (1993).

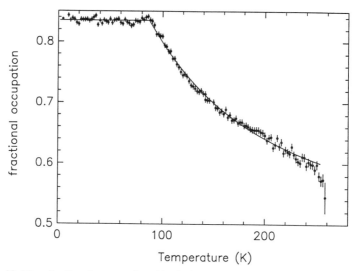

FIG. 19. The fractional occupation P of the more energetically favorable (major) orientation as a function of temperature. Relative populations of major and minor orientations are frozen at $\sim$85 K. The curved line corresponds to Eq. (14.1) with an energy difference $\Delta \bar{V} = 11.0$ meV. From David et al.[39]

where $\Delta \bar{V}$ is the energy difference between the two orientations ω_1 and ω_2, $\bar{V}(\omega_1) - \bar{V}(\omega_2)$. From a least squares fit to the diffraction data[39] it was found that $\Delta \bar{V} \approx 11$ meV, in good agreement with an analysis of thermal conductivity measurements.[77] Because of the large potential energy barrier between the two configurations, the transition rate reduces dramatically with temperature. At about 85 K, the rate is estimated to be on the order of 10^{-3} sec^{-1},[58] and thermodynamic equilibrium cannot be achieved except in extremely long-time-scale experiments. Since no further hopping occurs in a reasonable time in a nonequilibrium state, the system is orientationally frozen and P does not change below this temperature (Fig. 19). This state can then be considered an orientational glass, and 85 K can be taken as the approximate glass transition temperature. The height of the barrier $\bar{V}_a$ separating the ground state orientation and the defect orientation can be estimated by the Arrhenius equation

$$\frac{1}{\tau_g} = \nu_{lib} e^{-\bar{V}_a / k_B T_g}, \qquad (14.2)$$

where ν_{lib} is the librational frequency, T_g is the glass transition temperature, and τ_g is the equilibration time at the glass transition. Using the values of $\nu_{lib} = 0.65$ THz,[53] $T_g = 85$ K,[39,43] and $\tau_g = 1000$ sec,[58] one finds that $\bar{V}_a \approx 250$ meV.

A rather simple picture of the low temperature dynamics in C_{60} emerges from these considerations. There are two nearly degenerate orientations, differing in energy by about 11 meV and separated by a barrier of about 250 meV. Between an orientational glass transition at ~85 K and the orientational melting transition at 260 K, the molecules execute thermally activated jumps between these two orientations. Below the glass transition temperature, the probability of a thermally activated jump over the barrier becomes so small that thermodynamic equilibrium cannot be attained, and the sample is frozen into an orientational glass state. The probability of quantum mechanical tunneling between these states is so small that it can be ignored.

This picture is supported by a wide variety of other measurements. The earliest were the NMR measurements in which the ^{13}C line showed significant motional narrowing that changes gradually with temperature below the orientational order–disorder transition.[75,76] A more detailed examination of the lineshape revealed that the temperature dependence of the width is consistent with a thermally activated motion with an activation energy of 181 meV[75] or 250 meV.[76] Thus, the observed motion is simply the hopping of the molecules between the two orientations. Further evidence for the presence of a "glassy" state can be obtained by realizing that the apparent glass transition temperature depends on the time scale of the measurements; i.e., for an ac technique the transition temperature should be frequency dependent. This has been investigated by measuring the frequency dependence of the glass transition by ultrasonic attenuation, which shows that the glass transition occurs at 160 K at a frequency of 21.40 kHz, but shifts to 154 K at 10.73 kHz.[57] The rather small shift of only 6 K when the frequency changes by a factor of two is consistent with a barrier $\bar{V}_a$ of 240 meV. Frequency-dependent measurements of the glass transition have also been made using dielectric permittivity measurements. (Presumably the electric dipole moment probed in these measurements is induced by the orientational disorder.) The results again show a frequency-dependent glass transition temperature with $\bar{V}_a = 270$ meV. In addition, these results yield an estimate for the range of barriers present in solid C_{60} of ≈ 40 meV. The recent low frequency (~1 Hz) elastic response measurements by Schranz et al.,[73]

[75]C. S. Yannoni, R. D. Johnson, G. Meijer, D. S. Bethune, and J. R. Salem, *J. Phys. Chem.* **95,** 9 (1191).

[76]R. Tycko, R. C. Haddon, G. Dabbagh, S. H. Glarum, D. C. Douglass, and A. M. Mujsce, *J. Phys. Chem.* **95,** 518 (1991).

however, suggest that any spread in the distribution of $\bar{V}_a$ must be less than 10 meV. Further experiments are clearly required to determine the origin of this difference. A likely possibility is that it arises from the unintentional doping of one or both of the samples with dilute gaseous molecules such as N_2 or O_2.

One can also probe the glass transition by measuring the time dependence of an appropriate quantity near 85 K when the time scale is on the order of hundreds of seconds. This approach has been used in the measurements of heat capacity in Fig. 15 ($\delta C_p = 7$ J/K mole),[58] thermal expansion in Fig. 17 ($\delta\alpha = 10^{-5}$ K^{-1}),[40] and thermal conductivity.[77] In each case, the results are consistent with the presence of an orientational glass transition with the barrier heights varying between 235 and 288 meV. Figure 20 is an Arrhenius plot covering a variety of measurements. It is remarkable that experiments which span more than 10 decades in time yield essentially the same activation energy. Within the Angell classification[78] scheme for glass formers, this type of behavior is the extreme example of a strong glass former.

15. REORIENTATION MECHANISM AND THE ORIENTATIONAL POTENTIAL IN THE LOW TEMPERATURE PHASE

The foregoing observations strongly suggest that a single reorientation mechanism dominates below 260 K. Here we will infer this mechanisn based on the barrier height and the energy of the librations.

To relate the average energy of the librations to the barrier height, assume a potential $V(\Theta)$ of the form

$$\bar{V}(\Theta) = \bar{V}_a/2[1 - \cos(2\pi\Theta/\Theta_{\text{hop}})], \tag{15.1}$$

where Θ_{hop} is the angle through which the molecule jumps. Then

$$\bar{V}_a = E_{\text{lib}}^2 \left(\frac{\Theta_{\text{hop}}}{2\pi}\right)^2 \left(\frac{2I}{\hbar^2}\right), \tag{15.2}$$

where I is the moment of inertia of a C_{60} molecule ($\sim 1 \times 10^{-43}$ kg m^2)

[77]R. C. Yu, N. Tea, M. B. Salamon, D. Lorentz, and R. Malhotra, *Phys. Rev. Lett.* **68**, 2050 (1992).

[78]C. A. Angell, *J. Phys. Chem. Sol.* **49**, 863 (1988); *J. de Chemie Physique* **82**, 773 (1985).

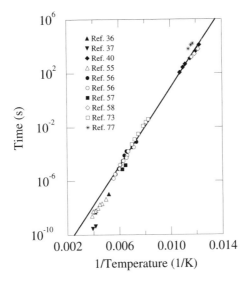

FIG. 20. Arrhenius plot of the equilibration time as a function of temperature for a wide variety of measurements including thermal conductivity,[77] dilatometry,[40] specific heat,[58] low frequency elastic response,[73] ultrasonic elastic response,[57] dielectric response,[56] and NMR.[36,37] The solid line represents an activation energy of 280 meV. The fact that the data can be described by a single activation energy over 12 orders of magnitude in time strongly suggests that a single reorientation mechanism predominates. Note that the NMR results represent the time between jumps rather than an equilibration time, and it is thus not surprising that they fall somewhat below the line. The slope, however, and therefore the activation energy, is consistent with the other measurements.

and E_{lib} is the measured average librational energy (2.5 meV).[53] It was originally speculated that the PF $\Leftrightarrow$ HF reorientation mechanism involved 60° rotations of a molecule about the single threefold molecular axis that is parallel to a threefold [111] direction in the crystal lattice, as illustrated in Fig. 21. The problem with this model is that this hopping angle implies an activation energy $\bar{V}_a$ of about 480 meV. Alternatively the molecules could execute ~42° hops about any one of the three twofold molecular axes that are normal to a body diagonal [111] direction in the lattice (Fig. 21). In this case the calculated activation energy is ~245 meV, in good agreement with experiment. The later model has the added merit that the spectrum of possible reorientations is almost isotropic; in the 60° jump model, a given molecule can only visit 3 of the 60 equivalent orientations, but the ~42° jump model permits access to all 60 equivalent orientations.

One can also use the preceding information to construct a schematic experimental version of the single particle rotational potential in the low temperature phase (Fig. 22). Here the angle of rotation $\theta = 0$

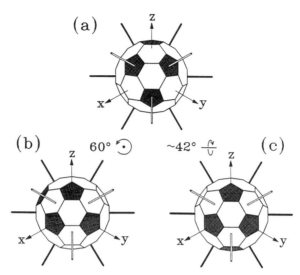

FIG. 21. Three possible orientations of a C_{60} molecule with respect to a fixed set of axes. The plane of each drawing is normal to a [111] direction. Thin shaded (unshaded) rods represent [110] directions that are normal (inclined at 35.26°) to the $\langle 111 \rangle$ direction. The 12 nearest neighbors of the molecule are in the 12 [110] directions. At (a), which represents the majority orientation, pentagons face neighbors in the "inclined" $\langle 1\bar{1}0 \rangle$ directions, and short double (6:6) bonds face neighbors in the "normal" $\langle 110 \rangle$ directions. At (b) and (c), which represent the minority orientation, hexagons have replaced pentagons in the "inclined" $\langle 110 \rangle$ directions. The transformation from (a) to (b) involves a 60° rotation about the [111] direction. The transformation from (a) to (c) involves a ~42° rotation about the $\langle 1\bar{1}0 \rangle$ direction. From Copley et al.[3]

corresponds to the ground state orientation of the molecule, the solid line is the energy required to rotate the molecule about a threefold [111] axis, and the dashed line is the energy required to rotate about a [110] axis, which roughly corresponds to the twofold molecular axis. First, consider the potential for rotations about a [111] axis. There is a barrier of ~500 meV between the ground state (majority) orientation and the defect (minority) orientation lying ~11 meV higher in energy. These orientations are separated by 60°. For rotations about the $\langle 1\bar{1}0 \rangle$ direction, the situation is somewhat more complicated. Here one sees the defect orientation at an angle of ~42° from the ground state orientation and that is separated from it by a barrier of ~250 meV. The other parts of this potential are more speculative, since there is no current low temperature experimental information upon which estimates of the potential can be made. Therefore, we are simply guided by the pattern of hexagons and pentagons on a molecule and how those appear as one rotates about the

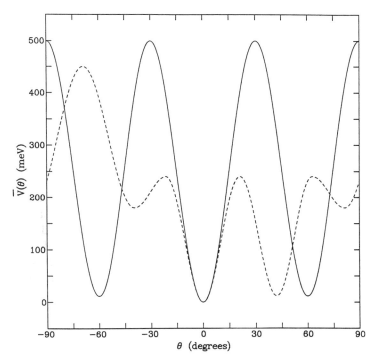

FIG. 22. Schematic diagram of the single particle potential $\bar{V}(\omega)$ in the low temperature Pa$\bar{3}$ phase. Here $\Theta = 0$ corresponds to the equilibrium position of the C$_{60}$ molecule, the solid line represents the potential for rotations of the molecule about the threefold axis aligned along a $\langle 111 \rangle$ direction, and the dashed line is the potential for rotations about the $\langle 1\bar{1}0 \rangle$ direction, which nearly corresponds to a twofold axis of the molecule. The defect orientation can be reached by a 60° rotation about [111] or a ~42° rotation about [1$\bar{1}$0].

twofold axis in Fig. 21. First, the two deepest dips are known to correspond to hexagons aligned along a [111] axis, which is required for cubic symmetry. These are the ground state and the observed defect orientations. Since the local relationships are not extremely different if a pentagon is aligned along the [111] direction, local minima are sketched in for these orientations in Fig. 22. Nevertheless, these minima must be very shallow, since x-ray diffraction results in the high temperature phase indicate that there is almost no probability of fivefold axes being aligned along the [111] direction in Fig. 7. The height of the barrier between these orientations is at present completely arbitrary. In spite of these uncertainties, Fig. 22 summarizes a considerable amount of experimental information on the single particle rotational potential in solid sc C$_{60}$. Any model that accurately reproduces the low temperature structural and dynamical properties in C_{60} must approximate the potential shown in Fig. 22 rather well. It should be reiterated, however, that for both the high temperature[17,18] and low temperature phases, the (mean field) rotational potential can be evaluated directly from Bragg data on the fcc set of peaks.

V. Phonons

16. Experimental Aspects

The rotational dynamics and orientational ordering phenomena of C_{60} are determined by the nonspherical part of the intermolecular potential ($l \neq 0$). The spherical ($l = 0$) part of the potential is reflected in the low energy intermolecular translation modes. Here we will refer to these modes as "phonons" in spite of the fact that the intramolecular bending and stretching modes and the librations are also momentum eigenstates of the system and therefore, in the strict sense of the term, are phonons as well. The justification for ignoring the intramolecular modes is that they occur above 33 meV[79-82] while the most energetic intermolecular modes are less that 7 meV. Thus, to an excellent approximation, these modes are completely decoupled, and for our purposes here the molecules can be considered as rigid objects. The distinction between librational modes (librons) and phonons is less obvious and is essentially for convenience, although some rationale for this artificial separation comes from lattice dynamical calculations that indicate the coupling between the rotational and translational degrees of freedom is rather small.[83-86] Information concerning the phonons can be obtained from a

[79]R. L. Cappelletti, J. R. D. Copley, W. A. Kamitakahara, F. Li, J. S. Lannin, and D. Ramage, *Phys. Rev. Lett.* **66,** 3261 (1991).

[80]W. A. Kamitakahara, J. R. D. Copley, R. L. Cappelletti, J. J. Rush, D. A. Neumann, J. E. Fischer, J. P. McCauley, and A. B. Smith, in *Novel Forms of Carbon, Mat. Res. Soc. Symp. Proc.* **270** (C. L. Renschler, J. J. Pouch, and D. M. Cox, eds.) Mat. Res. Soc., Pittsburgh (1992), p. 167.

[81]C. Coulombeau, H. Jobic, P. Bernier, C. Fabre, D. Schültz, and A. Rassat, *J. Phys. Chem.* **96,** 22 (1992).

[82]K. Prassides, T. J. S. Dennis, J. P. Hare, J. Tomkinson, H. W. Kroto, R. Taylor, and D. R. M. Walton, *Chem. Phys. Lett.* **179,** 181 (1991).

[83]T. Yildirim and A. B. Harris, *Phys. Rev. B* **46,** 7878 (1992).

[84]J. P. Lu, X.-P. Li, and R. M. Martin, *Phys. Rev. Lett.* **68,** 1551 (1992).

[85]X.-P. Li, J. P. Lu, and R. M. Martin, *Phys. Rev. B.* **46,** 4301 (1992).

[86]J. Yu, R. K. Kalia, and P. Vashishta, *Appl. Phys. Lett.* (in press); *Phys. Rev. B* (in press).

wide variety of measurements, including, for instance, the compressibility, the elastic constants, specific heat, various light scattering techniques including Brillouin and Raman scattering and IR absorption, and most importantly, inelastic neutron scattering. The earliest attempts to gain information on the radial part of the intermolecular potential involved measurements of the compressibility of powder samples of C_{60}.[87,88] The principal conclusion from those measurements was that the bonding is weak, consistent with the expected van der Waals nature of these interactions and similar to the c-axis bonding in graphite. An interesting by-product of these measurements, however, was the observation that the intensity of the (200) reflection increased continually with pressure. This Bragg peak is an fcc-allowed reflection whose absence is due solely to a zero in the spherical Bessel function $j_0(QR)$, which, by coincidence, occurs at the (200) position. On the application of pressure, because the lattice is considerably more compressible than the individual molecules, the (200) peak is shifted upward in Q and hence away from the zero in $j_0(QR)$;[88] i.e., $(d \ln R)/dp$ is much smaller than $(d \ln a)/dp$, where a is the lattice parameter.

Unfortunately, these techniques (along with Brillouin scattering) probe only the long wavelength acoustic modes. In addition, single crystals are necessary in order to measure the elastic constants C_{11}, C_{12}, and C_{44} rather than average polycrystalline quantities such as the bulk modulus. Thus, the compressibility, a single number, is not particularly sensitive to details of the interatomic potential, but does yield some information on the average intermolecular interactions.

Another approach is to use light scattering techniques that probe the zone center optic modes. For C_{60} the low temperature simple cubic phase must have 24 modes, since each of the four molecules in the primitive cell contributes three translations and three rotations to the total. The symmetries of these modes can be determined by taking the direct product of the characters for the molecular site symmetries of the four molecules $(A_g + T_u)$ with the characters for the vibrational modes for the C_{60} molecules in T_h^6 symmetry.[83,89–91] For the low energy modes, these are T_u (translations) and T_g (rotations). Then the symmetries of the 24 zone center modes can be expressed as $(T_g + T_u) \otimes (A_g + T_g) = A_g + E_g +$

[87]J. E. Fischer, P. A. Heiney, A. R. McGhie, W. J. Romanow, A. M. Denenstein, J. P. McCauley, and A. B. Smith, *Science* **252**, 1289 (1991).
[88]S. J. Duclos, K. Brister, R. C. Haddon, A. R. Kortan, and F. A. Theil, *Nature* **351**, 380 (1991).
[89]G. Dresselhaus, M. S. Dresselhaus, and P. C. Eklund, *Phys. Rev. B* **45**, 6923 (1992).
[90]M. Tinkham, "Group Theory and Quantum Mechanics," McGraw-Hill, New York, 1964.
[91]W. Que and M. B. Walker, *Phys. Rev. B* **48**, 13104 (1993).

TABLE III.   CHARACTER TABLE FOR THE T_h POINT GROUP SYMMETRY

T_h	E	$3C_2$	$4C_3$	$4C_3'$	i	$3\sigma_v$	$4S_3$	$4S_3'$	BASIC FUNCTIONS
A_g	1	1	1	1	1	1	1	1	$x^2+y^2+z^2$
E_g	1	1	ω	ω^2	1	1	ω	ω^2	$x^2+\omega y^2+\omega^2 z^2$
	1	1	ω^2	ω	1	1	ω^2	ω	$x^2+\omega^2 y^2+\omega z^2$
T_g	3	-1	0	0	3	-1	0	0	$(R_x,R_y,R_z),(yz,zx,xy)$
A_u	1	1	1	1	-1	-1	-1	-1	xyz
E_u	1	1	ω	ω^2	-1	-1	$-\omega$	$-\omega^2$	$x^3+\omega y^3\omega^2 z^3$
	1	1	ω^2	ω	-1	-1	$-\omega^2$	$-\omega$	$x^3+\omega^2 y^3+\omega z^3$
T_u	3	-1	0	0	-3	1	0	0	(x,y,z)

After Dresselhaus et al.[89]

$3T_g + A_u + E_u + 3T_u$. Since the translational modes in C_{60} are odd under inversion, they correspond to the "ungerade" modes $A_u + E_u + 3T_u$, while the rotations are even and therefore correspond to $A_g + E_g + 3T_g$. Of the translations, three are associated with the overall motion of the crystal and belong to the threefold degenerate T_u representation. These will become the acoustic branches at nonzero wavevector. Of the remaining four zone center phonons, inspection of the basis functions (i.e., the functions that transform as the symmetry elements) given in character tables (Table III) for T_h symmetry shows that the T_u modes are IR active, while the A_u and E_u modes are both IR and Raman inactive. Infrared measurements in the low frequency range necessary to observe these excitations are extremely difficult, and IR spectra of solid C_{60} typically show a profusion of features due to impurities in the sample. Recently, however, Fitzgerald and Sievers[63] have obtained exceptionally clean IR spectra by placing a pressed powder sample in an optically accessible vacuum cell, which was pumped for several hours at a pressure $<10^{-6}$ torr and a temperature of 100°C. After cooling the sample to 1.5 K, the IR spectrum displays only the two expected peaks at energies of 5.1 and 6.8 meV. Further evidence that these are the translational T_u modes is obtained by demonstrating that the peaks disappear above the transition to the orientationally disordered fcc phase. Upon exposing the sample to air, the profusion of peaks seen by other researchers was observed. This result demonstrates that some impurity atoms or molecules may have diffused from the ambient atmosphere into the large interstitial sites of the C_{60} structure. A similar inspection of the basis functions in the character table for T_h symmetry shows that modes with A_g, E_g, and T_g symmetry are Raman active. Thus, in principle all five zone-center librational modes are observable in a Raman scattering experiment. As difficult as the IR measurements are, however, the high symmetry of the C_{60} molecule will make the observation of these modes even more of an experimental tour-de-force.

The technique that yields the most detailed information on the phonon dispersion relations throughout the Brillouin zone is inelastic neutron scattering. The principal limitation of this technique is that rather large single crystals are required to obtain the necessary intensity for performing measurements in a reasonable length of time with reasonable resolution. This difficulty, which is always present in inelastic neutron scattering, is compounded by the molecular structure factor of the C_{60} molecule, particularly in the high temperature phase. For the purpose of making a simple estimate of the modulation of the Q dependence of the inelastic intensity, one can consider the molecule to be a spherical shell as was done in the initial structure determination. Then, for the translational modes, the intensity must be modulated by square of the $l = 0$ spherical Bessel function, $j_0^2(QR) = \sin^2(QR)/(QR)^2$. The dynamical structure factor[92] for these excitations, $S_T(\boldsymbol{Q}, \omega)$, is then given by

$$S_T(\boldsymbol{Q}, \omega) \propto j_0^2(QR) \left| \sum_{i,\, n} (\boldsymbol{Q} \cdot \hat{\varepsilon}_{i,\, n}) e^{-W_i} e^{i\boldsymbol{Q} \cdot \boldsymbol{R}_i} \right|^2, \tag{16.1}$$

where W_i is the Debye–Waller factor, $\boldsymbol{R}_i$ is the position of the ith molecule in the fcc unit cell, and $\hat{\varepsilon}_i$ is a unit polarization vector for the nth mode.

Since this expression contains a factor of Q^2, the Q dependence of the one-phonon cross section will be roughly given by $\sin^2(QR)$. The dynamical structure factor will therefore show a series of maxima separated by deep minima, making it imperative to choose an optimal position in reciprocal space to measure a particular phonon group.[93,94] Furthermore, since the phonons account for only 3 of the 180 degrees of freedom for a C_{60} molecule, the intensities must perforce be low, no matter where in reciprocal space one chooses to measure.

[92]G. L. Squires, "Introduction to Thermal Neutron Scattering." Cambridge University Press, Cambridge, 1978.

[93]L. Pintschovius, B. Renker, F. Gompf, R. Heid, S. L. Chaplot, M. Haluska, and H. Kuzmany, *Phys. Rev. Lett.* **69**, 2662 (1992).

[94]L. Pintschovius, S. L. Chaplot, R. Heid, M. Haluska, and H. Kuzmany, Proc. Int. Winterschool on Electronic Properties of Novel Materials, ed. H. Kuzmany, Springer Series on Solid State Science (in press).

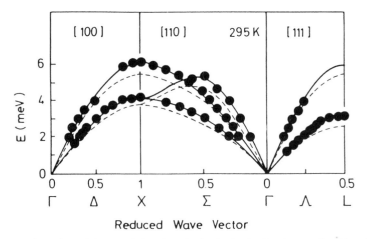

Fig. 23. Dispersion of the translational modes in C_{60} in the fcc phase. After Pintschovius et al.[94] The dashed line is the phonon dispersion for Xe. After Lurie et al.[95]

In spite of these difficulties, measurements of the phonon dispersion relations along the [100], [110], and [111] symmetry directions have been made using a single crystal with a volume of only 6 mm³ together with a particularly effective focusing geometry.[93,94] In the orientationally disordered phase at 300 K, the results in Fig. 23 display the classic fcc phonon dispersion. The solid line represents an axially symmetric two-neighbor Born–von Karman fit to the data, while the dashed line represents the phonon dispersion curves of the rare gas solid Xe.[95] As is readily apparent, the phonon branches in solid fcc Xe are very similar to those in the orientationally disordered phase of C_{60}. This is not particularly surprising in view of the nearly spherical molecular shape of C_{60}[17] and the predominantly van der Waals bonding in both cases. Therefore, a reasonable description of the translational modes in a C_{60} crystal might be obtained simply by treating the molecule as a large noble gas "atom."

This simple view is reinforced by the ability to fit the C_{60} acoustic dispersion with only two force constants. Furthermore, the second-neighbor force constant is only 10% of the nearest-neighbor one, indicating that the $l = 0$ part of the potential is indeed dominated by nearest-neighbor forces. The inclusion of second-neighbor forces is, however, necessary in order to reproduce some aspects of the measured dispersion. For instance, the ratio of the energy of the transverse phonon

[95]N. A. Lurie, G. Shirane, and J. Skalyo, Jr., *Phys. Rev. B* **9,** 5300 (1974).

at the X-point (100) to that at the R-point (1/2 1/2 1/2) of the Brillouin zone is significantly smaller in C$_{60}$ than would be expected for axially symmetric nearest neighbor forces alone, i.e., 1.2 compared with $\sqrt{2}$.[93] This rather low ratio is not reproduced by any theoretical calculation of the phonons reported to date,[83,84] and it is significantly smaller than 1.49, which is observed in Xe.[95] Thus, while describing the C$_{60}$ molecules as large noble gas "atoms" that interact via nearest-neighbor van der Waals forces provides a good first approximation to the lattice dynamics, there are subtle, but measurable, features of the phonon dispersion that require a more sophisticated model of the translational dynamics in the fcc phase.

Below the orientational order–disorder transition, the situation is considerably more complex, because there are no simple criteria that can be used to assign the observed peaks to translational or librational modes. Furthermore, many additional branches are present due to lifting of degeneracies by lowering the symmetry,[83] and the appearance of optic translational and librational modes. For instance the symmetry of the sc phase allows 10 distinct frequencies at the Γ-point, rather than just the $\omega = 0$ modes allowed in the fcc phase. In addition the reduction in symmetry removes all degeneracies of the branches along the "high" symmetry [010] direction, resulting in 24 distinct branches. In contrast, the fcc phase has only two nondegenerate translational branches, transverse and longitudinal along [010]. Inspection of Table IV and Fig.

TABLE IV. POINT SYMMETRIES AND THE NUMBER OF MODES OR BRANCHES AND THEIR DEGENERACIES AT VARIOUS HIGH SYMMETRY POINTS AND ALONG VARIOUS HIGH SYMMETRY DIRECTIONS OBTAINED FROM A GROUP THEORETICAL ANALYSIS OF THE LOW TEMPERATURE PHASE OF C$_{60}$

POINTS	q (π/a)	POINT SYMMETRY $G(q)$	NUMBER OF BRANCHES AND DEGENERACY
Γ	(0, 0, 0)	T_{h}	2(1) + 2(2) + 6(3)
X	(0, 1, 0)	$D_{2\mathrm{h}}$	12(2)
M	(1, 1, 0)	$D_{2\mathrm{h}}$	6(4)
R	(1, 1, 1)	T_{h}	6(4)
$\Delta(\Gamma X)$	(0, α, 0)	$C_{2v}(y)$	24(1)
$\Sigma(\Gamma M)$	(α, α, 0)	$C_3(z)$	24(1)
$\Lambda(\Gamma R)$	(α, α, α)	C_3	8(1) + 8(2)
$S(XR)$	(α, 1, α)	$C_3(y)$	12(2)
$Z(XM)$	(α, 1, 0)	$C_{2v}(x)$	12(2)
$T(MR)$	(1, 1, α)	$C_{2v}(z)$	6(4)

After Yildirim and Harris.[83]

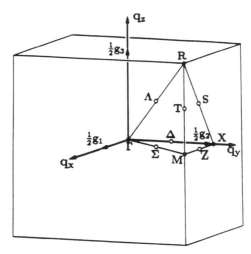

FIG. 24. Symmetry points and directions for a simple cubic lattice.

24 suggests that the simplest "high" symmetry directions in the Pa$\bar{3}$ structure are the [111] directions, where 16 branches are allowed, and along the Brillouin zone boundary between the R-point $(1/2, 1/2, 1/2)$ and M-point $(1/2, 1/2, 0)$, where only 6 branches are allowed.[83] Furthermore, the modes at the R-point separate into pure rotations and translations.

Figure 25 shows the best measurements to date together with current assignments along a [111] direction from Γ to R and along the zone boundary R to M. Here the closed (open) circles represent modes that are predominantly phonons (librations) and the solid (dotted) lines are to guide the eye.[94] The diamonds show the energies of the two T_u modes observed in IR measurements.[63] At Γ, six of the nine expected nonzero energy modes are observed. Of these, the four lowest energy peaks are assigned to librations and the two highest are assigned to phonons. The excellent correspondence between these two modes and the IR measurements indicates that they are almost certainly T_u modes. The only uncertainty is the assignment of the lower of these two modes, since all detailed lattice dynamical calculations performed to date[83,84] suggest that this T_u peak has very nearly the same energy as the E_u and A_u modes. Therefore, the peak in the neutron spectrum may contain all three.

The assignment of the librations is more problematic. While the excellent agreement between the lowest three modes and the density of states peak arising from librations[47,53] shows that these three excitations are undoubtedly librations, the assignment of these modes to particular

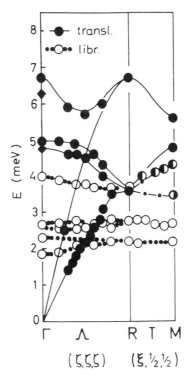

FIG. 25. Dispersion of the intermolecular modes of C_{60} in the orientationally ordered phase. The closed (open) circles are due to the translational (librational) modes[94] and the diamonds are the infrared active translational modes.[63] From Pintschovius et al.[94]

symmetries is not possible at this time. On the other hand, if the 4-meV mode is indeed a libration, it almost certainly has T_g symmetry since calculations to date[83,91] invariably give highest energy libration T_g symmetry. Further evidence for this assignment comes from the measurements at the M-point, where all of the six expected frequencies have been observed. While the modes do not separate into pure rotations and pure translations at this position in the Brillouin zone, the assignment of these lower energy mode frequencies as being principally librations and the higher energy modes as being principally phonons is certainly reasonable. Unfortunately, the six modes are not as well separated at the R-point, and the branch assigned to the highest energy "libration" has no measured points between R and M, leaving some uncertainty in tracing this back to the Γ-point.

This assignment of the 4-meV mode as a libration need not be in conflict with the powder measurements,[47,53] which show a single broad peak at ~2.3 meV, since less than a fourth of the total spectral weight is in this mode. Additionally, the presence of substantial dispersion in this branch compared with that of the other librations would reduce the apparent intensity at energies >3 meV since this scattering would be spread out over a considerable energy range. Also, the powder data has invariably been displayed as $S(Q, \omega)$ rather than being converted to a density of states. Conversion to a density of states, which directly reflects the spectral weight of the various modes, for the neutron energy loss side of the spectrum involves multiplying $S(Q, \omega)$ by $\omega/(n(\omega) + 1)$, where $n(\omega)$ is the Bose factor.[92] Thus, the higher energy features are de-emphasized in $S(Q, \omega)$ relative to their actual spectral weight. Finally, some indirect evidence for the assignment of the 4-meV mode to T_g symmetry comes from specific heat measurements.[96] Here it is found that the low temperature portion can be well described by a combination of a Debye term with a Debye energy of 3.2 meV and a spectral weight of 3 and two Einstein modes at 2.6 and 5.0 meV with spectral weights of 9 and 12, respectively. Then according to the interpretation presented here, the 3 modes corresponding to the Debye term would be assigned to T_u phonons, the 9 modes at 2.6 meV would be $A_g + E_g + 2T_g$ librations, and the 12 modes at 5.0 meV are $A_u + E_u + 2T_u + T_g$ modes.

While much is already known concerning the phonons, and therefore the $l = 0$ part of the intermolecular potential, a great deal of experimental work remains to be done in this area. For instance, some uncertainties remain in assigning peaks in the measured neutron spectra to specific excitations in the sc phase where the librational modes are present. There are several ways of addressing these uncertainties. The most obvious is simply to obtain more data in other directions, particularly along the $[\alpha, 1, 0]$ zone boundaries. To provide maximum confidence in the assignments, however, requires measuring the intensities as well as the positions of the phonon groups. Since the intensity of a particular peak depends on the eigenvector of the mode, an assignment can be directly checked against various lattice dynamical models. This is an iterative procedure, where improved measurements are used to refine a model whose predictions are then used to help assign more peaks, which will be used to further improve the model, etc. The Karlsruhe group[93,94] has embarked on this time consuming effort, which will prove very important for our understanding of both the dynamics and the underlying inter-molecular interactions.

[96]W. P. Beyermann, M. F. Hundley, S. D. Thompson, F. N. Diederich, and G. Grüner, *Phys. Rev. Lett.* **68,** 2046 (1992).

17. MODEL CALCULATIONS

The earliest models of phonons in C_{60} took advantage of the rather weak intramolecular forces and the similarity of these forces to the interplanar interaction in graphite. The molecules were treated simply as spherical shells that interact with their nearest neighbors via some average van der Waals interaction. The parameters for such models were based on the interlayer elastic constant, calculated interlayer binding energy, or any of several other properties that are determined by the interlayer forces in graphite.[97] While possessing the virtue of simplicity, models of this type have several unsatisfying features. First, they ignore the longer range interactions that are responsible for stabilizing the high temperature fcc structure and that are necessary to describe the measured phonon dispersion. Furthermore, these models simply ignore all components of the potential except $l = 0$. Since higher l components are responsible for the orientational ordering transition and rotational dynamics, a model of this type is incapable of describing them. Still in the high temperature fcc phase, the rotating molecules are remarkably close to spherical, and one would expect such models to have some correspondence to reality.

An example of this type of model is obtained by replacing the interatomic binding energy U between two molecules by the energy $\bar{U}$ corresponding to the interaction energy between two areas on adjacent molecules.[97] The paramaters of a modified Morse potential,

$$\bar{U}(r) = \tilde{D}_e[1 - e^{-\beta(r-r_e)})^2 - 1] + \tilde{E}_r e^{-\beta' r}, \qquad (17.1)$$

are then chosen to reproduce ab initio local density approximation calculations for the binding energy of graphite. Here r is the distance between atoms in adjacent layers of graphite, r_e is the equilibrium distance between atoms, $\tilde{D}_e$ is the effective equilibrium binding energy of these atoms, $\tilde{E}_r$ is an effective additional hard-core repulsion, and β and β' here describe the distance dependence of these interactions. After accounting for the curvature of the C_{60} molecule, the binding energy, and therefore the phonon dispersion relations, can be determined. This yields phonons energies that are on average $\sim30\%$ larger than those measured experimentally.

[97]Y. Wang, D. Tománek, and G. F. Bertsch, *Phys. Rev. B* **44**, 6562 (1991).

The first improvement one might envisage would be to model the interatomic van der Waals interactions with a Lennard–Jones 6–12 potential rather than using some average intermolecular interaction.[98] This greatly enhances the computational complexity, since one now must calculate $60 \times 60 = 3600$ interatomic potentials in order to obtain the interaction between a pair of molecules. The advantage, however, is that rotational components of the potential are now accessible. Molecular dynamics calculations performed with a Lennard–Jones potential do indeed display an orientational ordering transition.[99] Unfortunately it is accompanied by a tetragonal distortion.[100,101] Furthermore, the predicted librational density of states in the low temperature phase peaks at one-half of the measured energy,[47,53] indicating that the orientational potential is four times too small. On the other hand, the density of states for the translational modes that reflect the $l = 0$ contribution have an energy cutoff at about 6 meV, in excellent agreement with the experimental value in the fcc phase (~6.2 meV). The molecular dynamics calculations, however, only yield a density of states rather the detailed dispersion relations. A comparison with the results shown in Fig. 25 is thus not possible.

In order to stiffen the rotational part of the potential and account for the possibility that the low temperature structure may be stabilized by a Coulomb attraction between the "high" electron density short carbon–carbon (6:6) bond and the "low" electron density pentagon on adjacent molecules, the next models involved placing charges at various points along the bonds. Charge neutrality is maintained by positive charges located along the bonds, at the atoms, in the pentagon and/or hexagon centers, or in the center of the molecule. Several versions of this model have been proposed,[16,84,102] but there have been detailed calculations[83] of the intermolecular modes in the low temperature phase for only two. The first involves placing effective charges of q and $-2q$ ($q = 0.27e$) at the centers of the single and double bonds, respectively.[84,85] While this model reproduces the correct equilibrium orientation of the C_{60} molecule and yields defect (minority) orientations, the lowest energy defect orientation is not that observed experimentally (compare Fig. 26 and Fig.

[98]L. A. Girafalco, *J. Phys. Chem.* **96**, 858 (1992).
[99]A. Cheng and M. L. Klein, *J. Phys. Chem.* **95**, 6750 (1991).
[100]Y. Guo, N. Karasawa, and W. A. Goddard III, *Nature* **351**, 464 (1991).
[101]A. Cheng and M. L. Klein, *Phys. Rev. B* **45**, 1889 (1992).
[102]M. Sprik, A. Cheng, and M. L. Klein, *J. Phys. Chem.* **96**, 2027 (1992).

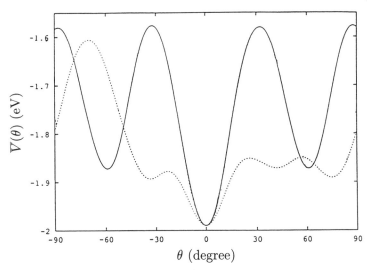

FIG. 26. Calculated single particle potential $\bar{V}(\omega)$ for the point charge model of Lu *et al.* The solid line represents the potential for rotations about the $\langle 111 \rangle$ direction, while the dashed line is for rotations about the $\langle 1\bar{1}0 \rangle$ direction. Compare with the schematic experimental version shown in Fig. 22. From Lu *et al.*[84]

22). In fact, it corresponds to having a fivefold axis along [111], and, therefore, while the three lattice constants are the same, the average symmetry of the system is not cubic. In addition, the glass transition temperature can be estimated using the energy difference between the lowest energy defect orientation and the lowest barrier between this orientation and the ground state, which is seen to be roughly 25 meV in Fig. 27. Using this value for V_a and Eq. (14.2), the glass transition temperature is estimated to be ~8 K rather than 85 K. Furthermore the highest calculated energy for any translational mode is 5.7 meV, which is somewhat lower than the observed energy of 6.8 meV. Thus, the $l = 0$ part of the potential is roughly 40% too small. In addition, while the rotational part of the potential proposed by Lu and coworkers[84,85] is somewhat stiffer than that obtained for the Lennard–Jones 6–12 potential used by Cheng and Klein,[99] the calculated librational energies lie between 1.3 and 2.5 meV, compared with the observed range of 1.8 to 4.0 meV, and therefore the $l = 0$ contribution to the potential is still more than a factor of 2 too soft.

The second model involves placing a charge of $q = -0.35e$ on the double bond and $-q/2$ on the atoms.[102] As with the other bond–charge

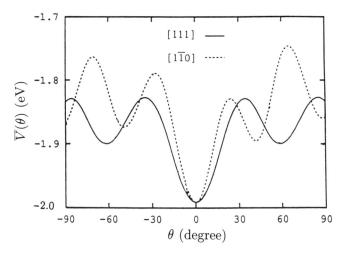

FIG. 27. Calculated single particle potential $\bar{V}(\omega)$ for the point charge model of Sprik *et al.*[102] The solid line represents the potential for rotations about the [111] direction, and the dashed line for rotations about [1$\bar{1}$0]. Compare with the schematic experimental version shown in Fig. 22. From Yildirim and Harris.[83]

models, this potential reproduces the experimentally observed equilibrium orientation of the C_{60} molecule. Additionally, it reproduces the correct defect orientation (compare Fig. 27 and Fig. 22). As before, one can estimate the glass transition temperature. In this case, the barrier between the defect and the ground state orientation is ~60 meV and Eq. (14.2) yields an estimated glass transition temperature of only ~20 K, compared with 85 K. While this model has obvious deficiencies, it is the best currently available for which detailed dispersion curves have been calculated, and therefore it will be discussed in somewhat more detail.

The dispersion curves,[83] based on the model of Sprik *et al.*,[102] are shown along relevant high-symmetry directions in Fig. 28 (1 meV = 8.0655 cm^{-1}). Because of the higher degree of degeneracy, the most detailed current experimental data in Fig. 25 are taken from Γ to R to M in Fig. 24. The first thing one notices when comparing this calculation[83,102] with the experimental results is the energy scale of the excitations; in the calculation, the highest energy phonon is only about 5.3 meV, compared with the observed value of 6.8 meV. Thus, the $l = 0$ part of the potential is too soft in this calculation as well. The disagreement for the librational branches, however, is greater still. Here the calculated librational energies are roughly between 0.9 and 2.2 meV, while the observed peaks currently assigned as librations are between 1.8 and 4 meV. This is also seen in the calculation of the density of states,

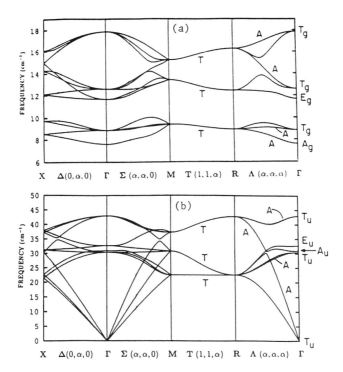

FIG. 28. Dispersion of the intermolecular modes in the orientationally ordered phase of C_{60} calculated by Yildirim and Harris[83] using the point charge model of Sprik *et al.*[102] (a) shows the librational bands and (b) shows the translational bands in the absence of rotation–translation coupling. The symmetries of the zone center (Γ) modes are given on the right. From Yildirim and Harris.[83]

which shows the rotational peak at about 1.6 meV, compared with the observed low temperature value of 2.7 meV.[47,53] In spite of the potential being too soft, there are many similarities, particularly for the translational modes. For instance, the calculated ratios of the energies of the two Γ-point phonons and the R-point phonons are reproduced well by the calculation. One possibility for the discrepancy in the energy scale between the model and the experiment could be the existence of defect orientations not included in the calculation. Calculations for ordered arrays of the ground state and defect orientations, however, show that the potential becomes softer rather than harder as the defects are included.[86]

In general, phenomenological calculations of this type are inadequate to reproduce all of the structural and dynamical properties together. For a given set of parameters, one can reproduce the ground state orientation, the transition temperature, and even the defect orientation. More

TABLE V.   THE $m = 0$ MULTIPOLES OF C_{60} (IN UNITS OF $|e|)q_l^0$
OBTAINED FROM THE CALCULATED QUANTUM MECHANICAL CHARGE DENSITY COMPARED
WITH TWO POINT CHARGE MODELS.

q_l^m	YILDIRIM ET AL.[103]	LU ET AL.[84]	SPRIK ET AL.[102]	$\dfrac{\text{LU ET AL.}^a}{\text{YILDIRIM ET AL.}}$
q_6^0	0.295	3.801	1.598	12.88
q_{10}^0	3.550	1.863	0.202	0.52
q_{12}^0	−0.751	−7.320	−3.637	9.75
q_{16}^0	−2.614	1.214	−0.854	−0.47
q_{18}^0	22.895	−0.031	4.790	−0.00135

After Yildirim *et al.*[103] This table demonstrates that these point charge models do not satisfactorily describe the microscopic charge density.

sophisticated models are required, however, to reproduce the activation energy, the lattice dynamics, the glass transition temperature, and the energy difference between the ground state and defect orientations. Additionally, these models require a rather large charge (~$0.2e$ or more) in order to stabilize the Pa$\bar{3}$ phase at low temperature. To determine whether or not these potentials reflect the actual Coulomb interaction between molecules, Yildirim *et al.*[103] have calculated the electrostatic multipole components of C_{60} using the quantum mechanical charge distribution obtained from ab initio calculations. These multipoles can then be compared with those obtained by the point charge models discussed earlier. Inspection of Table V shows that the multipoles obtained by the ab initio calculation are an order of magnitude smaller than those of the phenomenological models. In order to address this problem, Yildirim *et al.*[103] have developed a point charge model that reproduces the ab initio multipole moments for $l = 6$ through $l = 18$. This requires placing charges along all the bonds, at the centers of the pentagons and hexagons, and at the center of the molecule. This can be done in a variety of ways that reproduce the multipole moments up to $l = 18$, which is the minimum requirement to describe accurately the Coulomb part of the orientational potential. The potential obtained in this way does indeed stabilize the Pa$\bar{3}$ structure, albeit with the defect orientation rather than the experimentally observed ground state.

The unmistakable conclusion from the theoretical work to date is that potentials combining Coulomb interactions arising from charges placed on the molecule with a standard Lennard–Jones potential are inadequate to describe the structure and dynamics of C_{60} in a consistently

[103]T. Yildirim, A. B. Harris, S. C. Erwin, and M. R. Pederson, *Phys. Rev. B* **48**, 1888 (1993).

satisfactory way. Perhaps this is due to the inadequacy of the Lennard–Jones potential in describing the short range interaction. Evidence that the repulsive part of the potential is crucial in the orientational ordering transition comes from the large pressure dependence of the orientational ordering transition temperature[62,69] and from the sensitivity of the transition temperature to the choice of Lennard–Jones parameters.[41] Alternatively, the charge distribution on an isolated C_{60} molecule could differ significantly from that in the solid, requiring a careful treatment of the overlap integrals of the wavefunctions on neighboring molecules.[103]

VI. Summary

18. GENERAL REMARKS

Much of our discussion in this chapter has concentrated on the intermolecular potential in crystalline C_{60} as deduced through the combination of scattering methods and statistical mechanics. Molecular rotator functions and their thermodynamic averages $\bar{U}_{l, \gamma}$ provide the necessary parameters for describing the orientational dynamics and equations of state, and while the formalism is complex, it is correspondingly rich in possibilities. Both neutron and x-ray measurements, together with a variety of related techniques that probe the underlying orientational dynamics, have provided us with a matching richness of data, which must finally be understood from a consistent theoretical viewpoint. It has been our ambition here to assemble this in a critical fashion to highlight both the successes and failures of current theoretical understanding. For example, in Section 9, Fig. 8 we present a portion of the single particle potential $\bar{V}(\omega)$, extracted from single crystal Bragg intensities, which places some constraints on orientational potential calculations. Section 17 develops this theme further to illustrate how the theoretical state of affairs, as of this writing, wants further consideration.

It is quite interesting that the C_{60} molecule, while seemingly coupled rather weakly to its neighbors, nonetheless experiences at room temperature a potential barrier height to free rotation of approximately 600 K or 50 meV. This is an appreciable activation barrier, and it does not permit a linearized treatment of the orientational distribution function $f_i(\omega_i)$. Below the orientational order–disorder transition, this barrier increases to ~250 meV and the rotations are further restricted to jump reorientation among selected orientational minima. The nature of the phase transition and its description via a Landau treatment reveals a complex behavior due to rotational localization. Fortunately, this can be sorted out in scattering experiments, since the order parameters $\bar{U}_{l,\gamma}$ appear more or less separately at selected reciprocal lattice points. Much more structurally related work on this material remains to be done and includes additional studies of the orientational pair potential via x-ray scattering and single crystal dynamical studies. While an enormous amount of static and dynamical information has been obtained using orientationally averaged powder samples, the more detailed neutron determination of $S(Q, \omega)$ awaits larger crystals. In addition to obtaining information on the orientational pair potential and dynamic orientational correlations, this would allow an examination of the possible roles that rotational–translational coupling and librational mode softening play in the sc–fcc phase transformation. Currently, the only evidence that these effects are important is rather indirect, coming both from the jump in lattice constant at the orientational order–disorder transformation and the presence of a co-existence regime and from the precursor effects observed below the transformation. Also, the nature of the "glassy" low temperature behavior awaits further clarification. Finally, we look forward to the development of a similar body of information on the higher fullerenes, in particular crystalline C_{70}, which exists in a plastic crystal phase (dynamically disordered slightly above room temperature).

ACKNOWLEDGMENTS

We wish to express our appreciation to our many colleagues at Brookhaven, Houston, and NIST both for continuing collaborations and stimulating discussions, and for permission to incorporate currently unpublished results in this review. In particular, we thank P. C. Chow, P. Wochner, and G. Reiter at Houston, J. R. D. Copley at NIST, and R. K. McMullan and J. C. Hanson at Brookhaven. We would also like to thank G. B. Alers for providing some of the data shown in Fig. 20 and S. A. Fitzgerald for providing the energies of the intermolecular IR modes at 200 K in Fig. 25. This work was supported at Brookhaven by the U.S. Department of Energy, Division of Materials Sciences under contract no. DE-AC02-CH00016. At Houston, support was through the NSF, Division of Materials Research on DMR-9208450, and by the State of Texas through the Texas Center for Superconductivity at the University of Houston (TCSUH). This manuscript was expertly and patiently typed by Ms. Sandi White at the University of Houston, to whom we are deeply indebted.

Glossary of Symbols

ω_i	set of Euler angles α_i, β_i, γ_i defining the orientation of the ith molecule
$f_i(\omega_i)$	molecular orientational distribution function
F, F'	crystal and molecular (local) coordinate frames, resp.; this convention will carry through to Euler and polar angles, i.e., $(\omega, \omega'; \Omega, \Omega')$
$Y_{lm}(\Omega)$	spherical harmonic polynomial of rank l
$D^l_{m,m'}$	Wigner rotation matrix of rank l
$T^A_l(\Omega)$	the linear combination of spherical harmonics of order l that transforms into itself under all symmetry operations of an icosahedron, where $A = A_{1g}$
$T_{l,\gamma}(\Omega)$	as before, but that transforms irreducibly under the symmetry operations of the cubic point group
$U_{l,\gamma}(\omega)$	molecular rotator functions: linear combinations of the Wigner rotation matrices that transform irreducibly under the symmetry operations of cubic point group
$\bar{U}^i_{l,\gamma}$	thermal average over (ω) of $U_{l,\gamma}(\omega)$ for the ith molecule; this takes on the meaning of an orientational order parameter of rank l
g_l	2^l-th pole moments of the atomic distribution on a molecule, occasionally called a *molecular form factor*, defined in Eq. (4.3)
$C_{l,\gamma}$	coefficient of the cubic harmonic in the expansion of the atomic number density on a C_{60} molecule in the crystal frame; these coefficients are directly extracted from the crystallographic data
$\bar{\rho}_i(\Omega_i)$	angular part of the atomic number density averaged over the ith orientations of the molecule
$\bar{\rho}_0, \bar{\rho}_X, \bar{\rho}_Y, \bar{\rho}_Z$	decomposition of $\bar{\rho}_i(\Omega_i)$ into irreducible Fourier amplitudes
η_X, η_Y, η_Z	Landau order parameters describing the Fm3m to Pa3̄ phase transition
$\bar{V}_i(\omega_i)$	effective single particle potential, i.e. the potential experienced by the ith C_{60} molecule in the average crystal field of its neighbors; Copley and Michel[4] refer to this as V_{CF}
$\dfrac{d\sigma}{d\Omega_Q}$	conventional differential scattering cross section, which is proportional to the scattering intensity at wavevector Q.

$F(\boldsymbol{Q})$ molecular structure factor, defined in Eq. (7.8)

$f_c(Q)$ atomic scattering (form) factor

$j_l(QR)$ spherical Bessel function, where R = molecular radius

$P(\boldsymbol{\omega}, \boldsymbol{\omega}_0, t)$ probability that a molecule has an orientation ω at time t, if it has ω_0 at $t = 0$

$S_R(Q, \omega)$ rotational diffusion scattering law, defined in Eq. (10.5)

$J_{\gamma, \gamma'}(\boldsymbol{q})$ Fourier transform of the pair potential $v^{i,j}_{\gamma, \gamma'}(T_{2g})$ for $l = 10$, where $\gamma = (X, Y, Z)$ the triply degenerate sub-representations of T_{2g}. (This quantity may be extracted from diffuse scattering measurements above the Fm3m to Pa$\bar{3}$ transition.)

$\bar{V}_a$ activation barrier between majority and minority orientations for C_{60} below 260 K

SOLID STATE PHYSICS, VOLUME 48

Electrons and Phonons in C$_{60}$-Based Materials

WARREN E. PICKETT

Complex Systems Theory Branch
Naval Research Laboratory
Washington, D.C.

Copyright © 1994 by Academic Press, Inc.
All rights of reproduction in any form reserved.
ISBN 0-12-607748-7
ISBN 0-12-606048-7 (pbk.)

I. Introduction

1. BACKGROUND

The identification of the C_{60} molecule by Kroto *et al.*[1] and the invention of a method of isolation of bulk quantities by Krätschmer *et al.*[2] has led to intense study of the resulting materials. The similarity of the C_{60} structure of the geodesic domes of R. Buckminster Fuller[3] has given rise to the name *fullerenes* to C_n molecules, and to *buckminsterfullerene* and *buckyball* to the C_{60} molecule in particular.

Perhaps the earliest discussion of a fullerene-like structure of carbon atoms was given by Jones,[4] who discussed closed graphite sheets as hollow molecules that could provide a form of matter intermediate in density between that of common gasses and standard condensed matter. His initial suggestion was that the graphite sheet could be folded over on itself by foreign atoms or molecular units; only later did he note[5] that carbon pentagons could provide the "defects" necessary to close the graphite surface. An early suggestion in the chemistry literature that the C_{60} molecule might be stable was by Osawa.[6]

One of the earliest discussions (probably independent of Jones and Osawa) of the electronic structure of the buckyball was Bochvar and Gal'pern,[7] who used the Hückel method to demonstrate the closed shell

[1] H. W. Kroto, J. R. Leath, S. C. O'Brien, R. F. Curl, and R. E. Smalley, *Nature* **318**, 162 (1985).

[2] W. Krätschmer, L. D. Lamb, K. Fostiropoulos, and D. R. Huffman, *Nature* **347**, 354 (1990).

[3] R. B. Fuller, "Inventions—The Patented Works of Buckminster Fuller," St. Martins, New York, 1983.

[4] D. E. H. Jones, *New Scientist* **32**, 245 (1966).

[5] D. E. H. Jones, "The Inventions of Daedalus," W. H. Freeman, Oxford, 1982, pp. 118–119.

[6] E. Osawa, *Kagaku (Kyoto)* **25**, 854 (1970).

[7] D. A. Bochvar and E. G. Gal'pern, *Dokl. Akad. Nauk SSR* **209**, 610 (1973) [*Proc. Acad. Sci. USSR* **209**, 239 (1973)].

nature of the C_{60} electronic structure and investigated deviations from σ and π bonding due to the spherical geometry of the molecule. Stankevich et al.[8] presented an informative and provocative review of crystalline carbon systems that included discussion of the fullerene molecule. Davidson[9] derived analytically the simple Hückel eigenvalues for the "hypothetical molecule C_{60}, a spheroidal, polyhedral oligomer of carbon."

Accounts of the various discoveries relating to C_{60} have been presented in entertaining detail elsewhere. Several reviews[10–21] and collections of overviews[22–27] of various extent and intent also have appeared. Loktev[28] presented an excellent early general review of the properties of C_{60} solids. The present chapter is intended to present an overview of many of the properties, and to serve as a useful guide to the literature, of C_{60} and its compounds.

[8]I. V. Stankevich, M. N. Nikerov and D. A. Bochvar, *Uspekhi Khimii* **53**, 1101 (1984) [*Russian Chem. Rev.* **53**, 640 (1984)].

[9]R. A. Davidson, *Theoret. Chim. Acta (Berlin)* **58**, 193 (1981).

[10]H. Kroto, *Science* **242**, 1139 (1988).

[11]R. F. Curl and R. E. Smalley, *Science* **242**, 1017 (1988).

[12]W. Krätschmer, L. D. Lamb, K. Fostiropoulos, and D. R. Huffman, *Nature* **347**, 354 (1990).

[13]W. Krätschmer, *Z. Phys. D* **19**, 405 (1991).

[14]R. F. Curl and R. E. Smalley, *Scientific American,* October 1991, p. 54.

[15]R. F. Curl and R. E. Smalley, *Science* **242**, 1018 (1988).

[16]H. Kroto, *Science* **242**, 1140 (1988).

[17]D. R. Huffman, *Physics Today* **44**, 22 (1991).

[18]H. W. Kroto, A. W. Allaf, and S. P. Balm, *Chem. Rev.* **91**, 1213 (1991). Contains 263 references.

[19]M. A. Wilson, L. S. Pang, G. D. Willett, K. J. Fisher, and I. G. Dance, *Carbon* **30**, 675 (1992). Contains 280 references.

[20]W. Weltner, Jr., and R. J. van Zee, *Chem. Rev.* **89**, 1713 (1989). With 390 references.

[21]R. C. Haddon, *Accts. Chem. Res.* **25**, 127 (1992).

[22]"Clusters and Cluster-Assembled Materials," MRS Symp. Proc. **206**, eds R. S. Averback, J. Bernholc, and D. L. Nelson, MRS, Pittsburgh, 1991.

[23]"Fullerenes—Synthesis, Properties, and Chemistry of Large Carbon Custers," ACS Symp. Ser. **481**, eds. G. S. Hammond and V. J. Kuck, Amer. Chem. Soc., Washington DC, 1992.

[24]"Buckminsterfullerenes," eds. W. E. Billups and M. A. Ciufolini, VCH Publishers, New York, 1992.

[25]"Proc. First Italian Workshop on Fullerenes: Status, Perspectives and Perspectives," eds. C. Taliani, G. Ruani, and R. Zamboni, World Scientific, Singapore, 1992.

[26]*J. Phys. Chem. Solids,* Vol. **53**, No. 11 (1992).

[27]*Carbon,* Vol. **30**, No. 8 (1992).

[28]V. M. Loktev, *Fiz. Nizk. Temp.* **18**, 217 (1992) [*Sov. J. Low Temp. Phys.* **18**, 149 (1992)].

The excitement and extreme activity in this field of research arises from several novel aspects:

(1)   the discovery of a novel molecule of unparalleled symmetry and beauty, and great stability;

(2)   the production of other fullerene molecules C_m and related graphene tubules;

(3)   the encapsulation of atoms inside these fullerene molecules;

(4)   the making of a new form of solid carbon (besides graphite and diamond) in the condensed form of C_{60}; and

(5)   the discovery of metallicity, and then relatively high temperature superconductivity,[29] in fulleride compounds.

The discovery[30,31] of weak ferromagnetism in an organic fulleride TDAE-C_{60} has further heightened interest. TDAE is tetrakis(dimethylamino)ethylene, $C_2N_4(CH_3)_8$. A recent study[31a–31c] suggests the magnetic moments are localized on the C_{60} molecules. Like the high temperature copper oxide superconductors, research has been strongly multidisciplinary,[32] involving chemists, physicists, and materials scientists.

This chapter will focus on solid materials based on C_{60}. Encapsulations, tubules, and larger (or smaller) carbon molecules will not be discussed. The study of C_{60} compounds has already become an extensive area of research, with some questions seemingly resolved but many other questions remaining open.

The purpose of this chapter is to assemble a representative compilation of the available information on C_{60} compounds, to discuss the primary

[29]A. F. Hebard, M. J. Rosseinsky, R. C. Haddon, D. W. Murphy, S. H. Glarum, T. T. M. Palstra, A. P. Ramirez, and A. R. Kortan, *Nature* **350,** 600 (1991).

[30]P. M. Allemand, K. C. Khemani, A. Koch, F. Wudl, K. Holczer, S. Donovan, G. Grüner, and J. D. Thompson, *Science* **253,** 301 (1991).

[31]P. W. Stephens, D. Cox, J. W. Lauher, L. Mihily, J. B. Wiley, P. M. Allemand, A. Hirsch, K. Holczer, Q. Li, J. D. Thompson, and F. Wudl, *Nature* **355,** 331 (1992).

[31a]K. Tanaka, A. A. Zakhidov, K. Yoshizawa, K. Okahara, T. Yamabe, K. Yakushi, K. Kikuchi, S. Suzuki, I. Ikemoto, and Y. Achiba, *Phys. Rev. B* **47,** 7554 (1993).

[31b]K. Tanaka, A. A. Zakhidov, K. Yoshizawa, K. Okahara, T. Yamabe, K. Yakushi, K. Kikuchi, S. Suzuki, I. Ikemoto, and Y. Achiba, *Phys. Lett. A* **164,** 221 (1992).

[31c]R. Seshadri, A. Rastogi, S. V. Bhat, S. Ramesesha, and C. N. R. Rao, *Solid State Commun.* **85,** 971 (1993).

[32]F. Wudl and J. D. Thompson, *J. Phys. Chem. Solids* **53,** 1449 (1992).

topics of current study, and perhaps to focus on certain areas that may be of particular importance to resolve. The choice of topics and references is no doubt somewhat subjective, but an attempt has been made to present the research objectively.

It is important to establish the terminology that is to be used here, and which has become fairly standard. Early references to the shape of the C_{60} molecule led to some use of the terminology *footballene* (European) or *soccerballene* (U.S.). These have mostly given way, initially to the unwieldy *buckminsterfullerene* and now very widely to simply *fullerene* to denote the C_{60} molecule and in fact the class of closed C_m molecules with $m \sim 60$ or greater. The crystalline C_{60} solid is called *fullerite*, while the crystalline compounds $A_n C_{60}$ (and offshoots) are called *fullerides*.

2. GENERALITIES

There are some general considerations concerning these unusual molecular ionic solids that were alluded to previously in a brief note.[33] The copper oxide high temperature superconductors are quasi-two-dimensional materials, whereas previous conventional superconductors had been three dimensional. The organic superconductors are also quasi-two-dimensional systems. It is reasonable first of all to consider the corresponding question for the fullerene solids, especially the superconducting fullerides.

Certainly in the gross sense they are three-dimensional systems; in fact, they generally have cubic crystal structures. As will be seen in the discussion, however, many of the properties will be directly ascribable to the molecule, which is itself a zero-dimensional material. Then again, the electronic properties and forces are commonly discussed in terms of the relationship to graphite, a two-dimensional material. It seems, at least, that one-dimensional behavior can be ruled out, but that at various stages the discussion or description of properties will wander among the zero-(0D), two- (2D), and three- (3D) dimensional analogies.

This general question of dimensionality, I believe, is beginning to emerge as a basic problem to be addressed and resolved, although it is rarely expressed in this language. The conundrum emerges indirectly:

• How large is the intermolecular hopping energy (a 3D quantity) compared with the intramolecular coulomb repulsion (a 0D quantity,

---

[33]W. E. Pickett, *Nature* **351**, 602 (1991).

renormalized by solid state [3D] polarization)? The situation can be approached as the true nanoscale limit of the problem of Coulomb blockade in a lattice of quantum dots.

• How dominant is the intramolecular phonon coupling (0D) compared with that of the extramolecular phonons involving cations and librational coupling (3D)?

• How important is the orientational disorder (a 3D effect) for the electronic properties and especially superconductivity? The fullerene solids are certainly a collection of molecules (and cations), but to how much of an extent are they more than that?

One can even take this bit of philosophizing further. The molecule in question is a remarkable, particularly attractive one, and it is tightly bound and spherical. Is it possible, at least for certain purposes, to consider it as an entity, a "superatom," say, with chemical symbol Bf (buckminsterfullerium). It is "super" in size [diameter ~ 10 Å], in weight [$60 \times 12 = 720$ amu], in stability [ionization potential of 7.3 eV], in electronegativity [accepts at least six extra electrons in solution], and in symmetry and comeliness. As a superatom, it poses intriguing questions, such as whether Hund's rules hold, and how it behaves in electric and magnetic fields. Resolving these and related questions will be an integral part of the scientific puzzle posed by fullerene solids.

Fowler[34] has observed that the C_{60} molecule is notable for its lack of properties. It is chemically inert, which makes it difficult to manipulate and also to detect chemically. It lacks any magnetic dipole moment, which makes it invisible to magnetic resonance probes. It even has no quadrupole moment. The equivalence of its 60 atoms means there is no informative NMR splitting pattern. It has only two optically allowed absorption lines below 5 eV. It does have clear Raman and infrared signals below 0.2 eV that make it easy to identify, however, and these have become of central importance in identifying fullerene materials and for monitoring solid state effects.

Chemically, there is some ambiguity over whether the C_{60} molecule should be designated as "organic." A conservative definition requires that an organic molecule have C–H bonds. C_{60} is often said to be a completely conjugated system, however, which can be considered as an

[34]P. Fowler, *Nature* **350**, 20 (1991).

aromatic hydrocarbon without hydrogen.[35] In practice, most chemists seem to accept it as an organic molecule.

II. Electrons on the C_{60} Molecule

3. STRUCTURE AND SYMMETRY

The symmetry of the C_{60} molecule is described by that of the regular icosahedron: the icosahedral-reflection group $I_h = I \times C_i$, the products of the icosahedral rotation group I (60 rotations) and the (twofold) reflection C_i. As a consequence the irreducible representations are labeled by those of I (a, t_1, t_2, g, and h, with degeneracies 1, 3, 3, 4, and 5, respectively) and those of C_i (g, for gerade = even, and u, for ungerade = odd). Polynomial representations of the irreducible representations of I have been discussed by Fässler and Schwarzenback.[36] The manner in which spherical harmonics provide realizations of the irreducible representation of I_h are given in Table I. The specific representations in terms of real spherical harmonic have been given up through $L = 6$ by Troullier and Martins.[37] To obtain a representation of the one-dimensional A_u requires a linear combination of the $31 = 2l + 1$ special harmonics for $l = 15$.

This point group contains the most members (120) of any noncyclic group in three dimensions, and as a result it is the closest possible discrete approximation to spherical symmetry. In the C_{60} molecule (of a given radius) the geometry is specified by the position of a single carbon atom, which is invariant under one of the reflection operations and is rotated into each of the other carbon atoms by the operations in I.

The C_{60} molecule is characterized by 12 regular pentagons and 20 hexagons. There are two distinct types of bonds, one that separates two hexagons (b_{hh}) and one that separates a pentagon and a hexagon (b_{ph}). Only for equal bond lengths are the hexagons regular. The symmetry of the molecule is independent of whether these two lengths are equal, and a number of spectroscopies have shown that they are equal to

[35]E. Brendsdal, B. N. Cyvin, J. Brunvoll, and S. J. Cyvin, *Spectrosc. Lett.* **21**, 313 (1988).
[36]Von A. Fässler and H. Schwarzenbach, *Zeitschrift f. angew. Math. Phys. (ZAMP)* **30**, 190 (1979).
[37]N. Troullier and J. L. Martins, *Phys. Rev. B* **46**, 1754 (1992).

TABLE I. RELATIONSHIP BETWEEN
THE SPHERICAL HARMONICS OF QUANTUM
NUMBER L AND THE REPRESENTATIONS OF THE
SYMMETRY GROUP I_h OF THE C_{60} MOLECULE

L	REPRESENTATIONS[a]
0	a_g
1	t_{1u}
2	h_g
3	$t_{2u} + g_u$
4	$g_g + h_g$
5	$t_{1u} + t_{2u} + h_u$
6	$a_g + t_{1g} + g_g + h_g$
7	$t_{1u} + t_{2u} + g_u + h_u$
8	$t_{2g} + g_g + 2h_g$
9	$t_{1u} + t_{2u} + 2g_u + h_u$
10	$a_g + t_{1g} + t_{2g} + g_g + 2h_g$
11	$2t_{1u} + t_{2u} + g_u + 2h_u$
12	$a_g + t_{1g} + t_{2g} + 2g_g + 2h_g$
13	$t_{1u} + 2t_{2u} + 2g_u + 2h_u$
14	$t_{1g} + t_{2g} + 2g_g + 3h_g$
15	$a_u + 2t_{1u} + 2t_{2u} + 2g_u + 2h_u$

[a] In each row the representation labels indicate
the transformation properties of the $2L + 1$
harmonics. Note that it is necessary to go to
$L = 15$ obtain a representation of a_u.
From P. W. Fowler and J. Woolrich, *Chem.
Phys. Lett.* **127**, 78 (1986).

1.40 ± 0.01 Å and 1.45 ± 0.01 Å, respectively. Gas phase electron
diffraction studies of Hedberg *et al.*[38] led to the values 1.458 ± 0.006 Å
and 1.401 ± 0.010 Å. The weighted average of these bond lengths is
1.439 Å, which can be compared with the 1.421 Å bond length of
graphite, which is also threefold coordinated carbon. The molecule has 90
bonds, 60 b_{hh} and 30 b_{ph}. In terms of the bond lengths r_{hh} and r_{ph}, the
diameter of the molecule is given by $D = [\tau^2(2r_{hh} + r_{ph})^2 + r_{ph}^2]^{1/2}$, where
$\tau = (\sqrt{5} + 1)/2$ is the golden ratio. The number of interatomic distances
for a collection of 60 atoms, $60(60 - 1)/2 = 1770$, is reduced by symmetry
to 23.

C_{60} has some analogy as a spherical version of graphite. One of the
C–C–C bond angles is 108° rather than 120°, however. Moreover, each C
atom does not sit in the plane defined by its three neighbors, but

[38] K. Hedberg, L. Hedberg, D. S. Bethuen, C. A. Brown, H. C. Dorn, R. D. Johnson, and
M. de Vries, *Science* **254**, 410 (1991).

$\sim 0.2r \sim 0.3$ Å above it. This feature is much like the case of corrannulene, $C_{20}H_{10}$, which consists of a carbon pentagon surrounded by five carbon hexagons, with the exterior carbons bonded to hydrogen. Corrannulene is almost half of a buckyball.

It may be useful to recall the contribution of Euler to fullerene science. For a molecule of v threefold coordinated atoms (each at a vertex), the number of edges is $e = 3v/2$. The number of faces f is the sum of the numbers of hexagonal and pentagonal rings: $f = h + p$. The number of edges is $e = (5p + 6h)/2 = 3v/2$, and so $3v = 5p + 6h$. Euler's contribution[39] is the theorem that the number of faces plus the number of vertices is equal to the number of edges plus two: $e + 2 = v + f = v + p + h$. Combining the equations immediately gives $p = 12$: There must be 12 pentagons to get a closed threefold coordinated structure. The number of hexagonal rings is $h = (v/2) - 10$. Only an even number v of atoms is possible. Dodecahedrane, simply consisting of 12 pentagons of carbon atoms, corresponds to $h = 0$; C_{60} corresponds to $h = 20$.

If the Cartesian axes are chosen to extend from the origin through the centers of b_{hh} bonds lying on the two fold axes, the C_{60} point group can be generated from the three group operations

$$R_5 = \frac{1}{2} \begin{pmatrix} \tau^{-1} & 1 & \tau \\ -1 & \tau & \tau^{-1} \\ \tau & \tau^{-1} & 1 \end{pmatrix},$$

$$P = \begin{pmatrix} 0 & 1 & 0 \\ 0 & 0 & 1 \\ 1 & 0 & 0 \end{pmatrix}, \qquad (3.1)$$

$$I = \begin{pmatrix} -1 & 0 & 0 \\ 0 & -1 & 0 \\ 0 & 0 & -1 \end{pmatrix}.$$

The operation R_5 is a rotation by $2\pi/5$ about a fivefold axis, P is a cyclic permutation of the axes, and I is the inversion operation.

The C_{60} buckyball is often discussed[40] in the context of the Platonic and Archimedean solids. Platonic solids are those whose faces are formed by identical regular polygons and whose vertices are congruent (symmetry

[39]L. Euler, "Elementa Doctrinæ Solidorum" (1758).
[40]A. D. J. Haymet, *J. Am. Chem. Soc.* **108**, 319 (1986).

equivalent), and there are five: tetrahedron, cube (hexahedron), octahedron, icosahedron, and dodecahedron. Archimedean solids are semiregular polyhedra, each of whose faces is some regular polygon and whose vertices are all congurent. Archimedean solids can be obtained by truncating the Platonic solids. There are 13 nontrivial such solids,[40] one of which is the truncated icosahedron whose vertices correspond to the positions of carbon atoms of the C_{60} molecule when the two inequivalent bonds have equal length. The truncated icosahedron is pictured in Fig. 1 for $r_{hh} = r_{ph}$, $r_{hh} > r_{ph}$, and $r_{hh} < r_{ph}$.

4. ELECTRONIC STRUCTURE

a. *Spherical Approximation*

The first inclination of the theoretician is to simplify to the lowest possible level, and for the spherical C_{60} molecule this leads to the model of electrons confined to a spherical shell. With energy levels of angular momentum l denoted by E_l, Auerbach and Murthy[41] note that the noninteracting Hamiltonian would be

$$H_0 = \sum_{s=\pm} \sum_{l=0}^{\infty} \sum_{m=-l}^{l} E_l C_{lms}^+ C_{lms}. \qquad (4.1)$$

For particles of mass m^* constrained to a sphere of radius R, the kinetic energy is given by $E_l = [\hbar^2/2m^*R^2]l(l+1)$. The number of states with energy $E \leq E_l$ is $2(l+1)^2$, including spin degeneracy. Considering only the 60 π electrons of the C_{60} molecule as done by Ozaki and Takahashi,[42] the states through $l = 4$ would be filled (requiring 50 electrons), and the remaining 10 electrons would be distributed within the 22-fold degenerate $l = 5$ shell. Such an open shell situation is not a reasonable approximation to the correct molecular orbital description (as will be seen later), which makes it clear that the atomic potentials of the C atoms must be considered from the start for any realistic treatment. In fact, simply introducing I_h symmetry leads to $(l=5) \rightarrow t_{1u} + t_{2u} + h_u$, and filling the fivefold orbitally degenerate h_u subshell leads to a closed (sub)shell. Although not presented in the context of fullerenes, Kim et al.[43] have presented the effect of a magnetic field on the eigenvalue spectrum of noninteracting electrons on spherical shell.

[41]A. Auerbach and G. N. Murthy, *Europhys. Lett.* **19**, 103 (1992).
[42]M. Ozaki and A. Takahashi, *Chem. Phys. Lett.* **127**, 242 (1986).
[43]J. H. Kim, I. D. Vagner, and B. Sundaram, *Phys. Rev. B* **46**, 9501 (1992).

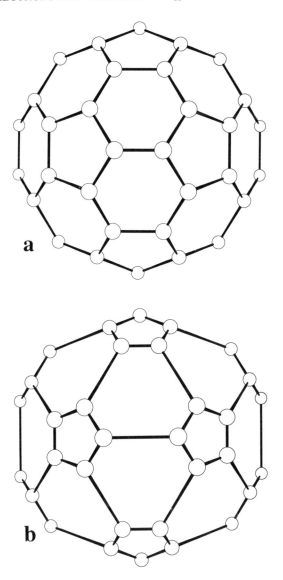

FIG. 1. C_{60} molecules, normal (a) and distorted (b) and (c). Each molecule has full I_h symmetry. The example with small pentagons (b) allows one to visualize the truncation of the icosahedron vertices, which leads to the pentagonal faces. Figures courtesy of S. C. Erwin.

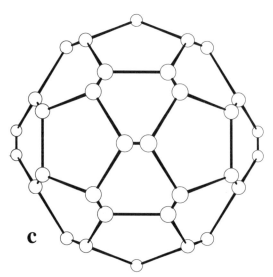

c

FIG 1.—(continued)

Although the eigenvalue spectrum of the spherical shell is not realistic, it is useful to note one further quantity that follows from this model: the energy of interaction of two uncorrelated electrons confined to such a shell. This energy is $U_{sph} = (1/\sqrt{2})e^2/R$, which for $R \sim 3.5$ Å leads to $U_{sph} \sim 2.9$ eV. In another simple estimate, taking the electrons as equal point charges on each C atom with an intraatomic Coulomb interaction integral $F_0 = 12$ eV, Antropov et al.[44] obtain a value of 3.7 eV. If the electrons were perfectly correlated, so they stayed always on opposite sides of the spherical shell, the potential energy would be $e^2/2R \sim 2$ eV. These values provide an estimate of the scale of the energy of interaction. The interaction energy of two electrons on the actual C_{60} molecule will be discussed further. A related continuum approximation to the C_{60} molecule was discussed by Gonzàles et al.[45]

Although the spherical model fails straightaway, Savina et al.[46] have investigated a perturbed spherical model of the C_{60} molecule. A perturbation of icosahedral symmetry splits the $l = 5$ into $h_u + t_{1u} + t_{2u}$ levels, and a physically based crystal field potential leads to occupation of the h_u level, giving a filled subband. By adjusting the crystal field strength and effective mass, the low energy optical spectrum can even be reproduced.

[44]V. P. Antropov, O. Gunnarsson, and O. Jepsen, *Phys. Rev. B* **46,** 13647 (1992).
[45]J. Gonzàles, F. Guinea, and M. A. H. Vozmediano, *Phys. Rev. Lett.* **69,** 172 (1992).
[46]M. R. Savina, L. L. Lohr, and A. H. Francis, *Chem. Phys. Lett.* **205,** 200 (1993).

b. *Local Density (and Related) Approximation*

(1) Neutral molecule   As strong as the inclination to simplify is, it is the duty of the theoretician not to oversimplify, and so the discrete nature of the C_{60} molecule must be addressed. The electronic structure of the neutral C_{60} molecule was first investigated by a number of semiempirical calculations,[47-52] by Hartree–Fock,[53] and then by local-density-based methods, a non-self-consistent calculation by Satpathy[54] and a self-consistent one by Hale.[55] These have been superseded by a number of detailed self-consistent calculations within the local density approximation (LDA),[56] in which the many-body effects that are collected into the exchange-correlation energy are evaluated using a known function of the density.

Much insight was provided by the early Hückel calculations, originally treating only the 60 π electrons and later extended to include all 240 $2s + 2p$ electrons of the 60 carbon atoms. These and more recent ab initio calculations indicated a distribution of the 240 valence s, p electrons into only 32 eigenlevels. Since the highest occupied molecular orbital (HOMO) level is completely filled, no symmetry breaking Jahn–Teller instability is anticipated. The HOMO, of h_u symmetry, is fivefold orbitally degenerate.

The lowest unoccupied MO (LUMO) is a threefold t_{1u} state. The character of this state is not adequately understood simply from the knowledge that its three degenerate members transform as a vector (x, y, z), because it has additional nodal structure. Haddon[21] has illustrated the nodal structure (three inequivalent lines of nodes on the molecule) resulting from Hückel calculations. Both this state and the next highest one (t_{1g}) are strongly bound (local density eigenvalues[57] of -4.26 and -3.15 eV, respectively), and Haddon, Brus, and Raghavachari[48] suggested that this molecule might be induced to accept up to 12 electrons. In the free state only the singly and doubly charged molecules have been detected.[58]

[47]A. D. J. Haymet, *Chem. Phys. Lett.* **122**, 421 (1985).
[48]R. C. Haddon, L. E. Brus, and K. Raghavachari, *Chem. Phys. Lett.* **125**, 459 (1986).
[49]R. L. Disch and J. M. Schulman, *Chem. Phys. Lett.* **125**, 465 (1986).
[50]P. W. Fowler and J. Woolrich, *Chem. Phys. Lett.* **127**, 78 (1986).
[51]S. Larsson, A. Volosov, and A. Rosén, *Phys. Chem. Lett.* **137**, 501 (1987).
[52]J. M. Schulman, R. L. Disch, M. A. Miller, and R. C. Peck, *Chem. Phys. Lett.* **141**, 45 (1987).
[53]H. P. Lüthi and J. Almlöf, *Chem. Phys. Lett.* **135**, 357 (1987).
[54]S. Satpathy, *Chem. Phys. Lett.* **130**, 545 (1986).
[55]P. D. Hale, *J. Am. Chem. Soc.* **108**, 6087 (1986).
[56]R. O. Jones and O. Gunnarsson, *Rev. Mod. Phys.* **61**, 689 (1989).
[57]B. I. Dunlap, *Intl. J. Quant. Chem. QC Symp.* **22**, 2571 (1988).
[58]R. L. Hettich, R. N. Compton, and R. H. Ritchie, *Phys. Rev. Lett.* **67**, 1242 (1991).

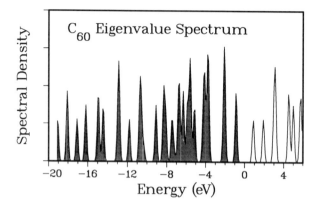

FIG. 2. The spectrum of one-electron eigenvalues in C_{60}, slightly broadened to give an impression of the spectral weight, of the neutral C_{60} molecule in the valence region, illustrating the distribution of states. The shaded states are occupied, and unshaded states are unoccupied. For this figure the zero of energy is taken at the center of the gap. Eigenvalues are from the local density approximation calculations of M. R. Pederson and A. A. Quong.

Self-consistent LDA calculations lead to the orbital eigenvalues in the region of primary interest shown in Fig. 2. The complete $s-p$ valence states span a range of 18.6 eV, comparable with but slightly less than the valence bandwidths of graphite and diamond. The atomic contribution to states in the HOMO–LUMO region have been presented by Guo et al.[59] The calculated ionization potential $I = E(C_{60}) - E(C_{60}^+)$ and electron affinity $A = E(C_{60}^-) - E(C_{60})$ are ~7.6 and ~2.8 eV (see the following). Measured values are $I = 7.6 \pm 0.2$ eV by Lichtenberger et al.[60,61] and $I = 7.6 \pm 0.1$ eV by Zimmerman et al.,[62] refined to $7.56 \le I \le 7.62$ eV by de Vries et al.[63] From photodetachment of cold C_{60}^-, Wang et al.[64] obtained $A = 2.65 \pm 0.05$ eV. The corresponding values from the one-electron eigenvalues, i.e., the (negatives of the) eigenvalues of the

[59]J. Guo, D. E. Ellis, and D. J. Lam, Chem. Phys. Lett. 184, 148 (1991).

[60]D. L. Lichtenberger, K. W. Nebesny, C. D. Ray, D. R. Huffman, and L. D Lamb, Chem. Phys. Lett. 176, 203 (1991).

[61]D. L. Lichtenberger, M. E. Jatcko, K. W. Nebensy, C. D. Ray, D. R. Huffman, and L. D. Lamb, in "Clusters and Cluster-Assembled Materials," MRS Symp. Proc 206, eds. R. S. Averback, J. Bernholc, and D. L. Nelson, Mat. Res. Soc., Pittsburgh, 1991, p. 673.

[62]J. A. Zimmerman, J. R. Eyler, S. B. H. Bach, and S. W. McElvany, J. Chem. Phys. 94, 3556 (1991).

[63]J. de Vries, H. Steger, B. Kamke, C. Manzel, B. Weisser, W. Kamke, and I. V. Hertel, Chem. Phys. Lett. 188, 159 (1992).

[64]L.-S. Wang, J. Conceicao, C. Jin, and R. E. Smalley, Chem. Phys. Lett. 182, 5 (1991).

HOMO and LUMO, respectively, are[65] 5.94 and 4.26 eV. Comparing the LUMO–HOMO eigenvalue difference of 1.7 eV with the measured value of $I - A \sim 5$ eV reflects the size of electron interactions (~ 3.3 eV) for the isolated C_{60} molecule. This value is quite close to the result obtained from LDA total energy calculations, discussed in more length later. The second ionization potential of C_{60} has been measured by McElvaney *et al.*[66] to be 9.7 eV.

There are several levels of sophistication in the estimates of these quantities. Within LDA, which is known to overestimate the binding energy of such molecules, the cohesive energy[57,66-68] of the molecule is 8.50–8.54 eV per atom. A treatment of exchange-correlation energy beyond LDA using the generalized gradient approximation (GGA) of Perdew and Wang,[69] which gives much improved binding energies for molecules involving light atoms, leads to a binding energy of 7.48 eV per atom given by Pederson *et al.*[66] or 7.24 eV as calculated by Kobayashi *et al.*[68] Considering that the GGA is known to overestimate binding energies of hydrocarbon molecules by 0.1–0.2 eV/atom, Pederson *et al.* estimate a binding energy of 7.25 eV/atom for C_{60}, which is rather close to that of graphite (7.38 eV) and diamond (7.35 eV).

Although as already noted the spherical description of the C_{60} molecule cannot be taken very far, Troullier and Martins[37] have observed that much can be learned from a spherical analysis of various properties of the molecule. They obtain the amount of character of angular momentum index l (at any distance $r = |\mathbf{r}|$ from the center of the molecule) of any function $f(\mathbf{r})$ defined on the molecule by projecting onto the spherical harmonics of that l:

$$f_l(r) = \sum_m \left| \int Y_{lm}^*(\mathbf{r}) f(\mathbf{r}) \, d\Omega \right|^2. \tag{4.2}$$

Troullier and Martins find that the molecular wavefunctions $[f(\mathbf{r}) \rightarrow \psi(\mathbf{r})]$

[65]M. R. Pederson, K. A. Jackson, and L. L. Boyer, *Phys. Rev. B* **45**, 6919 (1992).

[66]S. W. McElvaney, M. M. Ross, and J. H. Callahan, in "Clusters and Cluster-Assembled Materials," MRS Symp. Proc. **206**, eds. R. S. Averback, J. Bernholc, and D. L. Nelson, Mat. Res. Soc., Pittsburgh, 1991, p. 697.

[67]M. R. Pederson, S. C. Erwin, W. E. Pickett, K. A. Jackson, and L. L. Boyer, in "Physics and Chemistry of Finite Systems: From Clusters to Crystals, Vol. II," eds. P. Jena, S. N. Khanna, and B. K. Rao, Kluwer, Dordrecht, 1992, p. 1323.

[68]K. Kobayashi, N. Kurita, H. Kumahora, K. Tago, and K. Ozawa, *Phys. Rev. B* **45**, 13690 (1992).

[69]J. P. Perdew and Y. Wang, *Phys. Rev. B* **33**, 8800 (1986).

are invariably identified with a specific single value of l, in the sense that one value of f_l is an order of magnitude larger than any other, and they give examples showing that the spherical nodal behavior of a wavefunction is strikingly similar to that of the corresponding spherical harmonics. They also note that it is easy to identify radial nodes of the wavefunctions $\psi(r)$, and thereby assign them to σ type, with no radial node, or to π type, with a radial node that roughly corresponds to a sphere passing through all C atoms.

The corresponding characters of the states of most interest, within $\sim 4\,\mathrm{eV}$ of the HOMO and LUMO, are reproduced from Troullier and Martins in Table II. These results are specific for the fullerite crystal (with some crystal field splittings being evident), but in the energy range of interest they are essentially the same for the isolated molecule. The HOMO orbital is seen to be a π state with $l = 5$ character ("π_5"), as is the LUMO orbital, while the next higher lying unoccupied level is a π_6 state.

The corresponding spherical analysis of the self-consistent potential is revealing. Troullier and Martins[37] find it to be, in addition to a spherically symmetric part with a "valley" lying at the radius of the molecule and

TABLE II. EIGENVALUES, WAVEFUNCTION TYPE (σ, π, OR INTERSTITIAL ι), DOMINANT ANGULAR MOMENTUM CHARACTER OF THE WAVEFUNCTION, AND THE IRREDUCIBLE REPRESENTATION OF THE TETRAHEDRAL GROUP T_h, AT THE Γ-POINT FOR ORIENTATIONALLY ORDERED Fm3 (T_h^3) FULLERITE

ENERGY[a] (eV)	CHARACTER	L	T_h IRREP
-2.77	σ	9	e_u
-2.76	σ	9	t_u
-2.63	π	3	t_u
-1.85	π	4	e_g
-1.62	π	4	a_g
-1.08	π	4	t_g
-0.97	π	4	t_g
-0.23	π	5	e_u
0.00	π	5	t_u
1.71	π	5	t_u
2.37	π	6	t_g
3.47	π	5	t_u
4.05	π	6	e_g
4.10	π	6	t_g
4.22	ι	0	a_g
4.97	π	7	e_u

[a] Only states within $\sim 4\,\mathrm{eV}$ of the enter of the gap are given, and the zero of energy has been taken at the energy of the HOMO.
From Table III of N. Troullier and J. L. Martins, *Phys. Rev. B* **46**, 1754 (1992).

arising from the spherical average of the 60 atoms, primarily an $l = 10$ function that arises from the I_h symmetry. The $l = 6$ term allowed by symmetry is almost two orders of magnitude smaller. The splitting between the HOMO and LUMO, which as noted are both of π_5 type, arises from the $l = 10$ part of the potential.

Spectral properties   Spectral properties of the C_{60} molecule and solids are addressed in the lowest approximation in terms of the one-electron eigenvalues of Hückel, Hartree–Fock, or LDA calculations. For a collection of 60 atoms, the spectrum of eigenvalues is remarkably sparse, due of course to the high symmetry of the molecule. There are, however, a number of serious complications. First of all, there are the various electronic configurations, labeled by molecular symmetry and total spin (S) angular momentum. The corrections from simple eigenvalue differences can be of the order of an electron volt. Then there are the vibrational sidebands to the electronic transitions, which for single vibrational modes range up to 0.2 eV. Some data (such as luminescence) may also involve geometric relaxation of the molecules. Finally, in the solid there are the added effects due to banding, with a typical bandwidth being ~0.5 eV. Interpreting the experimental data and building a consistent theoretical picture form a continuing process.

As a simple approximation to the spectrum of low energy excitations of C_{60}, simple differences of (unrelaxed) LDA eigenvalues of Dunlap et al.[57,70] are presented in Table III. In Section 2 it was noted that the high symmetry of the C_{60} molecule leads to a sparse spectrum of eigenvalues. The restrictions arising from symmetry rules is even more striking for the optically allowed transitions. Within the one-electron approximation there are only two allowed transitions below 5 eV in energy. The lowest optically accessible configuration, $(h_u)^9(t_{1g})^1$, is 15-fold degenerate, but again the high symmetry makes only the T_{1u} combination allowed in one-photon absorption.[48] In the anion, where the t_{1u} level is partially occupied, there is an additionally allowed $t_{1u} \rightarrow t_{1g}$ transition at lower energy (1.11 eV, using Dunlap's eigenvalues).

A more justified approximation to the excited states of the neutral molecule has been provided by Negri et al.,[71] who evaluated the energies of several excited configurations using semiempirical methods that include some correlation effects. The least energetic of these results (up to roughly 4 eV), for both singlet and triplet excitations, also are presented in Table III. Their prediction is that the triplet T_{2g} state is

[70]B. I. Dunlap, D. W. Brenner, J. W. Mintmire, R. C. Mowrey, and C. T. White, *J. Phys. Chem.* **95**, 5763 (1991).

[71]F. Negri, G. Orlandi, and F. Zerbetto, *Chem. Phys. Lett.* **144**, 31 (1988).

TABLE III APPROXIMATIONS TO THE EXCITED STATES OF C_{60}

ΔE (LDA)[a]		SINGLETS[b]		TRIPLETS[b]	
$h_u \rightarrow t_{1u}$	1.68	T_{2g}	2.58	T_{2g}	2.06
$h_u \rightarrow t_{1g}$	2.79 ![c]	T_{1g}	2.61	T_{1g}	2.40
$g_g \rightarrow t_{1u}$	2.86	G_g	2.70	G_u	2.58
$h_g \rightarrow t_{1u}$	2.98 !	T_{2u}	2.89	H_g	2.59
$h_u \rightarrow t_{2u}$	3.69	H_u	3.15	T_{2u}	2.59
$h_u \rightarrow h_g$	3.87	H_g	3.15	T_{1u}	2.61
$g_g \rightarrow t_{1g}$	3.97	G_u	3.21	G_g	2.66
$h_g \rightarrow t_{1g}$	4.09	H_u	3.89	H_u	2.98
$g_g \rightarrow t_{2u}$	4.87	G_u	3.90	T_{2u}	3.66
$h_g \rightarrow t_{2u}$	4.99	T_{1u}	4.08 !	H_u	3.81
		T_{2u}	4.12	G_u	3.84
		G_u	4.37	G_g	4.26
		H_u	4.39	H_g	4.28
		T_{2g}	4.41	G_u	4.29
		T_{1u}	4.53 !	T_{2g}	4.32

[a] B. I. Dunlap, *Int. J. Quant. Chem. QC Symp.* **22**, 2571 (1988).
[b] F. Negri, G. Orlandi, and F. Zerbetto, *Chem. Phys. Lett.* **144**, 31 (1988).
[c] Exclamation points indicate an optically allowed transition/state. Note that the high symmetry of the molecule leads to an extreme sparsity of allowed transitions.

lowest, 0.5 eV lower in energy than the lowest singlet, which has the same symmetry. There are nine optically forbidden states below the first optically allowed state, of T_{1u} symmetry at 4.08 eV. This state is dominated by the $h_u \rightarrow t_{1g}$ configuration, and hence it corresponds directly to the corresponding lowest allowed one-electron transition, at 2.79 eV using the LDA eigenvalues (which are expected to underestimate the energy). The next optically allowed excitation, again of T_{1u} symmetry, at 4.53 eV corresponds to the $h_g \rightarrow t_{1u}$ configuration (at 2.98 eV, from LDA eigenvalues). László and Udvardi,[72] who presented configuration interaction results based on the Pariser–Parr–Pople model, pointed out that when electronic transitions of C_{60} are combined with its vibrations (Herzberg–Teller interactions), every symmetry-forbidden transition becomes allowed.

Using optical spectroscopy of molecular beams, Haufler *et al.*[73] reported the energy of the lowest triplet excitation to be 1.7 eV. From

[72] I. László and L. Udvardi, *Chem. Phys. Lett.* **136**, 418 (1987).
[73] R. E. Haufler, L.-S. Wang, L. P. F. Chibante, C. Jin, J. Conceicao, Y. Chai, R. E. Smalley, *Chem. Phys. Lett.* **179**, 449 (1991).

C_{60} in methylcyclohexane solutions, Wang[74] obtained the value of 1.5 eV for the triplet state, and a difference between singlet and triplet states of 0.24 eV. An absorption peak of unknown origin at 1.15 eV (followed by phonon sidebands) was reported by Kato et al.[75] The optical spectrum has been examined in much more detail in solids and will be discussed in Section 17.

Electric polarizability The linear electric polarizability and the hyperpolarizability of the C_{60} molecule are properties of widespread interest. Since the C_{60} solid is a collection of van der Waals–bonded molecules (a molecular solid), it is reasonable to consider the polarizability of the solid to arise from to a lattice of point dipoles and use the Clausius–Mossotti relationship between the crystal dielectric constant ε and the molecular polarizability α:

$$\alpha = \frac{3}{4\pi N} \frac{\varepsilon - 1}{\varepsilon + 2}, \tag{4.3}$$

where N is the number density of dipoles in the crystal, equal to $4/a^3 = 1.40 \times 10^{-3} \, \text{Å}^{-3}$ for an fcc lattice of lattice constant $a = 14.2 \, \text{Å}$. Hebard et al.[76] report $\varepsilon = 4.4 \pm 0.2$, implying an experimental value of $\alpha = 90.6 \, \text{Å}^3$.

There have been a number of calculations for isolated molecules. Bertsch et al.[77] obtained 37 Å^3 using a parametrized tight binding method. Using Hartree–Fock plus configuration interaction, Fowler et al.[78] obtained 66 Å^3, which they considered to be a rigorous lower bound (i.e., improving the basis set would increase the result). Due to the extensive computational requirements of this method (10 days on a Cray XMP/48 was quoted), they assumed bond lengths[79] of 1.376 and 1.465 Å for b_{hh} and b_{ph}, respectively. Using semiempirical quantum chemical methods, Shuai and Brédas[80] calculated a large value of 154 Å^3.

There have been two LDA calculations of the C_{60} molecule polarizability. Using a large basis set and calculating the density self-consistently in

[74] Y. Wang, *J. Phys. Chem.* **96**, 764 (1992).

[75] T. Kato, T. Kodama, T. Shida, T. Nakagawa, Y. Matsui, S. Suzuki, H. Shiromaru, K. Yamauchi, and Y. Achiba, *Chem. Phys. Lett.* **180**, 446 (1991).

[76] A. F. Hebard, R. C. Haddon, R. M. Fleming, and A. R. Kortan, *Appl. Phys. Lett.* **59**, 2109 (1991).

[77] G. F. Bertsch, A. Bulgac, D. Tomanek, and Y. Wang, *Phys. Rev. Lett.* **67**, 2690 (1991).

[78] P. W. Fowler, P. Lazzaretti, and R. Zanasi, *Chem. Phys. Lett.* **165**, 79 (1990).

[79] J. M. Schulman, R. L. Disch, M. A. Miller, and R. C. Peck, *Chem. Phys. Lett.* **141**, 45 (1987).

[80] Z. Shuai and J. L. Brédas, *Phys. Rev. B* **46**, 16135 (1992).

an applied field, Pederson and Quong[81] find an LDA prediction of 82.6 Å^3. For comparison, Pederson and Quong reported calculations for the carbon atom that correspond to ~68 Å^3 for 60 isolated atoms. This comparison indicates that the polarizability is enhanced by the extension of the valence states over the C_{60} molecule. Matsuzawa and Dixon[82] used the commercial DMol LDA package and obtained $\alpha = 78$ Å^3, quite close to the result of Pederson and Quong. The difference between the "experimental" value of the C_{60} polarizability, 90.6 Å^3, and apparently the best theoretical value, 80 ± 2 Å^3, could be due to the local density approximation or to the inexactness of the Clausius–Mossotti relation, but is also possibly due to the banding in the solid state. Ching et al.[83] have reported calculations of the frequency-dependent dielectric constant in the crystal that are discussed later.

Nonlinear optical response   The size and carbon bonding of the C_{60} molecule suggest the possibility of unusual nonlinear optical response. The experimental data is expressed alternatively in terms of the third-order optical susceptibility $\chi^{(3)}$, a property of the material, and second hyperpolarizability γ, a property of the molecule. They are related by

$$\chi^{(3)} = \frac{N\gamma}{[(\varepsilon + 2)/3]^4} \tag{4.4}$$

where ε is the dielectric constant of Eq. (4.3). Published results are not in full agreement, and Flom et al.[84] have provided a recent overview of the situation.

Kafafi et al.[85] reported $\chi^{(3)} = 7 \times 10^{-12}$ esu and $\gamma = 30 \times 10^{-35}$ esu from studies of thick fullerite films at 1.064 μm. These values and the small absorption coefficient of 6 cm^{-1} suggest fullerite may be an attractive optical material. Wang and Cheng[86] reported $\gamma = 75 \times 10^{-35}$ esu from studies at 1.91 μm of C_{60} in solutions. A value of γ three orders of magnitude larger inferred by Blau et al.[87] from C_{60}–benzene solutions

[81]M. R. Pederson and A. A. Quong, Phys. Rev. B **46**, 13584 (1992).

[82]N. Matsuzawa and D. Dixon, J. Phys. Chem. **96**, 6872 (1992).

[83]W. Y. Ching, M. Z. Huang, Y. N. Xu, W. G. Harter and F. T. Chan, Phys. Rev. Lett. **67**, 2045 (1991).

[84]S. R. Flom, R. G. S. Pong, F. J. Bartoli, and Z. H. Kafafi, Phys. Rev. B **46**, 15598 (1992).

[85]Z. H. Kafafi, J. R. Lindle, R. G. S. Pong, F. J. Bartoli, L. J. Lingg, and J. Milliken, Chem. Phys. Lett. **188**, 492 (1992).

[86]Y. Wang and L.-T. Cheng, J. Phys. Chem. **96**, 1530 (1992).

[87]W. J. Blau, H. J. Byrne, D. J. Cardin, T. J. Dennis, J. P. Hare, H. W. Kroto, R. Taylor, and D. R. M. Walton, Phys. Rev. Lett. **67**, 1423 (1991).

has been criticized by Knize and Partanen,[88] by Kafafi et al.,[89] and by Flom et al.[84]

There have been several calculations of the nonlinear optical coefficients. Shuai and Brédas[80] calculated $\gamma = 20 \times 10^{-35}$ esu using a semiempirical method. Matsuzawa and Dixon[82] and Quong and Pederson[90] performed self-consistent LDA calculations in applied electric fields, finding static hyperpolarizabilities of $\gamma = 1.6 \times 10^{-35}$ and 0.70×10^{-35} esu, respectively. Using Eq. (4.4) these values convert to $\chi^{(3)} \approx 0.32 \times 10^{-12}$ and 0.14×10^{-12} esu, respectively, a factor of 25–50 less than the experimental value of Kafafi et al.[85] The differences among the theoretical results and with the experimental values reflect the sensitive nature of this property and indicate that important questions about the nonlinear response remain to be answered.

Magnetic susceptibility   The magnetic susceptibility χ of the neutral fullerene molecule was first addressed in terms of ring currents by Elser and Haddon[91–93] in terms of London's ring current theory of aromatic molecules. Using a Hückel-like electronic structure, they concluded that the contribution to the orbital magnetic susceptibility of the neutral molecule from the π electrons was unusually weakly diamagnetic. An expository discussion of the calculations and the related NMR experiment was given by Mallion.[94] The NMR chemical shift is a local property that can be significantly affected by ring currents that give little or no net contribution to the susceptibility. Pasquarello et al.[95] concluded that paramagnetic currents flow within the pentagons, and that weaker diamagnetic currents flow all around the molecule. The net affect is a nearly vanishing orbital susceptibility.

Fowler et al.[78,96] have calculated χ using Hartree–Fock plus configuration interaction methods. They demonstrate that the calculation is rather sensitive to the quality of the basis set. With their best basis set, they obtain strong cancellation: The paramagnetic contribution is 70% the size of the diamagnetic part, with a net orbital susceptibility of -0.44×10^{-6} au. The results of Fowler et al. were supported by the

[88]R. J. Knize and J. P. Partanen, Phys. Rev. Lett. 68, 2704 (1992).
[89]Z. H. Kafafi, F. J. Bartoli, J. R. Lindle, and R. G. S. Pong, Phys. Rev. Lett. 68, 2705 (1992).
[90]A. A. Quong and M. R. Pederson, Phys. Rev. B 46, 12906 (1992).
[91]V. Elser and R. C. Haddon, Phys. Rev. A 36, 4579 (1987).
[92]V. Elser and R. C. Haddon, Nature 325, 792 (1987).
[93]R. C. Haddon and V. Elser, Chem. Phys. Lett. 169, 362 (1990).
[94]R. B. Mallion, Nature 325, 760 (1987).
[95]A. Pasquarello, M. Schlüter, and R. C. Haddon, Science 257, 1660 (1992).
[96]P. W. Fowler, P. Lazzeretti, M. Malagoli, and R. Zanasi, Chem. Phys. Lett. 179, 174 (1991).

chemical considerations discussed by Schmalz.[97] The experimental value reported by Haddon et al.[98] is -0.26×10^{-3} cgs. There do not seem yet to be any reports of the application of LDA calculations to the magnetic susceptibility.

(2) Charged molecules C_{60}^{-n} Pederson and Quong[81] have reported LDA calculations on negatively charged C_{60}^{-n} molecules. Taking atomic relaxation into account (within I_h symmetry) and including spin polarization, but neglecting JT distortions, they find the energy versus additional charge Q can be expressed

$$E[Q] = E_0 + \tfrac{1}{2}U(Q - Q_0)^2, \qquad E_0 = -3.34 \text{ eV},$$

$$U = 3.02 \text{ eV}, \qquad Q_0 = 1.49, \tag{4.5}$$

and its derivative (from theorems in density functional theory[56]) gives the (highest occupied) t_{1u} eigenvalue. Since the occupation of the $\pi^* t_{1u}$ states adds more charge Q to the b_{hh} bonds, they stretch and weaken, reducing the calculated ratio of b_{hh} to b_{ph} bond lengths very slightly (from 1.037 for $Q = 0$ to 1.029 for $Q = 2$). Pederson and Quong find that the LDA value of the electron affinity, 2.98 eV, is corrected by GGA to 2.75 eV, in excellent agreement with the experimental value of 2.74 eV reported by Yang et al.[99] The value of $Q_0 = 1.49$ is midway between the singly and doubly charged states, reflecting that they have similar energy. From Pederson and Quong's LDA calculations, the doubly charged molecule is unbound by 0.03 eV. Correlation effects must be included to account for the existence of the doubly charged molecule detected by Hettich et al.[58] The evidence is that higher charge states than -2 are unbound. In solution, however, higher charges of the C_{60} molecule are produced. Wilson et al.[100] have detected charge states up to and including C_{60}^{-6} in a variety of solutions using cyclic voltammetry spectroscopy. They identify the ground states of the singly and doubly charged molecules to be spin

[97]T. G. Schmalz, Chem. Phys. Lett. **175**, 3 (1990).

[98]R. C. Haddon, L. F. Schneemeyer, J. V. Waszczak, S. H. Glarum, R. Tycho, G. Dabbagh, A. R. Kortan, A. J. Muller, A. M. Mujsce, M. J. Rosseinsky, S. M. Zahurak, A. V. Makhija, F. A. Thiel, K. Raghavachari, E. Cockayne, and V. Elser, Nature **350**, 46 (1991).

[99]S. H. Yang, C. L. Pettiette, J. Conceicao, O. Chesnovsky, and R. E. Smalley, Chem. Phys. Lett. **139**, 233 (1987).

[100]L. J. Wilson, S. Flanagan, V. Khabashesku, M. Alford, F. Chibante, M. Diener, C. Fargason, and E. Roche, Appl. Supercon. **1**, 913 (1993) ["Proc. 3rd World Congress on Superconductivity," eds. K. Krishen and C. G. Burnham, Pergamon, New York, 1993].

$S = 1/2$ and 1, respectively, apparently following a "molecular Hund's rule." The triply charged molecule was identified as $S = 1/2$, however, breaking the trend to maximize the value of S on the molecule.

de Coulon et al.[101] have reported LDA calculations of C_{60}^m molecules, $m = +1, 0, -1, -2, -3$, at the equilibrium geometry of the neutral molecule. Their calculated electron affinities are 7.74, 2.78, −0.26, and −3.26 eV for $m = +1, 0, -1, 2-$, respectively (a negative electron affinity indicates the ion will not accept another electron). The first two compare well with experimental values[102] of 7.6 ± 0.2 and[60,61] 2.65 ± 0.05 eV. Also, the $m = -2$ molecule has been detected,[58] and so this affinity is positive. The calculation of such weakly bound electrons will likely be sensitive to the treatment of correlation effects.

The enhancement of the polarizability due to added electrons on the C_{60} molecule has been studied by Pederson and Quong.[81] Adding electrons to the t_{1u} LUMO decreases contributions to α from virtual transitions into this state (because its occupation blocks such processes) but increases α by the contributions from virtual excitations from this level to higher lying states. These effects are pictured in Fig. 3. The

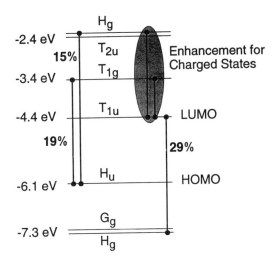

FIG. 3. Electronic states in the HOMO–LUMO region of the C_{60} molecule, with vertical lines indicating virtual transitions contributing to the polarizability. The transitions with percentages (fraction of the total polarizability) labeled contribute to both the neutral and charged molecules, while the transitions in the shaded region become available in the charged molecule. From M. R. Pederson and A. A. Quong, *Phys. Rev. B* **46**, 13584 (1992).

[101]V. de Coulon, J. L. Martins, and F. Reuse, *Phys. Rev. B* **45**, 13671 (1992).
[102]L.-S. Wang, J. Conceicao, C. Jin, and R. E. Smalley, *Chem. Phys. Lett.* **182**, 5 (1991).

transitions $h_g(\text{occ.}) \rightarrow t_{1u}$ and $t_{1u} \rightarrow h_g(\text{unocc.})$ give nearly canceling effects, but the $t_{1u} \rightarrow t_{1g}$ transition leads to a large enhancement of the polarizability. Their result is that α increases by ~7% for each added electron.

c. Electronic Interactions, Screening, and Correlation

The question of interelectronic interactions on the C_{60} molecule and in the solid has drawn considerable attention. The fact that several of the alkali and alkaline earth compounds superconduct and that these materials, being molecular crystals, are different from conventional superconductors raises the possibility that the pairing may be of an unconventional nature (mechanisms are discussed in Section IX). In conventional superconductors, pairing arises from coupling via phonons, and this topic will be covered in subsequent sections. The possibility of pairing arising from the Coulomb interaction has motivated many of the models and calculations covered in this section.

One fundamental question is formulated in terms of the binding energy of a pair of electrons: Given two "extra" electrons in an otherwise neutral C_{60} compound, will they prefer to separate or to bind together at some finite distance, or become unbound? The possibility most often addressed is whether they will reside on the same molecule or not; if not, it is assumed to be unlikely that they will prefer to reside on neighboring molecules rather than to be far apart and thereby reduce their Coulomb repulsion. This is typically addressed by considering the molecules to be uncoupled (i.e., intermolecular hopping is neglected) and at large separation, so that e^2/r repulsion is neglected if they are on separate molecules. In this limit, the binding energy is given by

$$\Delta_{\text{pair}} = -[E_{N+2} + E_N - 2E_{N+1}], \qquad (4.6)$$

where E_J is the energy of a C_{60} molecule with J electrons, and N is the average number of electron per molecule. (Here I use the convention that a positive value of the binding energy indicates the pair will bind.) In this isolated molecule limit, the same energy difference arises if one asks whether the solid with average electron number $N + 1$ will disproportionate into molecules of charge $N + 2$ and N.

Spherical approximation Murthy and Auerbach[103] considered

[103]G. N. Murthy and A. Auerbach, *Phys. Rev. B* **46**, 331 (1992).

the interaction on a spherical shell. They treated an interaction with variable strength and range, so that screening can be accounted for approximately by choosing appropriate values of the variables. Treating the interaction in second-order perturbation theory, pair binding was obtained for two electrons added above a closed angular momentum shell. The ground state can be singlet ($L = 0$) or triplet (magnetic $L = 1$) depending on the parameters. Gedik and Ciraci[104] have considered the complementary problem of two particles on a sphere interacting via an *attractive* potential. They find that a bound state exists no matter how weak the potential, analogous to the situation in one and two dimensions where arbitrarily weak attractive potentials result in a bound state.

Hubbard model for the C_{60} molecule   A carbon site Hubbard model for C_{60} has been studied from several viewpoints. Distinct hopping parameters t,t' for the two types of bonds are usually considered, along with the usual on-site repulsion energy U. The applicability of such a model has not been presented in any detail. The usual scenario is to deal only with the π electrons, whose states span ~ 10 eV in energy. One estimate of the on-site repulsion was given as $F_0 \sim 12$ eV by Antropov et al.[44] (F_0 is the Slater integral); this unscreened value would be screened substantially, giving perhaps $U \sim 4$–8 eV. If only an on-site interaction is used, the intersite interactions should be small. For C_{60}, e^2/r for $r = 1.4$ Å is ~ 10 eV; this might be screened by a factor of ~ 2, giving a nearest neighbor repulsion $V_1 \sim 5$ eV. The second neighbor is not much further away, perhaps $V_2 \sim 3$–4 eV. The effect of the on-site repulsion is reduced by these intersite interactions. These simple guesses suggest that intra-atomic interactions will not be dominant in the C_{60} molecule.

Chakravarty, Gelfand, and Kivelson[105,106] considered the weakly interacting case with only on-site repulsion using second-order perturbation theory. Electrons were added to the half-filled (60 electrons) case. Second-order perturbation theory in U/t results in pair binding, and a singlet ($S = 0$) ground state for $U < U_{pair} \sim 3t$. For larger values of U/t the magnetic $S = 1$ paired state is preferred. They extended their considerations up to six added electrons on the C_{60} molecule. For two, three, and four added electrons, the ground state can be either maximal spin or minimal spin depending on the size of U.

They also considered the interatomic Coulomb interaction as a "molecular charging" effect, in which case adding N electrons gives a

[104]Z. Gedik and S. Ciraci, *Phys. Rev. B* **45**, 8213 (1992).
[105]S. Chakravarty, M. P. Gelfand, and S. Kivelson, *Science* **254**, 970 (1991).
[106]S. Chakravarty and S. Kivelson, *Europhys. Lett.* **16**, 751 (1991).

repulsive contribution $E_{rep} = e^2 N^2/2C$, where C is the capacitance, $C \sim \varepsilon_\infty D/2$, where D is the molecular diameter. For $D \sim 7$ Å and $\varepsilon_\infty \sim 4$, this gives $E_{rep} \sim 1$–2 eV. This value would be reduced by metallic screening in the conducting solids. One can also consider the effect of electron–phonon coupling on pair binding: they find the effect of the symmetric A_g modes would be attractive, but in their picture the Jahn–Teller distortion would be pair breaking. Their model was extended by White et al.,[107] who found that pair binding arising from the correlation will arise only for intermediate coupling strength and on intermediate length scales.

A Gutzwiller treatment of the Hubbard model for the C_{60} model was considered by Joyes and Tarento.[108] They included in the energy also a repulsive capacitative charging term similar to that of Chakravarty et al. The charging term was found to be very important in reproducing the experimental ionization potentials and electron affinities for various charged molecules.

Goff and Phillips[109] considered the intramolecular electronic repulsion between two electrons in more detail. They addressed the extended Hubbard model, where a phenomenological repulsive screened Coulomb potential of the form

$$V_{ij} = U e^{-r_{ij}/\lambda_{sc}} \tag{4.7}$$

or of the Ohno form

$$V_{ij} = \frac{U}{[1 + (r_{ij}/\lambda_{sc})^2]^{1/2}}, \tag{4.8}$$

between all atomic sites i, j is added to the Hubbard model. The screening length λ_{sc} was varied from one-fourth to twice a bond length. For small U/t they find the maximal spin state to be the most stable, satisfying a molecular Hund's rule. As U/t increases beyond the range 1.5 (for small λ_{sc}) to 3.5 (for large λ_{sc}), lower spin states become favored. They find that the interatomic repulsion significantly reduces the ten-

[107]S. R. White, S. Chakravarty, M. P. Gelfand, and S. A. Kivelson, *Phys. Rev. B* **45**, 5062 (1992).
[108]P. Joyes and R. J. Tarento, *Phys. Rev. B* **45**, 12077 (1992).
[109]W. E. Goff and P. Phillips, *Phys. Rev. B* **46**, 603 (1992).

dency toward pair binding in second-order perturbation theory; for example, for pair binding to occur at $U/t = 4$, the screening length would have to be less than half the bond length.

Stollhoff and Häser[110] considered the ab initio "local ansatz" to evaluate correlation effects in the C_{60} molecule. In this method the many-body wavefunction is a Slater determinant of LDA one-electron wavefunctions with Gutzwiller-like modifications to represent charge and spin correlations. Intraatomic and interatomic correlations are found to be roughly equal in magnitude, and a weakly correlated picture of the neutral molecule was obtained. The binding energy of the fullerene molecule was calculated to be 0.45 eV/atom smaller than was found in graphite and diamond using similar methods.

Perhaps the most unusual suggestion of effects due to correlation was made by Friedberg et al.[111] Recognizing the relatively small splitting ("near degeneracy") of the t_{1u} and t_{1g} orbitals that are available for occupation in the charged C_{60}^{-n} molecule, they suggest that forming linear combinations of these one-electron states in a two-particle state may result in a gain in Coulomb energy that outweighs the loss in one-electron energy. For the doubly charged molecule there are low lying states of both even and odd parity. Their development of the implications of these "parity doublets" for properties of the fullerides will be discussed in Section 27. This aspect of correlation, which is unique to the charged C_{60} molecule, has not yet been followed up by a detailed computation, which seems entirely feasible.

Local density approximation calculations LDA methods, which have been found to give very reliable energy differences, have been applied to this question. It is standard to let the LUMOs (and of course other MOs) relax self-consistently in response to the added electrons, allowing the electrons to screen one another. Since the objective is to learn how the conduction electrons behave in the solid, solid state effects have been considered as well in arriving at the corresponding Coulomb repulsion U_{eff} that would be most appropriate in a Hubbard model representation of fullerite and fullerides. This has been done assuming the C_{60} molecules in the vicinity of a charge respond to the total electric field due to that charge and the induced dipoles on all other molecules. It is furthermore assumed that only the linear polarizability is important.

From total energy differences Pederson and Quong[81] (see Eq. (4.5))

[110]G. Stollhoff and M. Häser, Phys. Rev. B **45**, 13703 (1992).
[111]R. Friedberg, T. D. Lee, and H. C. Ren, Phys. Rev. B **46**, 14150 (1992).

and de Coulon *et al.*[101] obtained $U = 3.0$ eV. Antropov *et al.*[44] obtained $U = \delta\varepsilon_{t1u}(n)/\delta n = 2.7$ eV for 0, 1, and 2 added electrons in the t_{1u} LUMO. In the limit of infinitesimal charge difference δn (and identical calculational methods) the total energy difference and the eigenvalue derivative are identical within LDA. Using the LDA polarizability, Pederson and Quong[81] obtain a reduction due to the surrounding molecules of $\delta U^0_{xtal} = 1.74$ eV, similar to (but somewhat larger than) estimates of Antropov *et al.* Antropov *et al.* used a model screening function to estimate that the size of the molecules (i.e., non-point-dipole effects, which were not considered by Pederson and Quong) increases δU by 12%, which would increase the LDA value to $\delta U_{xtal} = 1.95$. Then the screened value is $U_{sc} = U - \delta U_{xtal} = 1.05$ eV.

Antropov *et al.* have then considered the next important solid state effect, the intersite interaction, that must be taken into account in a Hubbard model. The relevant energy, in the simplest picture in which only an on-site interaction is considered, is the energy of interaction when two electrons occupy the same molecule ("U") compared with the energy when they are close but do not occupy the same site ("V"). This latter, intersite repulsion can be obtained as the shift in the LUMO eigenvalue ε_{t1u} on molecule 1 due to an electron added to a neighboring molecule 2. This is the intersite repulsion $V_{sc} = e^2/R_{12} - \delta V$, where δV is the polarization effect arising from surrounding molecules. This interaction can be thought of as $V = e^2/\varepsilon R_{12}$; however, the screening constant ε should not be taken simply as the long wavelength value since local field effects will be important for a short range interaction. The value obtained by Antropov *et al.*, transformed to the LDA value[81] of the polarizability, gives $\delta V = 1.07$ eV, and $e^2/R_{12} = 1.43$ eV for an fcc lattice constant of 14.2 Å. The dielectric screening by the solid is seen to be substantial, and the net interaction is $V_{sc} = 1.43 - 1.07 = 0.36$ eV. Finally, the net cost of having two electrons on the same molecule rather than on neighboring molecules is $U_{eff} = U_{sc} - V_{sc} = 1.05 - 0.36 \approx 0.7$ eV. It is useful, both for clarifying the various terms and to allow revision as more information becomes available, to collect the various contributions and make the expression explicit:

$$U_{eff} = U_{sc} - V_{sc} = (U_0 - \delta U) - (V_0 - \delta V)$$

$$= \left[U_0 - (1 + \delta_{size})\frac{C_1}{\varepsilon a} \right] - \left[\frac{e^2}{\varepsilon R} - \frac{C_2}{\varepsilon a} \right], \tag{4.9}$$

where a is the lattice constant, R is the separation of molecular centers,

$\delta_{\text{size}} = 0.12$ is the correction for non-point-dipoles, and C_1 and C_2 are dipolar lattice sums. Antropov *et al.* estimated that reduced screening should increase U_{eff} at the surface by ~0.3 eV, and that such an increased value is what would arise in interpreting Auger and photoemission data (see Section VI).

III. Dynamics of the C_{60} Molecule and Solids

5. SYMMETRY AND THE MOLECULAR VIBRATIONS

An extensive rotational–vibrational symmetry analysis of icosahedral molecules has been provided by Harter and Weeks.[112,113] They provide the class structure of the I_h group and the multiplication table of the icosahedral group I. Their identification of the symmetries, multiplicities, and degeneracies of the eigenmodes of C_{60} are reproduced in Table IV. (The label T_2 that is used here is also sometimes called T_3 or F_2 in the literature.) There are 174 vibrational modes for the C_{60} molecule $(3 \times 60 -$ three rotational $-$ three translational). I_h symmetry reduces to 46 the number of distinct eigenfrequencies. A complete listing of

TABLE IV   THE SYMMETRY GROUP LABELS, MULTIPLICITIES, AND DEGENERACIES OF THE EIGENMODES OF THE C_{60} MOLECULE

I_h GROUP LABEL	MULTIPLICITY	DEGENERACY	ACTIVITY
A_g	2	1	Raman
T_{1g}	3	3	
T_{2g}	4	3	
G_g	6	4	
H_g	8	5	Raman
A_u	1	1	
T_{1u}	4	3	IR
T_{2u}	5	3	
G_u	6	4	
H_u	7	5	

From D. E. Weeks and W. G. Harter, *J. Chem. Phys.* **90**, 4744 (1989).

[112]W. G. Harter and D. E. Weeks, *J. Chem. Phys.* **90**, 4727 (1989).
[113]D. E. Weeks and W. G. Harter, *J. Chem. Phys.* **90**, 4744 (1989).

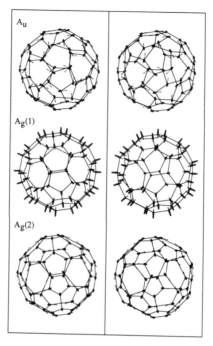

FIG. 4. Stereographic representation of the optically active normal modes (A_g, H_g, T_{1g}) and one silent mode (A_u) of the C_{60} molecule. The figures are adapted from the computer program "Buckyball Vibrations" of D. E. Weeks and W. G. Harter, and the normal mode displacements are given by a force constant model (see Refs. 116 and 117).

symmetry coordinates for the eigenmodes was presented by Brendsdal.[114] Because C_{60} is a 60-atom molecule, the resulting normal modes are far from intuitive, in spite of its high symmetry. Since there is rather little indication in the literature of what many of the normal modes are like, examples of the most widely discussed normal modes are pictured in Fig. 4. It should be recognized that displacements of the degenerate modes are not unique, and other choices of displacements patterns can be made.

A complication in the discussion and comprehension of vibrational spectroscopies is the matter of units, as meV, cm^{-1}, and K are all in common use. The experimental results will be reported here in the units used in the paper being quoted. The conversion is $10 \, \text{meV} = 80.1 \, \text{cm}^{-1} = 116 \, \text{K}$; $100 \, \text{cm}^{-1} = 144 \, \text{K} = 12.5 \, \text{meV}$; $100 \, \text{K} = 8.6 \, \text{meV} = 69 \, \text{cm}^{-1}$.

The Raman active modes consist of the two A_g and eight H_g

[114]E. Brendsdal, *Spectrosc. Lett.* **21**, 319 (1988).

$T_{1u}(1)$

$T_{1u}(2)$

$T_{1u}(3)$

$T_{1u}(4)$

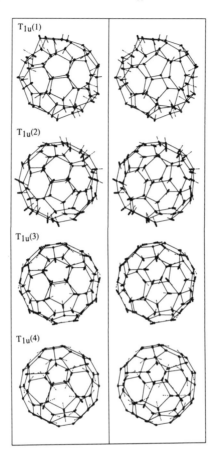

FIG. 4—(continued)

vibrations. The fully symmetric A_g modes are called the *breathing* mode, in which each C atom moves radially from the molecular center, and *pentagonal pinch* mode, in which the two distinct bond lengths are modulated and therefore the pentagonal face decreases/increases in area. Extreme distortions of the pentagonal pinch type lead to the distorted C_{60} molecules shown in Figs. 1(b) and 1(c).

Due to the large number of atoms in the C_{60} molecule and its unfamiliar symmetry, most modes other than the two A_g, fully symmetric modes are not trivial to visualize or to understand. The single A_u mode involves rotations of each of the pentagonal faces around the axis through its center and going through the center of the molecule. As a result all of the b_{hh} bonds that separate hexagonal faces rotate but remain unchanged in length to first order in the amplitude of vibration. All of the pentagon

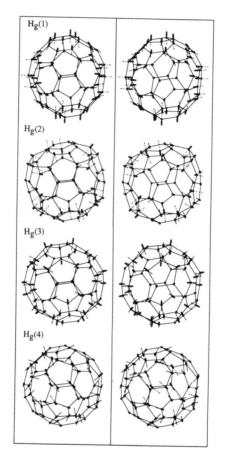

FIG. 4—(continued)

faces twist in phase. Since several of the bond lengths are almost unchanged, this should not be a high frequency mode.

The quadrupolar H_g modes, each fivefold degenerate, are distinguished by the retention of an axis of fivefold symmetry. The "squash" mode, in which pictorially the C_{60} "soccer ball" is set on a surface with a pentagonal face down and then squashed from above on the opposite pentagonal face, is a low frequency H_g mode. The four T_{1u} modes are the infrared active modes, each having three partners with symmetry characteristic of the vector (x, y, z) representation.

Several symmetry aspects adapted specifically to C_{60} crystals have been presented by Dresselhaus et al.[115] where a full character table for I_h

[115]G. Dresselhaus, M. S. Dresselhaus, and P. C. Eklund, Phys. Rev. B **45**, 6923 (1992).

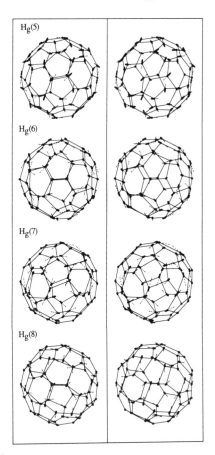

$H_g(5)$

$H_g(6)$

$H_g(7)$

$H_g(8)$

FIG. 4—(continued)

symmetry is also presented. Weeks and Harter[116,117] have considered
rotational–vibrational coupling in C_{60} molecules and presented stereographic figures of the T_{1u} infrared active rovibrational modes. A number of
results from symmetry analysis of fullerite crystals have also been
presented by Yildirim and Harris.[118] Eklund et al.[119] have presented a
much more comprehensive review of the optically active phonon modes
in C_{60}-based compounds than will be presented here.

[116]D. E. Weeks and W. G. Harter, *Chem. Phys. Lett.* **144,** 366 (1988).
[117]D. E. Weeks and W. G. Harter, *Chem. Phys. Lett.* **176,** 209 (1991).
[118]T. Yildirim and A. B. Harris, *Phys. Rev. B* **46,** 7878 (1992).
[119]P. C. Eklund, P. Zhou, K.-A. Wang, G. Dresselhaus, and M. S. Dresselhaus, *J. Phys. Chem. Solids* **53,** 1391 (1992).

6. FULLERITE C_{60}

a. Experimental Identifications

Some vibrational modes of C_{60} have been measured in the gas, solution, thin-film overlayer, and crystalline phases. Representative measured frequencies are listed in Table V. As can be seen from the table, there is agreement on the frequencies of the Raman active and infrared active modes. The A_g pentagonal pinch mode is hard, at 1470 cm^{-1} ($1500 \text{ cm}^{-1} = 186 \text{ meV} = 2160 \text{ K}$). The breathing mode is at one-third of that energy ($490-495 \text{ cm}^{-1}$). The eight H_g modes range in

TABLE V. REPRESENTATIVE MEASURED
VIBRATIONAL FREQUENCIES (cm^{-1}) OF C_{60} IN THE GAS, LIQUID,
AND SOLID PHASES

MODE	GAS	LIQUID	SOLID
A_g		491^a	496^c
A_g		1469^a	$1470,^c 1458^e$
H_g			$265,^a 273^c$
H_g			$434,^a 437^c$
H_g			$711,^a 710^c$
H_g			$773,^a 774^c$
H_g			$1100,^a 1099^c$
H_g			$1255,^a 1250^c$
H_g			$1427,^a 1428^c$
H_g			$1575,^a 1575^c$
T_{1u}	527^d		$528,^b 527^c$
T_{1u}	570^d		$577,^b 577^c$
T_{1u}	1170^d		$1183,^b 1183^c$
T_{1u}	1407^d		$1429,^b 1428^c$

[a] K. Sinha, J. Menendez, R. C. Hanson, G. B. Adams, J. B. Page, O. F. Sankey, L. D. Lamb, and D. R. Huffman, *Chem. Phys. Lett.* **186**, 287 (1991).
[b] W. Krätschmer, L. D. Lamb, K. Fostiropoulos, and D. R. Huffman, *Nature* **347**, 354 (1990).
[c] D. S. Bethune, G. Meijer, W. C. Tang, and H. J. Rosen, *Chem. Phys. Lett.* **174**, 219 (1990).
[d] C. I. Frum, R. Engelman, H. G. Hedderich, R. F. Bernath, L. D. Lamb, and D. R. Huffman, *Chem. Phys. Lett.* **176**, 504 (1991).
[e] T. Pichler, M. Matus, J. Kürti, and H. Kuzmany, *Phys. Rev. B* **45**, 13841 (1992).

energy from the very soft squash mode at $270\,\text{cm}^{-1}$ to the hard Jahn–Teller mode at $1575\,\text{cm}^{-1}$. Of the four infrared active modes, two are relatively soft at $527\,\text{cm}^{-1}$ and $577\,\text{cm}^{-1}$, and the others are progressively harder at $1183\,\text{cm}^{-1}$ and $1428\,\text{cm}^{-1}$. The silent modes (neither Raman nor infrared active) are trickier to assign, and there has been considerable reliance on theoretical calculations (given later). Since all atoms in fullerite are carbon, inelastic neutron scattering from powders[120] allows the direct identification of the vibrational density of states. The integrated intensity under a peak is a direct measurement of the number of modes, allowing the identification not only of frequencies but of degeneracies as well. Prassides et al.[121,122] have made tentative identifications of many of the peaks in the neutron scattering spectra in this way.

The vibrational density of states of fullerite is shown in the 10–110 meV range in Fig. 5 and in the 100–200 meV range in Fig. 6. The intramolecular modes extend from 30–200 meV ($240–1600\,\text{cm}^{-1}$). From the narrowness of the peaks beginning at 32 meV, it is clear that the dispersion of these modes in the solid is very small. The broader bands, e.g., the 90–100 meV range, contain several unresolved vibrations.

Another possibility for obtaining experimental information is to take advantage of the crystalline symmetry, which can make the crystalline analogs of the silent molecular vibrations become weakly Raman or infrared active. According to van Loosdrecht et al.,[123] all even modes become Raman active in the solid; and they detected several silent molecular modes that were activated and split by the crystal field. In the high temperature fcc phase, they detected 29 clear peaks (of 37 possible), while below the structural transition at 250 K they observed 37 peaks (of a possible 145). They also observed changes in some linewidths and frequencies at the structural transition, which provide information about the intermolecular interactions. They observed in addition Raman peaks at 56, 81, and $109\,\text{cm}^{-1}$ that may arise from the librations. Huant et al.[124] indicate that all odd modes become infrared active in the crystal, and

[120]R. L. Cappelletti, J. R. D. Copley, W. A. Kamitakahara, F. Li, J. S. Lannin, and D. Ramage, Phys. Rev. Lett. 66, 3261 (1991).

[121]K. Prassides, T. J. S. Dennis, J. P. Hare, J. Tomkinson, H. W. Kroto, R. Taylor, and . R. M. Walton, Chem. Phys. Lett. 187, 455 (1991).

[122]K. Prassides, J. Tomkinson, C. Christides, M. J. Rosseinsky, D. W. Murphy, and R. C. Haddon, Nature 354, 462 (1991).

[123]P. H. M. van Loosdrecht, P. J. M. van Bentum, and G. Meijer, Phys. Rev. Lett. 68, 1176 (1992).

[124]S. Huant, J. B. Robert, G. Chouteau, P. Bernier, C. Fabre, and A. Rassat, Phys. Rev. Lett. 69, 2666 (1992).

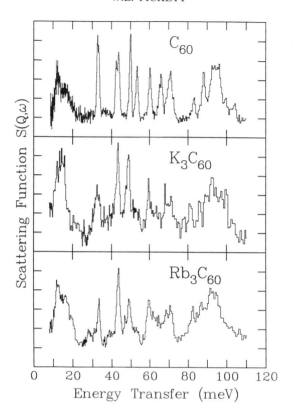

FIG. 5. High resolution inelastic neutron scattering spectra for (top) C_{60} at 25 K, K_3C_{60} (middle; summed data at 5 and 30 K), Rb_3C_{60} (bottom; summed data at 22 and 35 K). Note that the fullerite peaks at 54 meV (435 cm^{-1}) and 66 meV (530 cm^{-1}) are missing in the fullerides.
This figure is from J. R. D. Copley, D. A. Neumann, R. L. Cappelletti, and W. A. Kamitakahara, *J. Phys. Chem. Solids* **53**, 1353 (1992) and is a compilation of data. The sources of the data are C_{60}: C. Coulombeau, H. Jobic, P. Bernier, D. Schültz, and A. Rassat, *J. Phys. Chem.* **96**, 22 (1992); K_3C_{60}: K. Prassides, J. Tomkinson, C. Cristides, M. J. Rosseinsky, D. W. Murphy, and R. C. Haddon, *Nature* **354**, 462 (1991); Rb_3C_{60}: J. W. White, G. Lindsell, L. Pang, A. Palmisano, D. S. Sivia, and J. Tomkinson, *Chem. Phys. Lett.* **191**, 92 (1992).
Reprinted from *J. Phys. Chem. Solids*, Copyright 1992, with permission from Pergamon Press Ltd., Headington Hill Hall, Oxford OX3 0BW, UK.

they report several split modes corresponding to peaks in inelastic neutron scattering spectra. Narasimhan *et al.*[125] report a splitting and a

[125]L. R. Narasimhan, D. N. Stoneback, A. F. Hebard, R. C. Haddon, and C. K. N. Patel, *Phys. Rev. B* **46,** 2591 (1992).

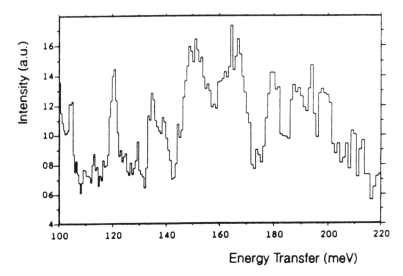

FIG. 6. Inelastic neutron scattering spectra of fullerite C_{60} in the high frequency range, taken at 25 K. Figure from J. R. D. Copley, D. A. Neumann, R. L. Cappelletti, and W. A. Kamitakahara, *J. Phys. Chem. Solids* **53**, 1353 (1992); the data was first presented by C. Coulombeau, H. Jobic, P. Bernier, D. Schültz, and A. Rassat, *J. Phys. Chem.* **96**, 22 (1992). Reprinted from *J. Phys. Chem. Solids*, Copyright 1992, with permission from Pergamon Press Ltd., Headington Hill Hall, Oxford OX3 0BW, UK.

frequency shift of the 1431-cm^{-1} T_{1u} mode at the structural transition at ~250 K, whereas the 1183-cm^{-1} T_{1U} mode showed only a slight narrowing in width.

The pressure dependence of the infrared active modes was reported by Klug *et al.*[126] for pressures up to 200 kbar at room temperature. The low frequency $T_{1u}(1)$ mode decreases in energy by ~1%, while the other three T_{1u} modes increase, slightly sublinearly, by ~2–3%. Klug *et al.* describe the $T_{1u}(1)$ mode as one in which the pentagonal faces at the top and bottom stay the same but oscillate (along the z axis) out of phase with the rest of the atoms, while the other modes involve a change in area of the pentagons on top and bottom. The reason for the various trends with temperature is not yet understood.

In the low temperature fullerite crystal with four molecules per cell, there are optic modes involving translations of the molecules that become optically active. Group theory indicates that from the 4 molecules per unit cell with 3 translational and 3 rotational degrees of freedom each, the spectrum consists of 12 translational modes $A_u + E_u + 3T_u$ and 12

---

[126]D. D. Klug, J. A. Howard, and D. A. Wilkinson, *Chem. Phys. Lett.* **188**, 168 (1992).

librational modes $A_g + E_g + 3T_g$. Taking account of the translational and rotational modes of the crystal as a whole leaves four optic modes (two being infrared active T_u) and four Raman active librations at $Q = 0$ (three of which may have been seen by van Loosdrecht et al.[123]). Huant et al.[124] assign infrared absorption peaks at 27 and 59 cm^{-1} to the optic modes.

The librational modes of C_{60} have been studied on powder samples by Neumann and coworkers.[127-129] They identify a single temperature-dependent peak that decreases linearly with increasing temperature, from ~2.7 meV near zero temperature to ~2 meV as the structural transition is approached at $T_s \approx 250$ K. The energy, linewidth, and signal intensity are shown in Fig. 7. Near T_s the width is comparable to the energy. The intensity does not vary appreciably from that expected of a harmonic oscillator.

The collective optic modes of single crystal C_{60} were measured by Pintchovius et al.[130] using inelastic neutron scattering. The vibrational optic modes are reminiscent of those in rare gas solids and extend up to 6 meV. They could be fit with a two-parameter force constant model, but the necessary value of the second neighbor force constant was surprisingly large for the C_{60} crystal (~10% of the first neighbor force constant). They also reported the librational modes and their Q dependence, finding rather flat branches (and therefore narrow peaks in the density of states) at 2.4, 3.6, and 4.6 meV.

Kim et al.[131] have employed a steady state photoexcitation technique to probe the effects of excited charge carriers on the infrared active modes. With their method, they expected to be measuring the effects of free electrons and free holes that have escaped from the recombination process, rather than bound excitons. They find the triple degeneracy of the $T_{1u}(1)$ mode is lifted by the charge carriers, and the $T_{1u}(2)$ mode is asymmetrically broadened below 30 K. The $T_{1u}(3)$ and $T_{1u}(4)$ modes are relatively unaffected. They attribute the splitting and line asymmetry to anharmonicity driven by the charge carriers, reflecting strong electron–lattice coupling. Although not named specifically by Kim et al., these

[127]D. A. Neumann, J. R. D. Copley, W. A. Kamitakahara, J. J. Rush, R. L. Cappelletti, N. Coustel, J. E. Fischer, J. P. McCauley, Jr., A. B. Smith III, K. M. Creegan, and D. M. Cox, *J. Chem. Phys.* **96**, 8631 (1992).

[128]K. Prassides, C. Christides, M. J. Rosseinsky, J. Tomkinson, D. W. Murphy, and R. C. Haddon, *Europhys. Lett.* **19**, 629 (1992).

[129]J. R. D. Copley, D. A. Neumann, R. L. Cappeletti, and W. A. Kamitakahara, *J. Phys. Chem. Solids* **53**, 1353 (1992).

[130]L. Pintchovius, B. Renker, F. Gompf, R. Heid, S. L. Chaplot, M. Haluska, and H. Kuzmany, *Phys. Rev. Lett.* **69**, 2662 (1992).

[131]Y. H. Kim, F. Li, and F. Diederich, *Phys. Rev. B* **45**, 10169 (1992).

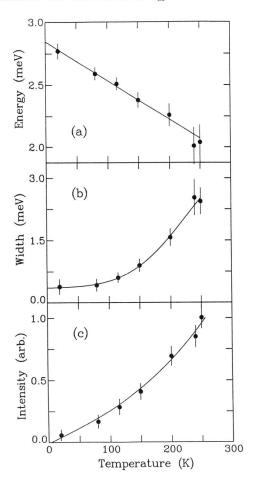

FIG. 7. Librational energy, linewidth, and scattering intensity versus temperature in the simple cubic phase of the C_{60} crystal. Solid lines in (a) and (b) are guides to the eye, while that in (c) is that predicted for a simple harmonic oscillator. From J. R. D. Copley, D. A. Neumann, R. L. Cappelletti, and W. A. Kamitakahara, *J. Phys. Chem. Solids* **53**, 1353 (1992). Reprinted from *J. Phys. Chem. Solids,* Copyright 1992, with permission from Pergamon Press Ltd., Headington Hill Hall, Oxford OX3 0BW, UK.

effects could be a rather direct result of the Jahn–Teller distortions that have received much attention. The Jahn–Teller question is discussed in Section 8.

van Loosdrecht *et al.*[131a] reported the effects on the Raman spectrum

[131a]P. H. M. van Loosdrecht, P. J. M. van Bentum, and G. Meijer, *Chem. Phys. Lett.* **205**, 191 (1993).

due to irradiation at 514 nm. They find that the electronic excitation introduces new modes just below the pentagonal pinch $A_g(2)$ (1468 cm^{-1}) by 3 cm^{-1} at low irradiance. At higher power, the spectrum becomes a single peak at 1458 cm^{-1}.

b. *Theoretical Calculations*

A classical force constant treatment of the dynamics has been pursued by several groups. The purpose of such models is to gain insight of the force constants, and Feldman *et al.*[132] have collected some general results related to the C_{60} molecule. First, nearest-neighbor force constants are not sufficient to stabilize a structure with a coordination of three, even if two distinct force constants (for b_{hh} and b_{ph}) are allowed. This result is understood by noting that such forces provide only $3N/2$ constraints on a molecule with $3N$ degrees of freedom. Second, the breathing A_g mode is unaffected, and the pentagonal pinch A_g mode almost so, by bond-bending forces. (A_u and A_g modes are affected very little by the puckering terms that will be discussed.) Third, it is found that the measured ratio of the two A_g modes cannot be reproduced by nearest-neighbor central forces.

Weeks and Harter[116] applied a model with two distinct bond-stretching spring constants and one or two bond-bending parameters. Onida and Benedek[133] fit a screened bond charge model, which incorporated in addition to central forces and bond bending a parameter representing a screened bond charge Z_{eff}^2/ε whose value was near unity. The 14 optically active modes were fit to within 3% of the experimental values, but this model predicts a large ($\sim$200 cm^{-1} wide) gap in the 1200–1400 cm^{-1} region that is contradicted by experiment. The model of Wu *et al.*[134] seems to be in conflict with other models.[116]

Brendsdal *et al.*[35,135] applied a five-parameter model determined from other aromatic hydrocarbon systems without variation of the parameters. The parameters were a bond-stretching force constant, a bond-bending constant, one for out-of-plane bending, and two torsional constants (one for each bond). All of the predicted frequencies were presented. Jishi *et al.*[136]

[132]J. L. Feldman, J. Q. Broughton, L. L. Boyer, D. E. Reich, and M. D. Kluge, *Phys. Rev. B* **46**, 12731 (1992).
[133]G. Onida and G. Benedek, *Europhys. Lett.* **18**, 403 (1992).
[134]Z. C. Wu, D. A. Jelski, and T. F. George, *Chem. Phys. Lett.* **137**, 291 (1987).
[135]S. J. Cyvin, E. Brendsal, B. N. Cyvin, and J. Brunvoll, *Chem. Phys. Lett.* **143**, 377 (1988).
[136]R. A. Jishi, R. M. Mirie, and M. S. Dresselhaus, *Phys. Rev. B* **45**, 13685 (1992).

and Ruoff and Ruoff[137] also applied force constant models based on bond-stretching and bond-bending interactions. Jishi *et al.* fit the observed Raman frequencies and found that the frequencies seen in infrared spectroscopies were reproduced well by the model.

Feldman *et al.*[132] used in addition to spring constants and bond-bending terms a "puckering" term, introduced by Lax, that represents a rotationally invariant form of elastic energy. This term is of the form $\delta \, |(\mathbf{r}_i - \mathbf{r}_k) \cdot (\mathbf{r}_i - \mathbf{r}_m) \times (\mathbf{r}_i - \mathbf{r}_n)|^2$; that is, it is proportional to the change in the (square of the) volume formed by the tetrahedron formed by the atoms i, k, m, n. Such terms have been useful in describing the lattice dynamics of graphite. This model produces optically active modes within 4% of the experimental values and does not show the unphysical high frequency gap. Feldman *et al.* have provided some discussion of the extent to which such force constant models can be used to fit and therefore interpret experimental data.

Calculations (without adjustment) began with the semiempirical methods (viz. MNDO) and have been extended to full LDA calculations. Newton and Stanton[138,139] used the MNDO method to calculate the frequencies; as an indication of its accuracy, the four infrared active T_{1u} modes are an average of 15% higher than the measured values.[140] Slanina *et al.*[141] used the AM1 method to obtain all frequencies.

LDA predictions of frequencies have been provided by several groups. Zhang *et al.*[142] carried out a plane-wave based ab initio molecular dynamics simulation of solid C_{60}, noting that intermolecular interactions are weak but not obtaining specific vibrational frequencies. Feuston *et al.*[143] performed similar simulations and used signal processing techniques to extract values of the four T_{1u}, two A_g, and six of the eight H_g modes. Subsequently, Kohanoff *et al.*[144] were able to determine all frequencies from this type of calculation. Adams *et al.*[145] carried out a local-orbital-based ab initio molecular dynamics simulation of the isolated molecule,

[137]R. S. Ruoff and A. L. Ruoff, *Appl. Phys. Lett.* **59**, 1553 (1991).

[138]M. D. Newton and R. E. Stanton, *J. Am. Chem. Soc.* **108**, 2469 (1986).

[139]R. E. Stanton and M. D. Newton, *J. Phys. Chem.* **92**, 2141 (1988).

[140]K. Raghavachari and C. M. Rohlfing, *J. Phys. Chem.* **95**, 5768 (1991).

[141]Z. Slanina, J. R. Rudzinski, M. Togasi, and E. Ósawa, *J. Mol. Struct. (Theochem)* **202**, 169 (1989).

[142]Q.-M. Zhang, J.-Y. Yi, and J. Bernholc, *Phys. Rev. Lett.* **66**, 2633 (1991).

[143]B. P. Feuston, W. Andreoni, M. Parrinelo, and E. Clementi, *Phys. Rev. B* **44**, 4056 (1991).

[144]J. Kohanoff, W. Andreoni, and M. Parrinello, *Phys. Rev. B* **46**, 4371 (1992).

[145]G. B. Adams, J. B. Page, O. F. Sankey, K. Sinha, J. Menendez, and D. R. Huffman, *Phys. Rev. B* **44**, 4052 (1991).

identifying all optically active modes. The frequencies of several of these groups, and others to be discussed, are presented in Table VI.

More standard frozen phonon and direct force constant calculations have been reported as well. Jones et al.[146] applied a local orbital, norm conserving pseudopotential method to obtain short range force constants and thereby all frequencies for the isolated molecule. Early all-electron local orbital results for the two A_g modes were reported by Pederson and Quong[81] (550 and 1500 cm^{-1}) and Dunlap[57] (710 and 1820 cm^{-1}), and Pederson and Quong also calculated the silent A_u mode frequency (871 cm^{-1}).

The most definitive theoretical predictions appear to be those of Wang et al.[146a] and Quong et al.[146b] The calculations use related but distinct local orbital basis methods, and their results are included in Table VI. With only a few exceptions, the predictions of these two groups are consistent to within 2–3%. The largest difference is for the $T_{2g}(2)$ mode (at 771 and 610 cm^{-1}, respectively), with $\Delta\omega/\bar{\omega} = 23\%$, and there are discrepancies for the $H_g(5)$ (7%), A_u (11%), and $G_u(2)$ (9%) modes. Wang et al. compare their calculated frequencies with the density of states measured for fullerite by Coulombeau et al.,[146c] obtaining excellent agreement for the 30–100 meV modes but some noticeable discrepancies in the 105–200 meV range.

It is instructive to contrast the carbon-based 60-atom molecule being considered here with the hypothetical silicon-based analog. The Si_{60} fullerene was considered by Khan and Broughton[147] using a tight-binding total energy scheme related to the one of Wang et al.[148] for C_{60}. The Si_{60} molecule relaxes to a puckered spheroid without I_h symmetry. It assumes this shape, rather than the spherical shape, apparently for the same reason that the graphitic Si layer is unstable to puckering: sp^2 bonding is much less favorable in Si than in C.

The intermolecular modes in fullerite have been studied by Yildirim and Harris[118] and by Li et al.[149] using classical intermolecular potentials. The resulting dispersion curves throughout the Brillouin zone calculated by Li et al. are shown in Fig. 8. The librational modes are found to

[146]R. Jones, C. D. Latham, M. I. Heggie, V. J. B. Torres, S. Öberg, and S. K. Estreicher, Phil. Mag. Lett. **65**, 291 (1992).

[146a]X. Q. Wang, C. Z. Wang, and K. M. Ho, Phys. Rev. B **48**, 1884 (1993).

[146b]A. A. Quong, M. R. Pederson, and J. L. Feldman, Solid State Commun. **87**, 535 (1993).

[146c]C. Coulombeau, H. Jobic, P. Bernier, D. Schültz, and A. Rassat, J. Phys. Chem. **96**, 22 (1992).

[147]F. S. Khan and J. Q. Broughton, Phys. Rev. B **43**, 11754 (1991).

[148]C. Z. Wang, C. T. Chan, and K. M. Ho, Phys. Rev. B **46**, 9761 (1992).

[149]X.-P. Li, J. P. Lu, and R. M. Martin, Phys. Rev. B **46**, 4301 (1992).

TABLE VI. CALCULATED VIBRATIONAL FREQUENCIES (cm^{-1}) OF
C$_{60}$, AND FOR COMPARISON, THE MEASURED OPTICALLY ACTIVE FREQUENCIES IN THE SOLID

MODE	SOLID[a]	ADAMS[b]	ANTROPOV[c]	JONES[d]	KOHANOFF[e]	NEGRI[f]	WANG[g]	QUONG[h]
A_g	496	537	458	582	455	513	483	478
	1458	1680	1463	1416	1365	1442	1529	1499
T_{1g}				661	547	597	566	580
				820	799	975	825	788
				1253	1211	1398	1292	1252
T_{2g}				514	527	637	550	548
				803	744	834	771	610
				1129	770	890	795	770
				1310	1186	1470	1360	1316
G_g				454	455	476	484	486
				517	560	614	564	571
				977	737	770	763	759
				1094	988	1158	1117	1087
				1326	1240	1450	1326	1297
				1466	1395	1585	1528	1505
H_g	273	249	281	288	246	258	263	258
	437	413	454	437	410	440	432	439
	710	681	753	787	689	691	713	727
	774	845	785	829	731	801	778	767
	1099	1209	1091	1106	1036	1154	1111	1193
	1250	1453	1290	1193	1140	1265	1282	1244
	1428	1624	1387	1395	1315	1465	1469	1444
	1575	1726	1462	1516	1484	1644	1598	1576
A_u				991	881	1206	947	850
T_{1u}	527	494		614	510	544	533	547
	577	643		647	534	637	548	570
	1183	1358		1127	1092	1212	1214	1176
	1428	1641		1403	1320	1437	1485	1461
T_{2u}				363	321	350	344	342
				850	730	690	717	738
				1009	833	999	987	963
				1164	1045	1241	1227	1185
				1494	1450	1558	1558	1539
G_u				374	332	358	356	357
				749	720	816	752	683
				936	751	832	784	742
				1115	886	1007	977	957
				1286	1216	1401	1339	1298
				1408	1316	1546	1467	1440
H_u				303	381	403	396	404
				510	509	531	534	539
				634	634	724	663	657
				909	725	812	742	737
				1203	1130	1269	1230	1205
				1317	1240	1469	1360	1320
				1507	1458	1646	1588	1565

[a]D. S. Bethune, G. Meijer, W. C. Tang, and H. J. Rosen, *Chem. Phys. Lett.* **174**, 219 (1990); D. S. Bethune, G. Meijer, W. C. Tang, H. J. Rosen, W. G. Golden, H. Seki, C. A. Brown, and M. S. de Vries, *Chem. Phys. Lett.* **179**, 181 (1991).
[b]G. B. Adams, J. B. Page, O. F. Sankey, K. Sinha, J. Menendez, and D. R. Huffman, *Phys. Rev. B* **44**, 4052 (1991).
[c]V. P. Antropov, O. Gunnarsson, and A. I. Liechtenstein, *Phys. Rev. B* **48**, 7651 (1993).
[d]R. Jones, C. D. Latham, M. I. Heggie, V. J. B. Torres, S. Öberg, and S. K. Estreicher, *Phil. Mag. Lett.* **65**, 291 (1992).
[e]J. Kohanoff, W. Andreoni, M. Parrinello, and E. Clementi, *Phys. Rev. B* **46**, 4371 (1992).
[f]F. Negri, G. Orlandi and F. Zerbetto, *Chem. Phys. Lett.* **144**, 31 (1988).
[g]X. Q. Wang, C. Z. Wang, and K. M. Ho, *Phys. Rev. B* **48**, 1884 (1993).
[h]A. A. Quong, M. R. Pederson, and J. L. Feldman, *Solid State Commun.* **87**, 535 (1993).

lie in the $10-20\,\mathrm{cm}^{-1}$ range, and the acoustic-optic modes extend up to $\sim 45\,\mathrm{cm}^{-1}$. The uncoupled vibrations are shown in Fig. 8(b), where no libration is allowed, and Fig. 8(c), where only librations are allowed. The reported measurements are similar, at $2.7\,\mathrm{meV} \approx 22\,\mathrm{cm}^{-1}$ by

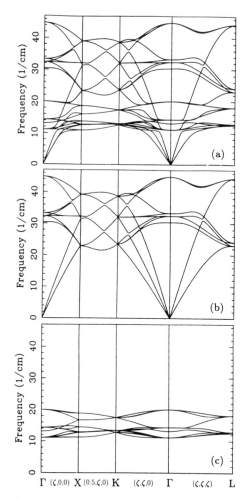

FIG. 8. Calculated dispersion curves for a C_{60} solid of rigid molecules. The structure is simple cubic Pa3 with four molecules per cell. The mixed curves (a) result from the interaction of the purely vibrational curves (b) and the purely librational curves (c). From X.-P. Li, J. P. Lu, and R. M. Martin, *Phys. Rev. B* **46,** 4301 (1992).

Neumann *et al.*[127-129] for the librational modes and up to 6 meV $\approx 50\,\text{cm}^{-1}$ by Pintchovius *et al.*[130] for the intermolecular translational modes.

7. FULLERIDES A_nC_{60}

Being molecular crystals, the fullerides (like fullerite) exhibit vibrational bands that can be related directly to those of the isolated molecule. Changes, relative to fullerite, that occur upon forming the ionic fulleride compounds give direct evidence of the strength of coupling of each mode to the conduction electrons that is so important in understanding the transport properties and superconductivity of the fullerides.

Differences in the lattice dynamics between the fullerides and fullerite can arise from a number of factors besides direct coupling to conduction (Fermi surface) electrons, however. An important factor is the charge transfer from the metal to the molecule, which will alter the bond-stretching and bond-bending force constants. The charge transfer can also change the structure, both by Jahn–Teller distortions (discussed in the following section) and by relaxation of the molecule. The introduction of metal ions also changes the crystal structure and symmetry and gives rise to new optic modes as well as altering the acoustic modes and librations. The charge transfer also introduces ionic forces and the possibility of LO–TO splittings in the intermolecular optic modes, which will be screened out if metallic screening occurs. The additional charge in the molecules also effects, sometimes drastically, the optical activity of the modes. Finally, the metallic fullerides have different screening properties than the nonmetallic ones. Thus, interpreting the differences between the vibrational spectra of fullerite and the fullerides may not be a straightforward matter.

Inelastic neutron scattering measurements of K_3C_{60} and Rb_6C_{60} were performed by Prassides *et al.*[122,128] They compare their vibrational density of states of K_3C_{60} directly with that of C_{60}, and both have sharp, relatively narrow peaks up to ~ 100 meV and rather broader features up to ~ 200 meV. They note strong changes in the 170–200 meV range[122] of the two highest H_g [$H_g(7)$ and $H_g(8)$] modes; however, other modes no doubt occur in this range as well. Changes in the 10–100 meV range can be seen in Fig. 5. In K_3C_{60} the $H_g(1)$ mode at 32 meV broadens, the $H_g(2)$ peak "disappears," either by broadening precipitously or by shifting downward by ~ 3–4 meV under another peak, and a peak at 67 meV (perhaps T_{1u}) "disappears." The resolution in the 75–100 meV range may not be sufficient to identify the changes in this region. The Rb_3C_{60} data are very similar to those K_3C_{60} (Fig. 5).

The librational modes of K_3C_{60}, which is metallic, and Rb_6C_{60}, an ionic insulator, were measured by Christides et al.[150] using inelastic neutron scattering. They reported a spectral peak for K_3C_{60} at 4.0 meV at 12 K that is softer by about 10% at room temperature. In Rb_6C_{60} the peak is at 6.1 meV at 150 K, decreasing to 5.6 meV at 350 K. (In fullerite, the librational peak[127-129] is at 1.7 meV.)

Optical spectroscopies, by detecting only certain optically allowed modes, provide fewer (especially in the metallic phases) but less ambiguous results. Wang et al.[151] have compared the Raman spectra of A_6C_{60}, $A = K$, Rb, and Cs, with that of fullerite. The Raman active modes are insensitive to the A atom, and therefore to the molecular volume. The high frequency, "tangential" modes are softer by ~60 cm^{-1} in the fullerides, which they attribute to charge-transfer-induced elongation of intramolecular bond lengths and softening of force constants. The breathing $A_g(1)$ mode, conversely, hardens by ~7 cm^{-1}, an effect that Jishi and Dresselhaus[152] suggest to be a consequence of the curvature of the molecule.

The Raman studies were extended[151,153,154] to the metallic K, Rb, and Cs fullerides as well. In the metals, where the penetration depth is reduced substantially, they observed only the two A_g modes and the lowest H_g mode. The pentagonal pinch $A_g(2)$ mode is softer by ~13 cm^{-1} in the fulleride; the width is 6 cm^{-1} narrower, probably due simply to the lack of orientational order in the fullerite rather than as any reflection of electron–phonon coupling in the fulleride. They propose a linear shifting of the pentagonal pinch $A_g(2)$ phonon downward by 5 cm^{-1} per electron in these fullerides.

Fu et al.[155] reported studies of infrared active modes in the A_6C_{60} compounds. They identified the high frequency $T_{1u}(4)$ mode at 1342 cm^{-1}

[150]C. Christides, D. A. Neumann, K. Prassides, J. R. D. Copley, J. J. Rush, M. J. Rosseinsky, D. W. Murphy, and R. C. Haddon, Phys. Rev. B **46**, 12088 (1992).

[151]K.-A. Wang, Y. Wang, F. Zhou, J. M. Holden, S.-L. Ren, G. T. Hager, H. F. Ni, P. C. Eklund, G. Dresselhaus, and M. S. Dresselhaus, Phys. Rev. B **45**, 1955 (1992).

[152]R. A. Jishi and M. S. Dresselhaus, Phys. Rev. B **45**, 6914 (1992).

[153]S. J. Duclos, R. C. Haddon, S. Glarum, A. F. Hebard, and K. B. Lyons, Science **254**, 1626 (1991).

[154]P. Zhou, K.-A. Wang, Y. Wang, P. C. Eklund, M. S. Dresselhaus, G. Dresselhaus, and R. A. Jishi, Phys. Rev. B **46**, 2595 (1992).

[155]K.-J. Fu, W. L. Karney, O. L. Chapman, S.-M. Huang, R. B. Kaner, F. Diederich, K. Holczer, and R. L. Whetten, Phys. Rev. B **46**, 1937 (1992).

in K_6C_{60}, a downward shift of $87\,cm^{-1}$ that translates into $14\,cm^{-1}$ per extra electron. The $T_{1u}(1)$ and $T_{1u}(2)$ modes at 526 and $576\,cm^{-1}$ in fullerite are 59 and $11\,cm^{-1}$ lower, respectively, in A_6C_{60}, while the $T_{1u}(3)$ mode at $1182\,cm^{-1}$ is unchanged. Fu *et al.* also report exceptionally large enhancements of the optical strength of the $T_{1u}(2)$ and $T_{1u}(4)$, by factors of 33 and 88, respectively. Pichler *et al.*[156] obtained similar frequency shifts, but enhancement factors of 70 and 20, respectively. These enhancements appear to be sensitive to sample condition but clearly indicate a strong electronic polarization contribution to these two vibrations.

Pichler *et al.*[156,157] have reported transmittance and reflectance measurements on K_3C_{60}, for which the $T_{1u}(1)$ mode is unobserved, the $T_{1u}(3)$ mode is unshifted and unenhanced, and the $T_{1u}(2,4)$ modes are shifted (by -6 and $-66\,cm^{-1}$, respectively) and enhanced (factors of 10 and 5, respectively). More unexpectedly, a number of other structures are observed that they relate to Raman modes of the C_{60} molecule. They fit their data with (1) a Drude intraband contribution ($\hbar\omega_p = 1500\,cm^{-1} = 0.19\,eV$, $\hbar/\tau = 400\,cm^{-1}$); (2) two electronic transitions within the t_{1u}-derived band complex; (3) three T_{1u} infrared active modes (one was neglected due to its apparent very weak activity), with the upper one split by crystal fields; and (4) Fano resonances with the six highest H_g modes and the high frequency A_g mode. They ascribe the appearance of the Raman active modes to a "charged phonon" effect, arising from a coupling of these modes to electronic excitations within the t_{1u} band complex. This "charged phonon" model is described in Section VII. This interpretation of the data requires extremely large softenings (up to $500\,cm^{-1}$) and broadenings (up to $140\,cm^{-1}$) of some of the H_g phonons that will have to be reconciled with the inelastic neutron scattering data.

An infrared transmission study of the K_xC_{60} and Rb_xC_{60} materials for $0 \le x \le 6$ was carried out by Martin *et al.*[157a] They confirm the occurrence of the distinct phases $x = 0$, 3, 4, and 6, and verify the strong enhancements and frequency shifts seen by Fu *et al.*[155] for $x = 6$. They reported quantitative fits to the data, with some of the notable findings being as follows:

(1)   The $T_{1u}(1)$ mode is discernible only in the $x = 0$ ($526\,cm^{-1}$) and $x = 6$ ($486\,cm^{-1}$) phases.

[156]T. Pichler, J. Kürti, and H. Kuzmany, *Cond. Matter and Mat. Commun.* (1993, in press).
[157]T. Pichler, M. Matus, and H. Kuzmany, *Solid State Commun.* **86**, 221 (1993).
[157a]M. C. Martin, D. Koller, and L. Mihaly, *Phys. Rev. B* **47**, 14607 (1993).

(2) The $T_{1u}(2)$ mode grows in strength with doping, with frequencies {576, 573, 570, 565} cm^{-1} for $x = \{0, 3, 4, 6\}$.

(3) The $T_{1u}(3)$ mode is discernible only in the $x = 0$ and $x = 6$ phases, and is unshifted at 1182 cm^{-1}.

(4) The $T_{1u}(4)$ mode shifts downward, $\omega = \{1428, 1393, 1363, 1340\}$ cm^{-1} for $x = \{0, 3, 4, 6\}$, with an 80-fold increase in oscillator strength and a width of 7.2 cm^{-1} for $x = 6$.

(5) The $T_{1u}(5)$ mode has a width of 21 cm^{-1} in the metallic $x = 3$ phase, interpreted in terms of strong electron–phonon coupling ($\lambda \sim 0.8$ using the normalization of Eq. (21.1) in Section VII). Notably, the (apparently) insulating $x = 4$ phase shows a similar linewidth for this mode.

(6) The spectra are almost identical for the K and Rb compounds.

(7) Eleven additional very weak modes become visible in the $x = 6$ phase, with many of the frequencies similar to those of the Raman active modes.

A different kind of study was undertaken by Mitch et al.[158,159] They examined ultrathin films, only a few monolayers thick, using interference-enhanced Raman scattering. They observed continuous frequency shifts of some lines, indicating lack of stoichiometric compound formation in these films. They also observed drastic broadenings of lines, especially the $H_g(8)$ mode, that have been invoked as indication of superstrong electron–phonon coupling. Due to the complexity of the surface region and the fact that the environment of the molecule is very different from that in the bulk, however, the line broadening may arise partially from inhomogeneity.

8. THEORY OF JAHN–TELLER DISTORTIONS

A Jahn–Teller (JT) distortion of the C_{60} molecule is expected if the six fold LUMO is partially occupied or if the HOMO has one or more holes.

[158]M. G. Mitch, S. J. Chase, and J. S. Lannin, Phys. Rev. Lett. **68**, 883 (1992).
[159]M. G. Mitch, S. J. Chase, and J. S. Lannin, Phys. Rev. B. **46**, 3696 (1992).

For several reasons (to be discussed) it is important to understand the character of, and energy gain due to, this JT distortion. Since the C_{60} LUMO, which will be the C_{60}^{n-} HOMO, is extended over the entire molecule, the distortion is not expected to be large. The resulting distortion has often been referred to loosely as a "polaron," a term that might better be reserved for any such associated charged excitations in the solid. Whether they survive in the solid is uncertain, since the electronic bandwidth is more than an order of magnitude larger than the JT distortion energy (to be discussed), and therefore the loss of kinetic energy in forming a polaron in the solid may dominate the gain in structural energy. This question has not been addressed in any detail for C_{60} compounds.

The JT theorem provides some general guidelines about instabilities of the charged C_{60} molecule, which have been discussed by Varma et al.[160] and Schlüter et al.[161-163] The more general question of Jahn–Teller splitting of molecules with icosahedral symmetry has been presented in some detail by Ceulemans and Fowler.[164,165] Since the added electron on C_{60} goes into one of the three t_{1u} orbitals that transform like a vector (x, y, z), group theory limits the linear coupling to molecular distortions of A_g, T_{1g} and H_g symmetry, which satisfy, in I_h symmetry,

$$t_{1u} \times t_{1u} = A_g + T_{1g} + H_g. \qquad (8.1)$$

Because the T_{1g} mode is asymmetric, there is no linear coupling to it. For A_g and H_g symmetry there are 2 and 8 distinct modes, respectively, of the C_{60} molecule (of unit and fivefold degeneracy, respectively). The A_g modes are fully symmetric, so coupling to them, although linear, will not lower the symmetry, and therefore they are not Jahn–Teller modes per se.

The linear coupling within the t_{1u} complex (labeled by indices i, j) is of the form

$$H_{ij} = E_0 \delta_{ij} + \sum_{S=A_g,H_g} \sum_{v=1}^{M_S} \sum_{\alpha=1}^{D_S} v_{ij}^{v\alpha}(S) Q_{v\alpha}(S) + H_{vib}, \qquad (8.2)$$

[160]C. M. Varma, J. Zaanen, and K. Raghavachari, Science **254**, 989 (1991).
[161]M. Schlüter, M. Lannoo, M. Needels, G. A. Baraff, and D. Tomanek, Phys. Rev. Lett. **68**, 526 (1992).
[162]M. Lannoo, G. A. Baraff, M. Schlüter, and D. Tomanek, Phys. Rev. B **44**, 12106 (1991).
[163]M. Schlüter, M. Lannoo, M. Needels, G. A. Baraff, and D. Tomanek, J. Phys. Chem. Solids **53**, 1473 (1992).
[164]A. Ceulemans and P. W. Fowler, Phys. Rev. A **39**, 481 (1989).
[165]A. Ceulemans and P. W. Fowler, J. Chem. Phys. **93**, 1221 (1990).

where S is the irreducible representation index, M_S is the number of eigenvalues of symmetry S, each with degeneracy D_S. The coupling coefficients are denoted by v, and Q is the normal coordinate. Within linear coupling, each mode v can be considered independently.

The simplest effect is from the coupling to A_g symmetry. Adding an electron affects both the spherical molecular breathing (br) mode and the pentagonal pinch (pp) mode. As mentioned in Section II, Pederson and Quong[81] note that occupation of the π^*t_{1u} LUMOs adds more charge to the b_{hh} bonds, causing them to stretch and weaken. Compared with the neutral molecule, double charging reduces the calculated ratio of b_{hh} to b_{ph} bond lengths from 1.037 to 1.029, which reflects the coupling to the pentagonal pinch A_g mode. The change in A_g frequencies was calculated to be nonlinear in charge, with $\omega_{br}(A_g) = \{550, 460, 458\}$ cm^{-1} and $\omega_{pp}(A_g) = \{1500, 1458, 1469\}$ cm^{-1} for charge $Q = \{0, -1, -2\}$, respectively. Experimental data indicates a linear shift in Raman frequencies with charge state, however (see the previous subsection).

The JT coupling to the vth mode of H_g symmetry is given by the coupling matrix[166]

$$\sum_\alpha v^{v\alpha} Q_{v\alpha} = \tfrac{1}{2} g_v \begin{pmatrix} Q_{v5} - \sqrt{3}\,Q_{v4} & -\sqrt{3}\,Q_{v1} & -\sqrt{3}\,Q_{v2} \\ -\sqrt{3}\,Q_{v1} & Q_{v5} + \sqrt{3}\,Q_{v4} & -\sqrt{3}\,Q_{v3} \\ -\sqrt{3}\,Q_{v2} & -\sqrt{3}\,Q_{v3} & 2Q_{v5} \end{pmatrix}, \quad (8.3)$$

where g_v is the energy per unit displacement of the vth mode. Since there are eight H_g modes in C_{60}, the task of obtaining the JT distortion of the ground state is far from simple.

Supposing that the distortion breaks symmetry in a minimal way, de Coulon et al.[101] have investigated within LDA distortions of D_{5d} symmetry (the largest subgroup of I_h). Within D_{5d} symmetry the LUMOs split according to $t_{1g} \rightarrow a_{2u} \oplus e_{1u}$. Varying the 10 structural parameters within D_{5d} symmetry amounts to probing one member of each of the eight H_g and two A_g modes. de Coulon et al. relaxed the energy with respect to two of these, the bond length around an equator perpendicular to a fivefold axis, and elongation of the molecule along this axis. They found an JT energy gain of 24 meV for this section of the energy surface (which of course is a lower bound on the true LDA value). Scherrer and Stollhoff[167] found that electronic correlations evaluated within the ab

[166]M. C. O'Brien, J. Phys. C 4, 2524 (1971).
[167]H. Scherrer and G. Stollhoff, Phys. Rev. B 47, 16570 (1993).

initio local ansatz enhances the coupling of electrons to the modes involving bond alternation, such as the pentagonal pinch mode, in C_{60}^{-n} molecules ($1 \le n \le 5$).

Friedman[168] and Harigaya[169,170] investigated a Su–Schrieffer–Heeger (SSH) model[171] involving a single orbital on each of the sites of the C_{60} molecule. The SSH model contains nearest-neighbor hopping t modulated by an "electron–phonon" coupling constant α, and a bond stretching/compressing spring constant κ, and the lattice was treated classically (adiabatic approximation). Without any restriction on the symmetry of the result, they obtain a JT distortion (sometimes referred to loosely as a polaron) at least qualitatively similar to that of de Coulon *et al.*: The primary structural change occurs around an equator perpendicular to a fivefold axis. The structural change was mostly a reduction in the degree of bond alternation (differing lengths of the b_{hh} and b_{ph} bonds) around this equator. Friedman found the energy gain to be $\sim 0.01t$; with $t \approx 2$ eV, this energy gain ~ 0.02 eV is consistent with de Coulon *et al.* Adding a second electron reinforced this distortion and resulted in an energy gain of $\sim 0.04t$. Friedman also found other metastable solutions $\sim 0.0002t$ (~ 5 K) higher in energy than this "string bipolaron." The SSH model has been applied by You *et al.*[171a] to calculate all the vibrational frequencies and to study electron–phonon coupling on the C_{60} molecule.

Hayden and Mele[172] earlier had obtained some of these results with a SSH-like model. They examined the cases of one added electron; one added hole; and an electron–hole pair. The distortion they obtained for one added electron was similar in character to that obtained by de Coulon *et al.* and Friedman, but the energy gain due to distortion (2.6 meV) was almost an order of magnitude less. The distortional energy for one hole was 7.7 meV. The relaxation in the case of the excited electron–hole pair was qualitatively similar to that for the single hole, but with a much larger energy gain, of 24 meV. The energy differences between Friedman's results and those of Hayden and Mele are apparently due to the differences in model parameters.

Harigaya[170] reported results of the SSH mode for the distortion for a wide range of C_{60}^n charge states, $-10 \le n \le +6$. The model was also used to study the changes in the electronic spectrum. With one or two extra

[168]B. Friedman, *Phys. Rev. B* **45**, 1454 (1992).
[169]K. Harigaya, *Phys. Rev. B* **45**, 13676 (1992).
[170]K. Harigaya, "Polaron excitations in fullerenes: Theory as π-conjugated systems." *Prog. Th. Phys. Suppl.* (in press).
[171]W. P. Su, J. R. Schrieffer, and A. J. Heeger, *Phys. Rev. Lett.* **42**, 1698 (1979).
[171a]W. M. You, C. L. Wang, F. C. Zhang, and Z. B. Su, *Phys. Rev. B* **47**, 4765 (1993).
[172]G. W. Hayden and E. J. Mele, *Phys. Rev. B* **36**, 5010 (1987).

electrons on the molecule, the t_{1u} LUMO splits into one- and twofold levels and the lowest state is singly degenerate. For three extra electrons, there are three distinct levels, with the total splitting ~0.3 eV. The effects of the Jahn–Teller distortion in the isolated molecule should be detectable by optical means. Because the LUMO splitting is only half of the bandwidth in the solid (and there are complications arising from crystal fields), the static JT distortion should not necessarily be expected to survive in the metal fullerides, and so far there is no structural evidence that it does.

The possibility of unusual dynamic behavior and strong electron–phonon coupling arising from the JT distortions has been suggested as a possible pairing mechanism by several groups. This work is described in Section 20.

IV. Crystal Structures

The crystal structures of fullerene solids cannot fail to be interesting, because of the contradictory symmetries of the C_{60} molecule (6 fivefold axes and 10 threefold axes) and of crystal lattices. The fivefold rotation axes of C_{60} and its 10 threefold axes are inconsistent with high symmetry three-dimensional space groups, suggesting that very low symmetry crystal phases could result. Since the C_{60} molecule has no fourfold axis, a cell with a C_{60} molecule at the origin can have a symmetry no higher than orthorhombic. This picture becomes even more complicated (or in a sense simplified) by the tendency of the nearly spherical C_{60} molecule to display merohedral (discrete orientational) disorder or complete (static or dynamic) orientational disorder.

9. Fullerite C_{60}

a. *Experimental Determinations*

Due to the availability of larger samples and the structural transitions that are observed as the temperature is lowered, more detailed structural studies have been carried out for fullerite than for the other solids. These transitions, which involve rotational dynamics and orientational ordering of the molecules, reflect directly the mutual interaction between the molecules in the solid that is one of the most basic aspects for solid state chemists and physicists to understand. Due to the low electron density of fullerenes, x-ray scattering should be less preferred to neutron scattering for elucidating structures. The presence of hydrogen in samples (remnants of the preparation procedure) is often at the 0.5–1 atomic percent level, however, which is roughly 1 H atom for every 2–4 C_{60} molecules. Because hydrogen is a very strong scatterer of neutrons, H contamination is a possible weakness of neutron scattering spectroscopy for these materials. This problem was recognized early on, however, and efforts to produce samples with low H content have been successful.

Here I will provide only an overview of some of the structural properties of C_{60} compounds. Considerably more extensive discussions of the structures and the procedures used in obtaining them have been given elsewhere by Prassides et al.,[173] Murphy et al.,[174] Heiney,[175] and Zhou and Cox.[176]

The fullerene solid C_{60} first produced by Krätschmer et al.[2] forms, when in the pure form, in the close-packed face-centered cubic (fcc) lattice[177] expected of a van der Waals collection of close-packed spherical molecules. At room temperature the molecules are dynamically orientationally disordered (i.e., they are rotating, or "constantly tumbling into different orientations"[129]), as shown by NMR measurements[178,179] and coherent quasi-elastic neutron scattering.[180] The (time and space average) symmetry is Fm3m (simple fcc) with lattice constant $a = 14.16 \pm 0.01$ Å at 300 K, and $a = 14.05$ Å at 110 K was reported by Liu et al.[181]

Lowering the temperature, a first-order transition to an orientationally

[173]K. Prassides, H. W. Kroto, R. Taylor, D. R. M. Walton, W. I. F. David, J. Tomlinson, R. C. Haddon, M. J. Rosseinsky, and D. W. Murphy, Carbon 30, 1277 (1992).

[174]D. W. Murphy, M. J. Rosseinsky, R. M. Fleming, R. Tycho, A. P. Ramirez, R. C. Haddon, T. Siegrist, G. Dabbagh, J. C. Tully, and R. E. Walstedt, J. Phys. Chem. Solids 53, 1321 (1992).

[175]P. A. Heiney, J. Phys. Chem. Solids 53, 1333 (1992).

[176]O. Zhou and D. E. Cox, J. Phys. Chem. Solids 53, 1373 (1992).

[177]J. E. Fischer, P. A. Heiney, A. R. McGhie, W. J. Romanow, A. M. Denenstein, J. P. McCauley, Jr., and A. B. Smith III, Science 252, 1288 (1991).

[178]C. S. Yannoni, R. D. Johnson, G. Meijer, D. S. Bethune, and J. R. Salem, J. Phys. Chem. 95, 9 (1991).

[179]R. Tycho, R. C. Haddon, G. Dabbagh, S. H. Glarum, D. C. Douglass, and A. M. Mujsce, J. Phys. Chem. 95, 518 (1991).

[180]D. A. Neumann, J. R. D. Copley, R. L. Cappelletti, W. A. Kamitakahara, R. M. Lindstrom, K. M. Creegan, D. M. Cox, W. J. Romanow, N. Coustel, J. P. McCauley, Jr., N. C. Maliszewskyj, J. E. Fischer, and A. B. Smith III, Phys. Rev. Lett. 67, 3808 (1991).

[181]S. Liu, Y.-J. Lu, M. M. Kappes, and J. A. Ibers, Sciences 254, 408 (1991).

ordered phase occurs at $T_s \approx 250$ K. This low temperature phase is a four-molecule supercell of the fcc lattice with simple cubic translational symmetry and lattice constant[182] $a = 14.04$ Å at 11 K. The decrease in volume at the transition is ~1%. The space group is Pa$\bar{3}$, with neighboring molecules oriented[183] such that the electron-rich short b_{hh} bonds about the electron-poor pentagon centers of neighboring molecules. This configuration can be achieved for all 12 nearest neighbors simultaneously, accounting qualitatively for the high ordering temperature, and this was the structure initially proposed.[175,183] Figure 9 provides

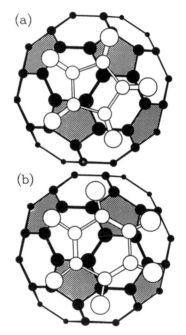

FIG. 9. Illustration of two favorable manners in which neighboring C_{60} molecules can be oriented. The view is along the axis joining their centers, and only 10–12 atoms of the nearer molecule are shown (white balls and sticks). In (a) a b_{hh} bond faces a pentagon on the nearer molecule, while in (b) the b_{hh} bond faces a hexagon on the nearer molecule. From J. R. D. Copley, D. A. Neumann, R. L. Cappelletti, and W. A. Kamitakahara, *J. Phys. Chem. Solids* **53**, 1353 (1992). Reprinted from *J. Phys. Chem. Solids,* Copyright 1992, with permission from Pergamon Press Ltd., Headington Hill Hall, Oxford OX3 0BW, UK.

[182]P. A. Heiney, J. E. Fischer, A. R. McGhie, W. J. Romanow, A. M. Denenstein, J. P. McCauley, Jr., A. B. SMith III, and D. E. Cox, *Phys. Rev. Lett.* **66**, 2911 (1991).
[183]W. I. F. David, R. M. Ibberson, J. C. Matthewman, K. Prassides, T. J. S. Dennis, J. P. Hare, H. W. Kroto, R. Taylor, and D. R. M. Walton, *Nature* **353**, 147 (1991).

pictures of how the electron-rich bonding regions on a molecule can align so as to abut the electron-poor pentagonal faces or the larger hexagonal faces. More details of the various fullerite structures can be found in the overviews by Heiney[175] and by Copley et al.[129]

X-ray diffraction of single crystals was used by Chow et al.[184] to determine whether the C_{60} moleucles are freely rotating at room temperature, or rather show indication of orientational correlations. They found that the crystal average of the charge density of the C_{60} molecule in the cubic (Fm3m) structure is *not* spherical. This finding indicates that the molecule is not freely rotating, but that it performs rotations with preferential orientations. The crystal-average molecule charge density has minima in the $\langle 111 \rangle$ directions and maxima located near (but somewhat off) the $\langle 110 \rangle$ directions, which are the directions of the neighboring molecules.

The transition itself was studied more closely by Heiney et al.[185] Using neutron and x-ray methods, they found the transition to be strongly first order, with a discontinuity of 0.044 Å in the lattice constant. There is phase coexistence over a range of ~5 K but negligible hysteresis. They also established that the thermal expansion coefficient $\alpha = 6.2 \times 10^{-5} \, \text{K}^{-1}$ is the same above and below the transition.

Evidence of a second transition at $T_{s'} = 90 \, \text{K}$ was presented by Heiney et al.[185] and David et al.[186] Superstructures have been reported in some samples.[187] David et al. suggest a two-state model in which two possible orientations are possible at $T < T_s = 250 \, \text{K}$ for any molecule, with unequal probabilities p and $1 - p$ for the orientations. The diffuse scattering was modeled in this way, allowing the molecules to librate isotropically[188] with r.m.s. amplitude $\Delta \theta$. Copley et al.[129] used a Monte Carlo method to obtain a low temperature value of $P = 0.83, \Delta \theta = 1.9°$, and showed that good agreement with the data could not be obtained if the librations were ignored.

[184]P. C. Chow, X. Jiang, G. Reiter, P. Wochnew, S. C. Moss, J. D. Axe, J. C. Hanson, R. K. McMullan, R. L. Meng, and C. W. Chu, *Phys. Rev. Lett.* **69**, 2943 (1992).

[185]P. A. Heiney, G. B. M. Vaughan, J. E. Fischer, N. Coustel, D. E. Cox, J. R. D. Copley, D. A. Neumann, W. A. Kamitakahara, K. M. Creegan, D. M. Cox, J. P. McCauley, Jr., and A. B. Smith III, *Phys. Rev. B* **45**, 4544 (1992).

[186]W. I. F. David, R. M. Ibberson, T. J. S. Dennis, J. P. Hare, and K. Prassides, *Europhys. Lett.* **18**, 219 (1992).

[187]G. Van Tendeloo, S. Amelinckx, M. A. Verheijen, P. H. M. van Loosdrecht, and G. Meijer, *Phys. Rev. Lett.* **69**, 1065 (1992).

[188]D. A. Neumann, J. R. D. Copley, R. L. Cappelletti, W. A. Kamitakahara, R. M. Lindstrom, K. M. Creegan, D. M. Cox, W. J. Romanow, N. Coustel, J. P. McCauley, Jr., N. C. Maliszewskyj, J. E. Fischer, and A. B. Smith III, *Phys. Rev. Lett.* **67**, 3808 (1991).

The transition at ~90 K was confirmed using high resolution dilatometry by Gugenberger et al.[189] They measured a change in the thermal expansion coefficient by a factor of 2–3 at the transition, which showed some hysteresis. Their data was interpreted in terms of a glass transition model with an activation energy for reorientation of 290 meV.

From the diffraction data discussed, the low temperature structure of C_{60} appears to be well ordered with four molecules per primitive cell. Hu et al.[190] have used pulsed neutron scattering to measure the atomic pair distribution (PDF) function of C_{60} at 10 K. The PDF measures the distribution of interatomic distances, which in an ordered crystal consists of a series of discrete peaks. This study, and the PDF data of Soper et al.,[191] confirmed very directly that the C_{60} bond lengths are 1.400 and 1.445 Å, to about 0.005 Å accuracy. Hu et al. found that the structural model that best fit the data involves deviations of the orientations of the molecules locally from the long range average determined in diffraction experiments. Their model predicts that 30–40% of the molecules have the sixfold face oriented toward the adjacent molecule.

Experimental studies of the rotational dynamics have been reported by Copley et al.[129] and Neumann et al.[127,180] All diffuse scattering can be modeled as diffusive librational/rotational motion. At low temperature, the measured librational vibration energy is 2.75 meV and softens to 2 meV at T_s. The width is ~0.4 meV at low temperature and increases to ~2.5 meV at T_s. The behavior of these quantities with temperature is shown in Fig. 7, along with the diffuse scattering intensity. $\Delta\theta$ increases to more than 7° at T_s, indicating a maximum librational amplitude ~10°. Since the angle subtended by near neighbor carbon atoms is 23°, a Lindemann picture (at melting, the atomic vibration amplitude becomes a large fraction of the interatomic distance) applies and the transition can be thought of as "orientational melting."

b. *Theoretical Treatments*

The substantial experimental data and analysis on the structure and dynamics of fullerite implicitly provides detailed information on the interaction of C_{60} molecules in the solid phase. Orientational ordering of

[189]F. Gugenberger, R. Heid, C. Meingast, P. Adelmann, M. Braun, H. Wühl, M. Haluska, and H. Kuzmany, *Phys. Rev. Lett.* **69**, 3774 (1992).

[190]R. Hu, T. Egami, F. Li, and J. S. Lannin, *Phys. Rev. B* **45**, 9517 (1992).

[191]A. K. Soper, W. I. F. David, D. S. Sivia, T. J. S. Dennis, J. P. Hare, and K. Prassides, *J. Phys.: Cond. Mat.* **4**, 6087 (1992).

the C_{60} molecules and the Fm3m→Pa$\bar{3}$ structural transition at T_s have been modeled in various ways. Molecular dynamics using classical potentials has been applied, with the potentials being adjusted by Cheng and Klein[192] and Sprik *et al.*[193] to produce the structural transition near 250 K. The basis for such a treatment is the closed-shell nature of C_{60} and the observation that interatomic distances between neighboring molecules are greater than 3 Å, much larger than the covalent bond length, and therefore the interaction is primarily van der Walls in nature. Sprik *et al.* introduced bond charges of $-0.35e$ and charged $(0.175e)$ carbon atoms to reproduce the structural transition and associated dynamical behavior. This model has been generalized by Lu *et al.*[194] to allow charge transfer among the bonds within a molecule in response to charges on a neighboring molecule. Their model predicts not only the transition temperature, but also its pressure dependence. Samara *et al.*[195] measured an increase of T_s with pressure of 10.4 K/kbar [equivalently, $d(\log T_s)/d \log(\text{volume}) = -7.5$] and a strong reduction of fluctuations with pressure. The model of Lu *et al.* predicts that below T_s there are many nearly degenerate orientations for each C_{60} separated by potential barriers of ~0.3 eV ~ 350 K. This is not far from the estimate of the rotational barrier of ~600 K by Chow *et al.*[184]

The Coulomb multipole moments of C_{60} were calculated directly from the LDA charge density by Yildrim *et al.*[195a] The $l = 6$ and $l = 10$ moments dominate, but many of the low order moments are about an order of magnitude smaller than given by the point charge models already discussed. They find the Coulomb energies to lead to two local configurations that are nearly degenerate: Pa3 space group with setting angles of 23° and 93°. They conclude that short range quantum mechanical repulsions are required to understand the observed ordering.

Ordering has been treated in a microscopic way by Gunnarsson *et al.*,[196] and in a more general, group theoretical way by Michel *et al.*[197,198] Gunnarsson *et al.* applied a tight-binding model of the energy, based on

[192]A. Cheng and M. L. Klein, *Phys. Rev. B* **45**, 1889 (1992).
[193]M. Sprik, A. Cheng, and M. L. Klein, *J. Phys. Chem.* **96**, 2027 (1992).
[194]J. P. Lu, X.-P. Li, and R. M. Martin, *Phys. Rev. Lett.* **68**, 1551 (1992).
[195]G. A. Samara, J. E. Schriber, B. Morosin, L. V. Hansen, D. Loy, and A. P. Sylvester, *Phys. Rev. Lett.* **67**, 3136 (1991).
[195a]T. Yildirim, A. B. Harris, S. C. Erwin, and M. R. Pederson, *Phys. Rev. B* **48**, 1888 (1993).
[196]O. Gunnarsson, S. Satpathy, O. Jepsen, and O. K. Andersen, *Phys. Rev. Lett.* **67**, 3002 (1991).
[197]K. H. Michel, *Z. Physik B* **88**, 71 (1992).
[198]K. H. Michel, J. R. D. Copley, and D. A. Neumann, *Phys. Rev. Lett.* **68**, 2929 (1992).

LDA calculations, to obtain energy differences between crystal structures involving different orientations of the molecules. They find that the chemical bonding between neighboring molecules, although rather weak, is an important factor in determining both the cohesive energy and the structural energy differences. They also raise the interesting possibility that the geometric centers of all C_{60} molecules may not lie exactly on the (average) lattice sites assumed in the structural refinements. Michel *et al.* derived a theory for the orientational dynamics of fullerite by expanding the molecular interaction potential and the crystal field in symmetry-adapted functions. These functions contained large $l = 6$ and $l = 10$ terms, with the molecular form factor leading to a dominance of the $l = 10$ term. Their analysis indicates that a T_{2g} symmetry multipolar mode drives the transition to the low temperature $Pa\bar{3}$ phase, and the transition is first order.

Mazin *et al.*[199] extended the procedure of Gunnarsson *et al.*, in large part by showing that the orientational energy is almost completely determined by the relative orientation of each pair of molecules, and therefore can be mapped onto a three-dimensional antiferromagnetic Ising model (AFIM), with interaction $J \sim 100$ K. The AFIM on an fcc lattice describes a frustrated "antiferromagnet," whose ground state[200] consists of antiferromagnetically ordered (001) layers. The energy is independent of how these layers are stacked, since for either of the two choices of arranging one such layer with respect to its neighboring layer each site has two ferromagnetic couplings and two antiferromagnetic couplings. The resulting ground state is infinitely degenerate with energy $-2J$ per molecule. Their prediction, then, is that the fullerite ground state will be bidirectionally ordered in-plane and disordered perpendicular to this plane.

Mazin *et al.* note several predictions that follow. The energy of a domain edge is $8J$ per lattice constant, and that of a point defect is $8J$. The number of point defects grows as $\exp(-8J/k_B T)$, and the in-plane order is lost at $T_N \approx 1.76J \approx 200$ K. They show that this picture is consistent with the diffraction data on crystalline C_{60}. A difference between fullerite and the AFIM is that there is a potential barrier B (they estimate $B \approx 4800$ K from NMR data) to rotation (flipping of "spins"), so that the orientational lifetime is $\tau_{or} \sim (1/\nu) \exp(B/k_B T)$ with

[199]I. I. Mazin, A. I. Liechtenstein, O. Gunnarsson, O. K. Andersen, V. P. Antropov, and S. E. Burkov, *Phys. Rev. Lett.* **70**, 4142 (1993).
[200]A. Danielian, *Phys. Rev. A* **133**, 1344 (1964).

$v \approx 10^{12} \sec^{-1}$ the frequency of rotational phonons. At the ordering transition, $\tau_{or}(T_N) \approx 1$ sec.

10. ALKALI FULLERIDES

Introduction of alkali metals (often referred to as doping or intercalation) leads to an equilibrium phase diagram that generally includes the line compounds A_nC_{60}, $n = 3$, 4, and 6. The compounds are respectively metallic and superconducing ($n = 3$), nonsuperconducting and (apparently) nonmetallic ($n = 4$), and insulating $n = 6$). In A_xC_{60} near $x \sim 0$ or $x \sim 6$ there is a small region of solid solution. A phase diagram proposed by Weaver et al.[201] is shown in Fig. 10. The alkaline atom and C$_{60}$ molecule positions in these fullerene alkalide compounds discussed here are illustrated in Fig. 11. (At somewhat elevated temperature an $x = 1$ phase, with only octahedral interstitial sites occupied, has been detected.[201a,201b])

The compound K$_3$C$_{60}$ was prepared by Hebard et al.[29] and found to be superconducting at $T_c = 18$ K (later improved by several groups to be

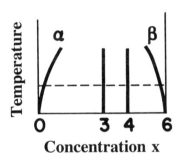

FIG. 10. The current understanding of the phase diagram of K$_x$C$_{60}$ (and Rb$_x$C$_{60}$), which contains small regions of solid solutions near $x \sim 0$ and $x \sim 6$ as well as the compounds at $x = 0$, 3, 4, and 6. Adapted from J. H. Weaver, P. J. Benning, F. Stepniak, and D. M. Poirier, *J. Phys. Chem. Solids* **53**, 1707 (1992). Reprinted from *J. Phys. Chem. Solids*, Copyright 1992, with permission from Pergamon Press Ltd., Headington Hill Hall, Oxford OX3 0BW, UK.

[201]J. W. Weaver, P. J. Benning, F. Stepniak, and D. M. Poirier, *J. Phys. Chem. Solids* **53**, 1707 (1992).

[201a]D. M. Poirier, T. R. Ohno, G. H. Kroll, P. J. Benning, F. Stepniak, J. H. Weaver, L. P. F. Chibante, and R. E. Smalley, *Phys. Rev. B* **47**, 9870 (1993).

[201b]D. M. Poirier and J. H. Weaver, *Phys. Rev. B* **47**, 10959 (1993).

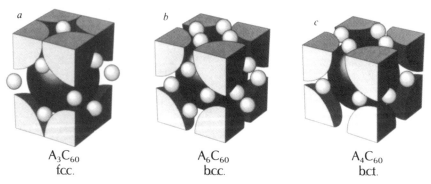

A_3C_{60} A_6C_{60} A_4C_{60}
fcc.   bcc.   bct.

FIG. 11. Crystal structure of A_nC_{60} compounds, $n = 3, 4, 6$. Large spheres represent the C_{60} molecules, and small spheres are the A cations. The fcc cell for A_3C_{60} has been depicted as an equivalent bct cell to emphasize the similarity to the other structures. C_{60} is like the $n = 6$ compound with the cations missing. From R. M. Fleming, M. J. Rosseinsky, A. P. Ramirez, D. W. Murphy, J. C. Tully, R. C. Haddon, T. Siegrist, T. Tycho, S. H. Glarum, P. Marsh, G. Dabbagh, S. M. Zahurak, A. V. Makhija, and C. Hampton, *Nature* **352,** 701 (1991). Reprinted with permission from *Nature*, Copyright 1991, MacMillan Magazines Limited.

somewhat about 19 K). X-ray diffraction data from K_3C_{60} at room temperature was fit by Stephens *et al.*[202] to a structure comprising an fcc array of C_{60} molecules ($a = 14.24$ Å) with both tetrahedral and octahedral interstices occupied by K ions. The C_{60} molecules take two orientations, apparently randomly, in such a way that cubic symmetry is restored to the solid. The result is the cryolite structure, space group Fm3m, a common structure for ionic solids. The K ions show large thermal factors in the structural refinements. This fcc structure seems to be common for the other alkali fullerides A_3C_{60} that have been made.

Information on the crystal structure at lower temperature was obtained by Barrett and Tycho[203] from NMR data. Below 220 K, the NMR linewidth displayed structure indicating a deviation from the behavior at higher temperature, which was motionally narrowed. A model fit to their data involved extensive molecular reorientation occurring as the result of small rotational jumps, with an activation energy of 460 ± 60 meV. This picture involves considerably more disorder than the high temperature structure of Stephens *et al.*[202]

[202]P. W. Stephens, L. Mihaly, P. L. Lee, R. L. Whetten, S.-M. Huang, R. Kaner, F. Deiderich, and K. Holczer, *Nature* **351,** 632 (1991).
[203]S. E. Barrett and R. Tycho, *Phys. Rev. Lett.* **69,** 3754 (1992).

The structure of K_6C_{60} and Cs_6C_{60} was elucidated by Zhou et al.[204] At room temperature, these compounds form a body-centered cubic (bcc) array of C_{60} molecules ($a = 11.39$ Å for K, 11.79 Å for Cs), with alkali atoms located at positions on the (unit cell) cube faces $(0, \frac{1}{2}, 0.28)$ and permutations allowed by the space group Im3. The lattice structure is shown in Fig. 11(b). The alkali atoms reside at distorted tetrahedral environments and form rhombic arrangements on the unit cube faces. In this crystal, the C_{60} molecules are fully oriented at room temperature and below, with two fold axes directed along the cubic axes, resulting in the lower-than-cubic symmetry.

The fullerides K_4C_{60} and Rb_4C_{60} were prepared by Fleming et al.,[205] who identified the structure as body-centered tetragonal (bct), with $a = 11.886$ Å, $c = 10.774$ Å for K; and $a = 11.962$ Å, $c = 11.022$ Å for Rb. The lattice structure is shown in Fig. 11(c). The space group is I4/mmm. As already noted, a simple tetragonal structure is incompatable with the C_{60} symmetry unless it is orientationally disordered. The detailed orientational behavior is not yet known. NMR experiments suggests this compound is insulating, and no superconducting signal has been seen. Stephens et al.[206] have made and reported the structures of Rb_nC_{60} compounds for $n = 3$, 4, and 6. A number of alkali-based $A_xB_yC_{60}$ compounds with $x + y \sim 3$ have been reported, and most are superconducting.

In addition to the those already mentioned, some more complex compounds have been reported. Rosseinsky et al.[207] reported studies of the Na_xC_{60} system, finding compounds at $x = 2$, 3, and 6 that could all be indexed on fcc unit cells. The phase diagram seems to allow all x between 0 and 6, unlike the larger alkalis, where only line compounds occur. In Na_2C_{60} the Na atoms occupy primarily the (smaller) tetrahedral sites. The composition $x = 6$ is quite different from other A_6C_{60} compounds, since it has a Na_4 cluster situated on the large octahedral interstitial site. Since fully ionized Na ions would strongly repel each other at such close distances, the observation of Na_4 clusters raises the possibility of partial covalency and therefore lack of complete charge transfer to the C_{60}

[204]O. Zhou, J. E. Fischer, N. Coustel, S. Kycia, Q. Zhu, A. R. McGhie, W. J. Romanow, J. P. McCauley, Jr., A. B. Smith III and D. E. Cox, Nature 351, 462 (1991).
[205]R. M. Fleming, M. J. Rosseinsky, A. P. Ramirez, D. W. Murphy, J. C. Tully, R. C. Haddon, T. Siegrist, R. Tycho, S. H. Glarum, P. March, G. Dabbagh, S. M. Zahurak, A. V. Makhija, and C. Hampton, Nature 352, 701 (1991); erratum, ibid. 353, 868 (1991).
[206]P. W. Stephens, L. Mihaly, J. B. Wiley, S.-M. Huang, R. B. Kaner, F. Diedrich, R. L. Whetten, and K. Holczer, Phys. Rev. B 45, 543 (1992).
[207]M. J. Rosseinsky, D. W. Murphy, R. M. Fleming, R. Tycho, A. P. Ramirez, T. Siegrist, G. Dabbagh, and S. E. Barrett, Nature 356, 416 (1992).

molecules in this $x = 6$ phase as is the case for other alkali fullerides. The $x = 3$ phase, not corresponding to a line phase, may tend to dispropor-tionate into the $x = 2$ and $x = 6$ phases.

This line of study was extended by Yildrim et al.,[208] who prepared Na_xC_{60} samples up to near $x = 11$. The structure refinement led to occupation of each of the two tetrahedral sites by one Na atom and occupation of the octahedral site by clusters of up to nine Na atoms (which consisted of one Na inside a cube of eight Na atoms). Their results are consistent with a solid solution behavior for $3 < x < 11$. No supercon-ductivity was detected for any x by either group.

This last example gives a glimpse of the complications to come in the class of metal-fullerene compounds. Recently Zhou et al.[208a] have reported the reaction of Na_2CsC_{60} ($T_c = 10.5$ K) with ammonia to form $(NH_3)_4Na_2CsC_{60}$ with an increased T_c of 29.6 K. In Table VII a collection of many of the reported compounds by the end of 1992 are provided, along with their lattice structure, lattice constants, and reported value of T_c.

11. ALKALINE-EARTH FULLERIDES; OTHERS

The first superconducting compound not based on alkaline ions was Ca_5C_{60}, reported by Kortan et al.[209] to be superconducting at $T_c = 8$ K. This compound is found to have a cluster of three Ca ions in each octahedral site and one Ca in each of the two tetrahedral sites of an fcc arrangement of C_{60} molecules. An Fm3m space group was suggested.

The compound Ba_6C_{60} was prepared by Kortan et al.[210] and found to be superconducting[211] at $T_c = 7$ K. Unlike all the other superconducing fullerides, the C_{60} molecules in this compound lie on a bcc lattice. The space group was identified as Im3, the same as the insulating alkali-based A_6C_{60} compounds. There appears to be a substantial range of solid solution in the bcc phase, but superconductivity occurs only at or very near $n = 6$.

[208]T. Yildirim, O. Zhou, J. E. Fischer, N. Bykovetz, R. A. Strongin, M. A. Cichy, A. B. Smith III, C. L. Lin, and R. Jelinek, *Nature* **360**, 568 (1992).

[208a]O. Zhou, R. M. Fleming, D. W. Murphy, M. J. Rosseinsky, A. P. Ramirez, R. van Dover, and R. C. Haddon, *Nature* **362**, 433 (1993).

[209]A. R. Kortan, N. Kopylov, S. Glarum, E. M. Gyorgy, A. P. Ramirez, R. M. Fleming, F. A. Thiel, and R. C. Haddon, *Nature* **355**, 529 (1992).

[210]A. R. Kortan, N. Kopylov, and F. A. Thiel, *J. Phys. Chem. Solids* **53**, 1683 (1992).

[211]A. R. Kortan, N. Kopylov, S. Glarum, E. M. Gyorgy, A. P. Ramirez, R. M. Fleming, O. Zhou, F. A. Thiel, P. L., Trevor, and R. C. Haddon, *Nature* **360**, 566 (1992).

TABLE VII.   REPORTED STRUCTURES, LATTICE PARAMETERS (a AND
c) AND SUPERCONDUCTING TRANSITION TEMPERATURES (T_c) OF M_nC_{60} SOLIDS

COMPOUND[i]	LATTICE	a (Å)	c (Å)	T_c	Ref.
$Na_{11-x}C_{60}$	fcc	14.59	—	—	c
$RbCs_2C_{60}$	fcc	14.56		33	a
$(NH_3)_4Na_2CsC_{60}$	fcc	14.47		30	h
Rb_2CsC_{60}	fcc	14.43		31	a
Rb_3C_{60}	fcc	14.38		29	a
KRb_2C_{60}	fcc	14.32		27	a
K_2CsC_{60}	fcc	14.29		24	a
Na_6C_{60}	fcc	14.25		—	b
K_2RbC_{60}	fcc	14.24		23	a
K_3C_{60}	fcc	14.24		19	a
Na_3C_{60}	fcc	14.19		—	b
Na_2C_{60}	fcc	14.19		—	b
Na_2CsC_{60}	fcc	14.13		12	a
Li_2CsC_{60}	fcc	14.12		12	a
Na_2RbC_{60}	fcc	14.03		2.5	a
Ca_5C_{60}	fcc	14.01		8	d
C_{60}	fcc	14.16		—	a
Ba_3C_{60}	bcc	11.34		—	f
Ba_6C_{60}	bcc	11.17		7	e
Cs_4C_{60}	bct	12.06	11.44	—	g
Rb_4C_{60}	bct	11.96	11.02	—	g
K_4C_{60}	bct	11.89	10.77	—	g

[a] K. Tanigaki, I. Hirosawa, T. W. Ebbesen, J. Mizuki, Y. Shimakawa, Y. Kubo, J. S. Tsai, and S. Kuroshima, *Nature* **356,** 419 (1992).
[b] M. J. Rosseinsky, D. W. Murphy, R. Fleming, R. Tycho, A. P. Ramirez, T. Siegrist, G. Dabbagh, and S. E. Barrett, *Nature* **356,** 416 (1992).
[c] T. Yildirim, O. Zhou, J. E. Fischer, N. Bykovetz, R. A. Strongin, M. C. Cichy, A. B. Smith III, C. L. Lin, and R. Jelinek, *Nature* **360,** 568 (1992).
[d] A. R. Kortan, N. Kopylov, S. Glarum, E. M. Gyorgy, A. P. Ramirez, R. M. Fleming, F. A. Thiel, and R. C. Haddon, *Nature* **355,** 529 (1992).
[e] A. R. Kortan, N. Kopylov, S. Glarum, E. M. Gyorgy, A. P. Ramirez, R. M. Fleming, O. Zhou, F. A. Thiel, P. L. Trevor, and R. C. Haddon, *Nature* **360,** 566 (1992).
[f] A. R. Kortan, N. Kopylov, R. M. Fleming, O. Zhou, F. A. Thiel, R. C. Haddon, and K. M. Rabe, *Phys. Rev. B* **47,** 13070 (1993).
[g] R. M. Fleming, M. J. Rosseinsky, A. P. Ramirez, D. W. Murphy, J. C. Tully, R. C. Haddon, T. Siegrist, R. Tycho, S. H. Glarum, P. Marsh, G. Dabbagh, S. M. Zahurak, A. V. Makhija, and C. Hampton, *Nature* **352,** 701 (1991).
[h] O. Zhou, R. M. Fleming, D. W. Murphy, M. J. Rosseinsky, A. P. Ramirez, R. B. van Dover, and R. C. Haddon, *Nature* **362,** 433 (1993).
[i] The fcc structure materials are arranged with decreasing lattice constant; note that this is correlated very strongly with decreasing value of T_c.

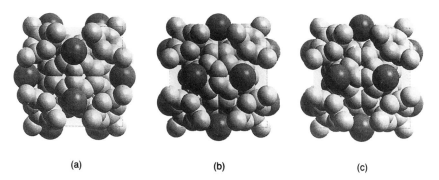

(a)   (b)   (c)

FIG. 12. Crystal structure of $A15$ structure Ba_3C_{60}, showing three possible orientations of the C_{60} molecule. From A. R. Kortan et al. Ref. 212.

A familiar superconducting crystal structure reappeared in the fullerides when Kortan et al.[212] reported the fabrication of Ba_3C_{60} in the $A15$ structure (viz. Nb_3Sn, V_3Si), with $a = 11.329$ Å. In the $A15$ structure of a compound A_3B, shown in Fig. 12, the B entities (C_{60} molecules in this case) form a bcc lattice, and the A atoms form three mutually orthogonal chains along the faces of the corresponding cube. (Recall that the $n = 3$ alkali fullerides form an fcc, rather than bcc, lattice of molecules.) The $Pm\bar{3}n$ space group that was favored by the structural refinement has the twofold axis of the body center molecule rotated by 90° with respect to the one at the origin. With this orientational order, there are two types of interstitial sites, one facing four hexagons on surrounding molecules and one facing four pentagons. The pentagon-surrounded interstices are fully occupied, while the others are empty.

The structure of the ferromagnetic fullerene compound TDAE-C_{60} discovered by Allemand et al.[30] was reported by Stephens et al.[31] TDAE is tetrakis(dimethylamino)ethylene, $C_2N_4(CH_3)_8$. The structure is a c-centered monoclinic cell ($a = 15.87$ Å, $b = 12.99$ Å, $c = 9.98$ Å), with space group C2 or possibly C2/m. This very low symmetry of the structure contrasts with the cubic symmetries of the superconducting phases; of course, the "intercalant" here is a low symmetry molecule. The fact that fullerene compounds can be ferromagnetic as well as superconducting is remarkable in itself; it is notable even more because the Curie temperature $T_m = 16.1$ K is the highest of any molecular organic ferromagnetic. The saturation moment at low temperature is 0.33 μ_B. The

[212]A. R. Kortan, N. Kopylov, R. M. Fleming, O. Zhou, F. A. Thiel, R. C. Haddon and K. M. Rabe, Phys. Rev. B **47**, 13070 (1993).

fact that the g-factor is relatively low, $g = 2.0008$, led to the suggestion that the magnetic moment lies primarily on the C_{60} molecule.

V. C_{60} Crystals: Local Density Calculations

The band structures of the various $A_n C_{60}$ solids have been calculated by a number of groups.[37,213–238] Considering the difficulties, particularly the large number of atoms (≥ 60) in the unit cell, the simplication (in several cases) of the actual C_{60} orientations, and the early stage of interpretation of the experimental data, the differences between the

[213] S. C. Erwin, in "Buckminsterfullerenes," eds. W. E. Billups and M. A. Ciufolini, VCH Publishers, New York, 1992.

[214] S. C. Erwin, M. R. Pederson, and W. E. Pickett, *Phys. Rev. B* **41**, 10437 (1990).

[215] S. Saito and A. Oshiyama, *Phys. Rev. Lett.* **66**, 2637 (1991).

[216] C. Pan, M. P. Sampson, Y. Chai, R. Hauge and J. L. Margrave, *J. Phys. Chem.* **95**, 2944 (1991).

[217] S. J. Duclos, K. Brister, R. C. Haddon, A. R. Kortan, and F. A. Thiel, *Nature* **351**, 380 (1991).

[218] S. C. Erwin and W. E. Pickett, *Science* **254**, 842 (1991).

[219] S. C. Erwin and M. R. Pederson, *Phys. Rev. Lett.* **67**, 1610 (1991).

[220] S. Satpaty, V. P. Antropov, O. K. Andersen, O. Jepsen, O. Gunnarsson, and A. I. Liechtenstein, *Phys. Rev. B* **46**, 1773 (1992).

[221] M. J. Puska and R. M. Nieminen, *J. Phys.: Cond. Mat.* **4**, L149 (1992).

[222] S. Ishibashi, N. Terada, M. Tokumoto, N. Ninoshita, and H. Ihara, *J. Phys.: Cond. Mat.* **4**, L169 (1992).

[223] Y. Lou, X. Lu, G. H. Dai, W. Y. Ching, Y.-N. Xu, M.-Z. Huang, P. K. Tseng, Y. C. Jean, R. L. Meng, P. H. Hor, and C. W. Chu, *Phys. Rev. B* **46**, 2644 (1992).

[224] J. L. Martins and N. Troullier, *Phys. Rev. B* **46**, 1766 (1992).

[225] W. Andreoni, F. Gygi, and M. Parrinello, *Phys. Rev. Lett.* **68**, 823 (1992).

[226] A. Oshiyama, S. Saito, N. Hamada, and Y. Miyamoto, *J. Phys. Chem. Solids* **53**, 1457 (1992).

[227] A. Oshiyama, S. Satio, Y. Miyamoto, and N. Hamada, *J. Phys. Chem. Solids* **53**, 1689 (1992).

[228] N. Hamada, S. Saito, Y. Miyamoto, and A. Oshiyama, *Jpn. J. Appl. Phys.* **30**, L2036 (1991).

[229] Y. Miyamoto, A. Oshiyama, and S. Saito, *Solid State Commun.* **82**, 437 (1992).

[230] A. Oshiyama and S. Saito, *Solid State Commun.* **82**, 41 (1992).

[231] S. Saito and A. Oshiyama, *Phys. Rev. B* **44**, 11536 (1991).

[232] W. Y. Ching, M.-Y. Huang, Y.-N. Xu, W. G. Harter, and F. T. Chan, *Phys. Rev. Lett.* **67**, 2045 (1991).

[233] M.-Z. Huang, Y.-N. Xu, and W. Y. Ching, *J. Chem. Phys.* **96**, 1648 (1992).

[234] Y.-N. Xu, M.-Z. Huang, and W. Y. Ching, *Phys. Rev. B.* **46**, 4241 (1992).

[235] M.-Z. Huang, Y.-N. Xu, and W. Y. Ching, *Phys. Rev. B* **46**, 6572 (1992).

[236] L. Ye, A. J. Freeman, and B. Delley, *Chem. Phys.* **160**, 415 (1992).

[237] S. Saito and A. Oshiyama, *Solid State Commun.* **83**, 107 (1992).

[238] S. C. Erwin and M. R. Pederson, *Phys. Rev. B* **47**, 14657 (1993).

various calculations are not important at present. It is more important to identify the trends and distinctions of the crystals, which can be done most easily with a given calculational method. For this reason the figures shown are (mostly) those of Erwin.[213] They are calculated with an accurate all-electron, full potential, local orbital method that is well documented.[214] In this section only results for the orientationally ordered structures will be discussed; the effects of merohedral disorder will be discussed in the section on transport properties.

It is important to understand some general features of the electronic structure of fullerene solids that arise simply from the large number of atoms in the unit cell. First, one cannot directly compare densities of states (DOS) per cell with those of conventional materials (such as graphite or diamond in the case of fullerite, or elemental metals in the case of the metallic fullerides). Rather it is essential to compare the DOS *per atom*, or per unit volume; otherwise the comparison is meaningless. Likewise, the lattice constant of the solids is approximately three times that of conventional (simple) solids, leading to the result that individual bandwidths will be ~3 times less than for a conventional material in which the band velocities (viz. the Fermi velocity) are the same, and the velocity may be the more relevant quantity for consideration. These considerations will enter the discussions in later sections.

12. FULLERITE

The equation of state (energy vs. volume) and the LDA band structure of fcc C_{60} was presented by Saito and Oshiyama[215] using a local orbital pseudopotential method and by Troullier and Martins[37] using a plane wave pseudopotential method. The calculated cohesive energy of the solid was $1.6\,eV/C_{60}$ molecule. The value reported by Pan *et al.*[216] is $\sim 1.7\,eV/C_{60}$ for the heat of sublimation/vaporization. Since each molecule has 12 neighbors, the energy per C_{60}–C_{60} "bond" is $1.7/6 = 0.28\,eV$. A typical C–C covalent bond energy is ~ 3–$4\,eV$, and so it is immediately clear that covalent bonding is not occurring; it is rather van der Waals cohesion resulting from mutual polarization and mild overlap of the molecular orbitals. The lattice constant has been predicted to be ~4% smaller[215] or 1–2% smaller[37] than the experimental value. This theoretical discrepancy and the difference from experiment are not such severe theoretical problems considering the softness of the C_{60} lattice, which implies a small difference in prediction of the energy.

The bulk modulus calculated by Troullier and Martin[37] is $B = 16.5$–$18.5\,GPa$, compared with the measured value[217] of $18.1 \pm 1.8\,GPa$. The

C$_{60}$ bandstructure

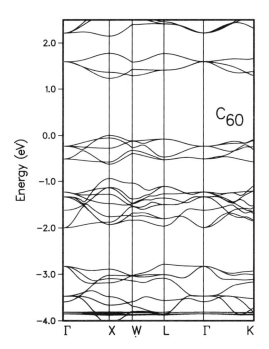

FIG. 13. Band structure of orientationally ordered fcc C$_{60}$ along high symmetry directions. The zero of energy lies at the top of the h_u-derived HOMO band. From S. C. Erwin, in *Buckminsterfullerenes,* eds. W. E. Billups and M. A. Ciufolini (VCH Publishers, New York, 1992). Reprinted with permission by VCH Publishers, Copyright 1192.

pressure derivative of B varied widely depending on the equation of state that was fit to, indicating that more extensive calculations are necessary to obtain this quantity accurately.

The energy bands of the solids, shown for the calculations of Erwin[213,218,219] for the potassium solids K$_n$C$_{60}$, $n = 0$, 3, 4,6 in Figs 13–16 with the corresponding densities of states in Fig. 17, are broadened versions of the molecular orbital (MO) energies. The band structures reflect (1) the degeneracy of the molecular level, (2) the symmetry of the molecular level, and (3) the type of overlap between the neighboring molecules of the C p orbitals that form the MOs. As has been done in most cases, the structure assumed for fullerite in the calculation leading to Fig. 13 (and for the studies of energetics already discussed) is an

K₃C₆₀ bandstructure

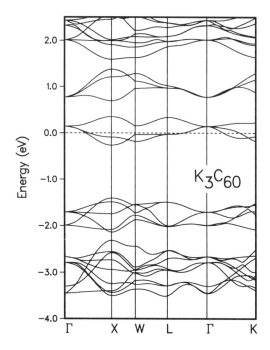

FIG. 14. Band structure of orientationally ordered fcc K$_3$C$_{60}$ along high symmetry directions. The Fermi level (dashed line) lies in the middle of the three t_{1u}-derived bands. Note the extreme similarity to the C$_{60}$ bands in Fig. 13. From S. C. Erwin, in *Buckminsterfullerenes*, eds. W. E. Billups and M. A. Ciufolini (VCH Publishers, New York, 1992). Reprinted with permission by VCH Publishers, Copyright 1192.

orientationally ordered structure with the highest symmetry consistent with both the icosahedral symmetry of C$_{60}$ and with the fcc lattice: Fm$\bar{3}$ (T_h^3).

The bands of most interest arise from the h_u-derived HOMOs and the t_{1u}-derived LUMOs, with calculated bandwidths of about 0.5–0.6 eV. The calculations for C$_{60}$ predict a semiconductor with a direct gap at the $X = (1, 0, 0)\pi/a$-point. The calculated gap of 1.2 eV is expected to be an underestimate by up to a factor of two, based on well-established behavior of (local) density functional theory[56] in underestimating bandgaps in both covalent and ionic solids. The next complex in the conduction band consists of the three t_{1g}-derived bands, separated from the t_{1u} complex by a 0.4-eV gap at the X-point.

K$_4$C$_{60}$ bandstructure

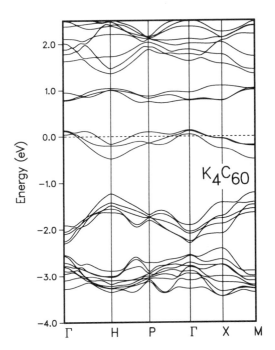

FIG. 15. Band structure of orientationally ordered fcc K$_4$C$_{60}$ along high symmetry directions. The Fermi level (dashed line) lies above the middle of the t_{1u}-derived bands which are quite different from the C$_{60}$ bands of Fig. 13 due to the different crystal structures. From S. C. Erwin, in *Buckminsterfullerenes*, eds. W. E. Billups and M. A. Ciufolini (VCH Publishers, New York, 1992). Reprinted with permission by VCH Publishers, Copyright 1192.

Troullier and Martins[37] have reported the band masses for the valence and conduction band states of fullerite. Since the band extrema occur at the X-point, there are both longitudinal (l) and transverse (t) masses, for both electrons (e) and holes (h). Their calculated values are: $m_h^l = 3.31$, $m_h^t = 1.26$, $m_e^l = 1.33$, $m_e^t = 1.15$. There are indications of a small region of solid solution for dilute concentrations of alkali atoms in fullerite, and such atoms are expected to donate their electrons easily to the conduction band states, leading to the possibility of an n-type semiconductor. Up to now, however, transport studies of such materials are not sufficiently detailed to compare with these predictions.

Troullier and Martins[37] have also reported on the appearance of

W.E. PICKETT

K$_6$C$_{60}$ bandstructure

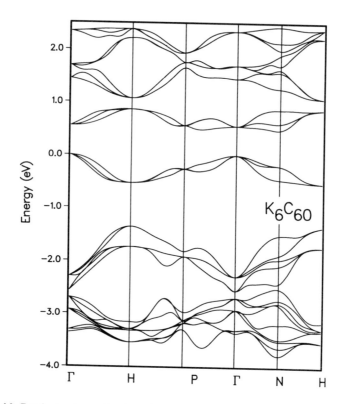

FIG. 16. Band structure of orientationally ordered fcc K$_6$C$_{60}$ along high symmetry directions. The zero of energy is at the top of the filled bands. The structural difference between bcc molecules (K$_6$C$_{60}$) and fcc molecules (C$_{60}$, K$_3$C$_{60}$) results in very different band dispersions. From S. C. Erwin, in *Buckminsterfullerenes*, eds. W. E. Billups and M. A. Ciufolini (VCH Publishers, New York, 1992). Reprinted with permission by VCH Publishers, Copyright 1992.

nonmolecular states in the fullerite conduction bands, related to flat regions of the potential in the interior of the C$_{60}$ molecule and in the interstitial sites. The lowest such state is an *s*-like state inside the C$_{60}$ molecule, at an energy of 4.2 eV above the valence band maximum. An octahedral site state lies at 6.5 eV, and two slightly split tetrahedral site states lie at 7.0 and 7.2 eV. These interstitial states will be the most strongly perturbed states when these intersticies are filled in the fulleride compounds.

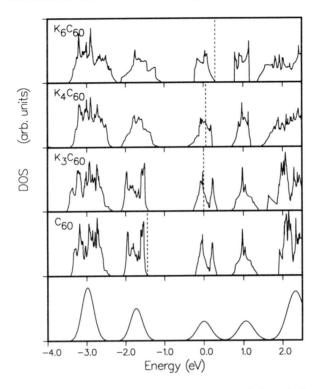

FIG. 17. Densities of states of the $K_n C_{60}$ crystals, $n = 0, 3, 4, 6$, from the bands picutred in Figs. 13–16. The lower panel is a broadened spectrum of the C_{60} molecule, illustrating the direct correspondence of bands in the solids with the molecular levels. From S. C. Erwin, in *Buckminsterfullerenes*, eds. W. E. Billups and M. A. Ciufolini (VCH Publishers, New York, 1992). Reprinted with permission by VCH Publishers, Copyright 1992.

The effect of other orientational orderings of the molecules in the fullerite crystal has been addressed by Satpathy et al.[220] They considered not only the idealized $Fm\bar{3}$ structure already discussed ("unidirectional" orientation, in their notation) but also "bidirectional" and "quadridirectional" structures with two and four molecules per cell, respectively. The bidirectional structure has space group $P4_2/mmm$ and corresponds to an idealization of the orientations occurring in the $K_3 C_{60}$ crystal in which the same two orientations were equally occupied, but in apparently a random manner. The quadridirectional structure is precisely that obtained by David et al.[183] for fullerite at low temperature. This structure, with space group $Pa\bar{3}$, also has the molecules confined to an fcc lattice but has four distinct orientations.

Although differences in the bands themselves are expected simply due to the differing symmetries, Satpathy *et al.*[220] find fairly striking differences in the bandwidths and the shapes of the DOSs. The t_{1u} bandwidths are calculated (from tight-binding Hamiltonians) to be 0.52, 0.64, and 0.44 eV in the uni-, bi-, and quadridirectional cases, respectively, with quite differently shaped DOS curves. The differences illustrate that the banding in the solid is dependent on the specific intermolecular atomic interactions, and the resulting band structures are sensitive to the relative orientation of the molecules.

The characteristics of the positron state in fullerite and various fulleride concentrations have been calculated by Puska and Nieminen,[221] Ishibashi *et al.*,[222] and Lou *et al.*[223] The positron lies in the large interstitial regions, most strongly in the larger the octahedral site, and is expelled from the interior of the C_{60} molecule. The positron density is strongly affected by K cations in the interstices. Positron lifetimes were also calculated by these authors and compared with the existing experimental data.

13. ALKALINE FULLERIDES

a. M_3C_{60}

The band structure of Erwin and Pickett[218] for orientationally ordered K_3C_{60} is shown in Fig. 14 along the same symmetry directions as for C_{60} in Fig. 12. The Fermi level lies in the middle of the t_{1u} complex of three bands. The h_u-, t_{1u}-, and t_{1g}-derived bands are quite similar to the corresponding bands in C_{60}, reflecting the fact (borne out by analysis of the charge and state density) that the alkali metal ions do not contribute to the states in this region. Rather, they donate their valence electrons to the C_{60} molecule, making this compound essentially an ionically bound metallic $(M^+)_3C_{60}^{3-}$ solid. Very similar results hold for Rb_3C_{60}, with differences being accounted for by the small increase in lattice constant. The calculated gaps between band complexes change little from C_{60} to M_3C_{60}, and most changes that occur can be accounted for by the difference in bandwidths arising from the variation in intermolecular overlap.

Martins and Troullier[224] report an enthalpy of reaction, given by the energy of K_3C_{60} minus the energies of the C_{60} crystal and that of potassium metal, to be $\Delta H = 1.7$ eV per K atom within LDA. Relaxation of the internal coordinates of the C_{60} molecules, which were not considered by Martins and Troullier, were found by Andreoni *et al.*[225] to be ~0.2 eV per K atom in K_6C_{60}, and Martins and Troullier suggest a

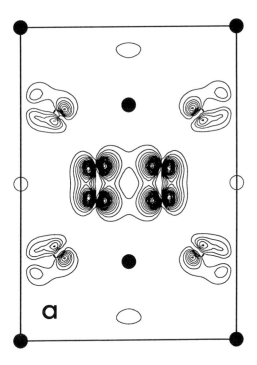

FIG. 18. Contour plots of valence charge densities, plotted in the (110) plane including halves of two adjacent molecules. In (a), the total charge density of C$_{60}$ is shown, with logarithmic spacing of the charge density contours. Panel (b) shows the density of the t_{1u}-derived states (at Γ) that become partially occupied in K$_3$C$_{60}$. In (c), the difference in valence density between K$_3$C$_{60}$ and C$_{60}$ is shown, illustrating the extreme similarity to that in panel (b). The filled circles in (a) give the position of K ions, and the contour separations are linear in (b) and (c). From J. L. Martins and N. Troullier, *Phys. Rev. B* **46,** 1766 (1992).

similar correction should be applied in K$_3$C$_{60}$, leading to an LDA prediction of $\Delta H \approx 1.9$ eV.

By comparing the charge density of the t_{1u}-derived orbitals with the difference in charge densities of K$_3$C$_{60}$ and C$_{60}$, Martins and Troullier[224] demonstrate explicitly that the K charge is donated to the fullerene states, and charge density of K$_3$C$_{60}$ and the density difference are pictured in Fig. 18. Troullier and Martins[37,224] also monitored the change in position of the conduction states that are not derived from MOs (discussed previously for fullerite). The state that is interior to the C$_{60}$ molecule is lowered by 0.68 eV, the octahedral site state is lowered by 0.44 eV, and the two tetrahedral site states are lowered by 0.37 eV (even combination) and 0.03 eV (odd combination) upon introducing K atoms

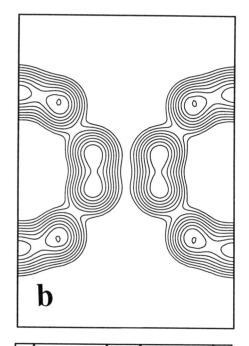

Fɪɢ. 18—(continued)

into the crystal. These states remain 2–5 eV above the Fermi level. They also report the change of t_{1u} bandwidth with volume: $d \log W / d \log V \approx -5$.

b. M_6C_{60}

For these compounds the C_{60} molecules lie on a bcc lattice instead of an fcc lattice. The bands calculated by Erwin and Pederson[219] for K_6C_{60} along symmetry lines are shown in Fig. 16. The parentage of the bands is analogous to that in M_3C_{60}, but the distinct band dispersions reflect the different symmetry of the bcc lattice. The gap between h_u- and t_{1u}-derived bands is now only 0.8 eV. The compound is insulating, and now the gap lies between the t_{1u}- and t_{1g}-derived complexes. The calculated gap is 0.5 eV and indirect, with the valence band maximum at Γ and the conduction band minimum at N.

Again, the bands on either side of the gap are simply broadened versions of the C_{60} MOs, without significant participation of alkali states. Thus again, there is essentially complete charge transfer from the metal atoms to the C_{60} molecules, leading to the ionic insulator $(M^+)_6C_{60}^{6-}$. The C_{60} molecule therefore accepts six extra electrons in this solid.

Andreoni et al.[225] have allowed the K and C atoms to relax and minimize the energy, obtaining ~1 eV energy gain per molecule compared with the frozen C_{60} molecule and K ions at their ideal positions. The energy gains come roughly equally from C_{60} relaxation and displacements of the K ions. The relaxation, carried out using ab initio molecular dynamics, showed that first the molecules distorted; this was followed by a slower displacement of the K ions. This indicates the presence of low energy K vibrational modes in the crystal. The resulting distortion preserves the T_h symmetry of the molecule.

c. M_4C_{60}

The compounds of this stoichiometry have C_{60} molecules arranged on a bct lattice. The band structure of orientationally ordered K_4C_{60} calculated by Erwin[213] is shown in Fig. 15. Again, the bands in the region of interest are simply broadened versions of the corresponding MOs, without any significant alkali state participation, leading to ionically bound $(M^+)_4C_{60}^{4-}$ solids. In this structure K_4C_{60} is predicted to be metallic, with the complex of three t_{1g}-derived bands being two-thirds filled. No indication of superconductivity was found in this compound, and the experimental information to date suggests that it is an insulator. Considering the case of K_3C_{60} and the crystalline symmetry, it seems that K_4C_{60} must possess merohedral disorder of the molecules. Disorder would affect the electronic states and may lead to the observed non-conducting behavior by making states near the band edges nonpropagating.

Densities of States

The electronic densities of states (DOS) of the K_nC_{60} crystals calculated by Erwin[213] are shown in Fig. 17 in the region of the bandgap or Fermi level. The extreme similarity of the $n = 0$ and $n = 3$ curves reflects the complete lack of participation of the K atom states in this region of energy (a difference can be seen in the region 1.6–1.9 eV in this figure). The differences between these two curves and those for $n = 4$ and $n = 6$ reflect the changes in crystal structure and, for example, number of nearest neighbors. The valence bandwidths in the K_6C_{60} crystal are 20–30% broader than in the fcc structure $n = 0$ and $n = 3$ crystals.

The reported values of $N(E_F)$, the DOS at the Fermi level, reflect the differences in calculational details, assumed lattice constants, and in some cases phenomenological broadenings that were applied. Erwin and Pickett[218] obtained $N(E_F) = 13.2$ states/eV-cell (both spins) at a lattice constant of 14.24 Å. Since the crystals do not have the precise orientational ordering assumed in the calculations, it is unrewarding to pursue the calculational differences. The variation with pressure is more relevant, since the relative variation may be similar in oriented and unoriented crystals. Erwin (private communication) obtained $d \log N(E_F)/d \log V \approx 4$, which can be compared with the value assuming $N(E_F) \propto 1/W$ from Martins and Troullier,[224] which gives $d \log(1/W)/d \log V = 5$. Thus, there is a ~20% deviation from a simple $N(E_F) \propto 1/W$ behavior due to internal rearrangement of the bands.

The Fermi surface calculated for the orientationally ordered structure of K_3C_{60} by Erwin and Pickett[218] is shown in Fig. 19. It consists of two inequivalent sheets. One is a closed, spheroidal surface centered at Γ with protrusions along the $\langle 111 \rangle$ directions. The other surface is a rather unusual, multiply connected surface. In fact, it consists of two distinct but symmetry-equivalent surfaces, arising from the fact that this surface contacts the boundary separating one Brillouin zone from its second neighbor, rather than the boundary with its first neighboring zone. The shape of this surface vividly reflects the lack of cubic symmetry of this oriented crystal: there are two necks connecting the surface along the $\langle 100 \rangle$ directions, rather than the four (or only one) that would occur if the $\langle 100 \rangle$ axes were fourfold axes. The velocities, discussed in Section VII, vary strongly over this piece of Fermi surface.

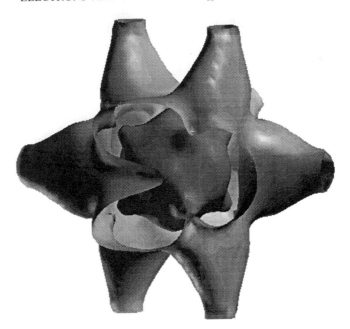

FIG. 19. Calculated Fermi surfaces of K$_3$C$_{60}$ in the ordered crystal structure. The multiply connected outer surface has two arms at the zone boundary along each of the crystal axes, reflecting the twofold (noncubic) symmetry. The surface inside is a closed hole surface. Figure courtesy of S. C. Erwin.

14. ALKALINE-EARTH FULLERIDES; OTHERS

a. Ca$_n$C$_{60}$

Satio and Oshiyama[237] have reported LDA calculations for the $n = 3$ solid. Although different from the $n = 5$ compound that is superconducting at 8 K, this example provides insight into the differences between alkali and alkaline earth C$_{60}$ compounds. For their studies, they considered Ca atoms placed in both of the tetrahedral sites of the fcc lattice, and the third Ca atom was placed on center, and also off center, in the octahedral site. They find that Ca $4s$ electrons transfer to the t_{1u} bands, leaving the Ca ionized, similar in this respect to the alkali compounds. Strong hybridization of the next higher lying t_{1g} bands with the Ca $4s$ band is found, however, quite unlike in the alkali compounds. They suggest then that in the $n = 5$ superconducting compound the t_{1u} bands will be completely occupied (with 6 of the 10 Ca valence electrons) and the hybridized $t_{1g} = $ Ca $4s$ bands will be the active ones for superconductivity in this compound.

b. Ba_6C_{60}

This compound was reported by Kortan et al.[210] to be superconducting at 7 K. The C_{60} molecules lie on a bcc Bravais lattice ($a = 11.17$ Å), and the structural refinement was carried out in orientationally ordered structure with $Im\bar{3}$ space group. A fully ionic picture of this compound would have the 12 Ba valence electrons transfering into the t_{1u}- and t_{1g}-derived bands, leading to an insulating compound.

Erwin and Pederson[238] have calculated the band structure of this compound, shown in Fig. 20. They obtain a band structure that differs considerably from the crystals with fcc Bravais lattice, and from the simple ionic picture. Indeed, there are bands that are primarily t_{1u}- and

bandstructure

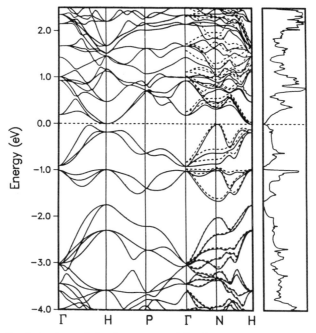

FIG. 20. Band structure of the fcc compound Ba_6C_{60} along high symmetry directions. Unlike the alkali fullerides, the band structure in this compound is strongly altered from broadened molecular orbitals, due to participation of Ba states. An accidental zero gap spectrum results. The dashed lines in the Γ–N–H panel include spin–orbit interactions and demonstrate that spin–orbit corrections do not alter the zero gap result. From S. C. Erwin and M. R. Pederson, Phys. Rev. B **47**, 14657 (1993).

t_{1g}-derived, but the separation between them is only ~0.2 eV. Moreover, the t_{1g}-derived bands are hybridized with bands from above, leading to a zero-gap semimetal. This gap is accidental, i.e., it does not arise from a filling of bands up to a symmetry-determined degenerate state at a high symmetry point, and so it is not necessarily precisely zero. The zero gap arises rather from what appears to be an anticrossing surrounding the H-point, which can be seen from the band structure along the Γ–H and H–P directions. The conduction band minimum occurs at the H-point; valence band maxima occur at N and at almost the same energy along the Γ–H line.

15. TIGHT-BINDING REPRESENTATIONS

A tight-binding representation of the electronic structure is useful for insight into the effects of symmetry and coordination as well as for application in model calculations. Manousakis[239] provided a simple picture of the states derived from π orbitals based on the fivefold symmetry of C$_{60}$. He began with a fivefold axis (think of it as north) running through the center of the molecule and the center of a pentagon. Near a great circle of "longitude" (which he called a "slice"), one can identify 12 atoms that generate all 60 atomic positions upon performing the $2\pi j/5$ rotations, $j = 0$, 4, around the axis. The electronic wavefunctions can be expressed in terms of their transformation properties under these rotations by writing

$$|\psi_m\rangle = \frac{1}{\sqrt{5}} \sum_{j=1}^{5} \exp\left(i\frac{2\pi}{5}jp\right)|\phi_p\rangle, \tag{15.1}$$

where the "basis" states are

$$|\phi_p\rangle = \sum_{n=1}^{12} a_{pn}c_n^+ |0\rangle. \tag{15.2}$$

The quantum numbers $m = 0$, ± 1, ± 2 identify a rotational character of the states, and Manousakis presented the corresponding eigenvalues and their degeneracy, parity, and m index.

[239]E. Manousakis, *Phys. Rev. B* **44**, 10991 (1991).

In terms of a single hopping parameter t_π, taken for simplicity as equal for b_{hh} and b_{bp} bonds, he determined the π bandwidth as $5.62t$ and the gap as $0.76t$. The degeneracies can be identified with the LDA eigenvalues. Not surprisingly, the fivefold h_g HOMO corresponds to $m = -2$, -1, 0, 1, $+2$, and both the t_{1u} and t_{1g} unoccupied states correspond to $m = -1$, 0, $+1$. Manousakis noted that it would require $t_\pi = 2.50$ eV to account for the calculated LDA gap. Also, the nearly degenerate g_g and h_g states just below the HOMO are accidentally degenerate when the hopping parameters of the inequivalent bands are taken to be equal in the tight-binding model, and so the splitting of these states is related to the unequal hopping across the two bonds.

Tight-binding calculations of one, two, and four molecule structures of fullerite were presented by Satpathy et al.[220] They adopted a fixed ratio of σ to π hopping parameters and parametrized the distance dependence as

$$t_\sigma = -4t_\pi = t_0 \frac{d}{d_0} \exp\left(-\frac{d - d_0}{L}\right), \qquad (15.3)$$

with $L = 0.505$ Å, $t_0 = 0.90$ eV, and $d_0 = 3.00$ Å. On the molecule only nearest-neighbor hopping was considered; these are -2.78 and -2.59 eV for the b_{hh} and b_{ph} bonds, respectively. For intermolecular hopping, they found that $L = 0.58$ Å reproduced better the LDA calculations for varying volumes. Satpathy et al. also considered the more specialized situation of a tight-binding basis consisting only of the t_{1u} orbitals, and they presented analytic expressions for the corresponding tight-binding Hamiltonian matrix elements.

The task of obtaining a tight-binding representation of the band structure of K_3C_{60} that is as accurate as possible was carried by Pickett et al.[240] They obtained both orthogonal and nonorthogonal fits of the complete valence and lower conduction band structure at normal and expanded volumes, starting from a fit for the C_{60} molecule itself. For the intermolecular hopping they assumed the distance dependence, for both the hopping amplitudes and the overlap parameters of Eq. (15.3) from Satpathy et al. Unlike others, they did not consider only nearest-neighbor intramolecular hopping (first and second neighbor distances, due to the different bond lengths), but also included the third and fourth neighbor intramolecular distances of ~ 2.5–2.9 Å, which are shorter than the intermolecular distances that are considered: 3.12, 3.45, and 3.56 Å.

[240]W. E. Pickett, D. A. Papaconstantopoulos, M. R. Pederson, and S. C. Erwin, J. Supercond. (1994, in press).

VI. Electronic Spectroscopy; Coulomb Interactions

The peculiarities of the fullerene compounds—the molecular crystal made of ideally spherical organic molecules, the metallic behavior in ionic compounds, the appearance of superconductivity at impressively high temperatures, the occurrence of ferromagnetism in at least one compound—has led to a great number of questions to be answered. Electronic spectroscopies afford an important means of addressing these questions.

16. Photoemission

Photoemission spectroscopy (PES) was applied to fullerenes, fullerite, and the fullerides as soon as they each became available in sufficient quantity. Photoemission is a powerful spectroscopy, but it is often hampered in answering questions about bulk solids because of its extreme surface sensitivity. It is important to consider the physical region and the likely properties that are being probed before interpreting the spectra.

Up until this time, the focus has been on trying to understand the electronic structure of the bulk crystal. In PES, the escape depth of the photoelectrons that reach the detector ranges from 5 to 25 Å, depending on its energy in the solid (and thus the incident photon energy). The separation of C_{60} molecues is ~ 10 Å, indicating most if not all of the photoelectrons are initiated in the first one or two molecular layers at the surface. There are somewhat different considerations for the fullerite (C_{60}) and the fullerides (A_xC_{60}).

Assuming a good surface (a flat, unreconstructed, truncated bulk lattice), for fullerite the difference between the electronic spectral density at the surface and in the bulk arises from the different coordination of the molecules at the surface. In the bulk of the fcc crystal, a molecule is coordinated with 12 nearest neighbors. For an ideally flat (111) surface, for example, the outer layer of molecules will be coordinated with only 9 other molecules (6 in the surface layer and 3 in the subsurface. As a result, the bandwidth of states on the surface layer is expected to be only $\sim 9/12 = 75\%$ of the value in the bulk. As noted in Section II, the effective on-site repulsion U_{eff} is also larger at the surface, due to reduced screening. Due to both the smaller bandwidth and the stronger interaction, the surface electronic structure will be less bandlike than for the bulk. The difference between surface and bulk spectra becomes magnified if the surface is not flat (so that some surface atoms have even lower coordination and more reduced screening) or if the surface is (100),

(110), (311), etc. Thus, one must first consider to what extent PES spectra reflect bulk phenomena.

For the metal fullerides, the complications are substantially greater. There is the same issue of C_{60}–C_{60} coordination, of course, from which the banding is expected to arise. In addition, the presence of the metal atoms and the charge transfer lead to additional surface effects. Taking K_3C_{60} as an example, the solid is understood as ionically bonded due to the essentially complete charge transfer to give K^+ and C_{60}^{-3} ions. At the surface the K stoichiometry and coordination will differ from the bulk, although indication of the $K:C_{60}$ stoichiometry within an escape depth can be obtained by comparing intensities of core levels. The charge state of the surface layer of C_{60} molecules is uncertain, and it will depend on the surface $K:C_{60}$ ratio. There is no reason a priori to expect the surface spectra to be representative of the bulk.[241] An important feature of the electronic structure of an ionic solid is the resulting ionic (Madelung) potentials at the various atomic sites. These can be quite different at a surface, even if the charge state does not vary from the bulk, and this change will raise or lower the surface C_{60} molecular orbitals with respect to those of the bulk. The filling of surface states could then differ from that of the bulk. The surface Madelung shifts, as well as surface disorder, will tend to broaden the spectrum, while the lower coordination will tend to narrow the spectrum. The net effect may in fact be quite dependent of the surface preparation technique.

Implications of the short escape depth have been discussed by Wertheim et al.[241] They quote escape depths of ~3–8 Å for electron kinetic energies of 20–50 eV, increasing to 10–20 Å for kinetic energies of ~1 keV. It is difficult to avoid the surface sensitivity by going to higher photon energies because the photoelectric cross section of the $C\,2p$ electrons decreases rapidly. Wertheim et al. suggest that such surface effects, rather than correlation effects, may be responsible for the weakness of the Fermi cutoff seen in many spectra.

a. *Fullerite C_{60}*

A recent discussion of the experimental data for fullerite C_{60} and its interpretation has been given by Weaver.[242] There is reasonable consensus on the data in this material. The photoemission data shows good agreement with the density of occupied states as obtained by LDA calculations, discussed in Section V. The measured peak positions

[241]G. K. Wertheim, D. N. E. Buchanan, E. E. Chaban, and J. E. Rowe, *Solid State Commun.* **83**, 785 (1992).
[242]J. H. Weaver, *J. Phys. Chem. Solids* **53**, 1433 (1992). With ~100 references.

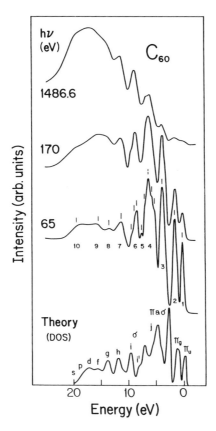

FIG. 21. Photoemission spectra from C$_{60}$ films at photon energies of 65, 170, and 1487 eV, compared with local density approximation calculations of the valence band density of states, broadened somewhat. The zero of energy is different for the calculation and the data. From J. H. Weaver, J. L. Martins, T. Komeda, Y. Chen, T. R. Ohno, G. H. Kroll, N. Troullier, R. E. Haufler, and R. E. Smalley, *Phys. Rev. Lett.* **66**, 1741 (1991).

correspond closely to the calculated ones, and even the widths are comparable. This comparison is shown in Fig. 21. The measured full width at half maximum of the HOMO is 0.65 eV. Band calculations give a (full) bandwidth of 0.6 eV. This apparent agreement may not be as simple as it appears. As previously noted, the lower coordination at the surface will narrow the states on the surface molecules compared with the bulk bands. Broadening due to vibronic effects, which are seen in the gas phase[60,61] and should occur in the solid as well, may add to the experimental width.

The gas phase photoemission spectrum of C$_{60}$ has been calculated from LDA eigenvalues and wavefunctions by Mintmire *et al.*[243] They illustrate clearly the photon energy dependence of the spectra: ultraviolet spectra (15–50 eV, say) emphasize the states with p character, while x-ray spectra allow the observation of only the s states. The spectrum calculated for 40.8-eV photon energy is in excellent agreement, both in energy and spectral weight, for the first four features, over ∼7 eV, with the gas phase data of Lichtenberger *et al.*[60,61]

The inverse photoemission data is also very similar to the calculated density of unoccupied states, with peaks up to 8 eV agreeing to within 0.5 eV. An interesting, but not unexpected, difference between experiment and LDA calculations is in the "gap" E_g^0, which is not the optical gap (which may be affected by exciton formation) but rather the difference in energy between a well-separated electron and hole, and the ground state. The direct/inverse photoemission data implies $E_{g,exp}^0 \sim 2.6$ eV (uncertainy arises because of difficulty in identifying the band edges), whereas the calculated eigenvalue difference is 1.5 eV. Such "discrepancies" of a factor of ∼1.5–2 occur quite generally in semiconductors and are reasonably well understood[56] as an "excited state effect" that LDA calculations do not include.

b. *Metal Fullerides*

Unlike the case for fullerite, in the metal fullerides there has been considerable disagreement in the *data*[244–249] even before the interpretation is considered. The situation may have been clarified somewhat in studies of K_xC_{60} by Weaver *et al.*[250] and Benning *et al.*[251] They provide

[243]J. W. Mintmire, B. I. Dunlap, D. W. Brenner, R. C. Mowrey, and C. T. White, *Phys. Rev. B* **43**, 14281 (1991).

[244]P. J. Benning, J. L. Martins, J. H. Weaver, L. P. F. Chibante, and R. E. Smalley, *Science* **252**, 1418 (1991).

[245]P. J. Benning, D. M. Poirier, T. R. Ohno, Y. Chen, M. B. Jost, F. Stepniak, G. H. Kroll, J. H. Weaver, J. Fure, and R. E. Smalley, *Phys. Rev. B* **45**, 6899 (1992).

[246]T. Takahashi, T. Morikawa, S. Hasegawa, K. Kamiya, H. Fujimoto, S. Hino, K. Seki, H. Katayama-Yoshida, H. Inokuchi, K. Kikuchi, S. Suzuki, K. Ikemoto, and Y. Achiba, *Physica C* **190**, 205 (1992).

[247]T. Takahashi, S. Suzuki, T. Morikawa, H. Katayama-Yoshida, S. Hasegawa, H. Inokuchi, K. Seki, K. Kikuchi, S. Suzuki, K. Ikemoto, and Y. Achiba, *Phys. Rev. Lett.* **68**, 1232 (1992).

[248]G. K. Wertheim, J. E. Rowe, D. N. E. Buchanan, E. E. Chaban, A. F. Hebard, A. R. Kortan, A. V. Makhija, and R. C. Haddon, *Science* **252**, 1419 (1991).

[249]C. T. Chen, L. H. Tjeng, P. Rudolf, G. Meigs, J. E. Rowe, J. Chen, J. P. McCauley, Jr., A. B. Smith III, A. R. McGhie, W. J. Romanow, and E. W. Plummer, *Nature* **352**, 603 (1991).

[250]J. W. Weaver, P. J. Benning, P. Stepniak, and D. M. Poirier, *J. Phys. Chem. Solids* **53**, 1707 (1992).

evidence that not only is there phase separation into $x = 0$, 3, 4, and 6, which is well accepted in the bulk, but that doping of the thin films occurs via nonequilibrium growth. When the overall stoichiometry corresponds to (say) $x = 3$, there are other phases present, and when off-stoichiometry there are more than the two (neighboring) phases that would occur under equilibrium conditions. So in addition to the surface complications already listed, any given surface may contain several phases, and the concentration of phases probably depends on the technique of preparation.

A critical interpretation of the various data that have been presented on metal fullerides is beyond the capabilities of this reviewer, so only a recent study of the Minnesota group will be presented. The data of Benning et al.[251] is shown in Fig. 22, for various average compositions of K determined from K $2p$ and C $1s$ intensities. Low energy electron diffraction (LEED) experiments were performed concurrently to monitor the condition of the surface. Their analysis indicates that indeed phase separation does occur, the $x = 3$ surface is metallic, and the $x = 4$ surface (which had not been identified previously) is insulating, as is the $x = 6$ surface. These results are consistent with bulk data. The $x = 3$ result differs from that of Takahashi et al.,[246,247] who interpreted their data in terms of a pseudogap at the Fermi level.

While the interpretation of the data is not at all simple, the valence band feature corresponding to $x = 3$ (see the data for $x = 3.5$ in Fig. 22) is ~ 1 eV wide and shows a peak at 0.7 eV, whereas the occupied part of the corresponding band in LDA calculations is only ~ 0.25 eV. Benning et al.[251] suggest that plasmon losses and vibrational losses may contribute to this discrepancy, but that the extra broadening more likely results from spectral weight shifts due to strong intramolecular interactions. They suggest that the data is consistent with an effective Hubbard U repulsion of ~ 1.5 eV. We return to the question of strong correlations later. The surface complications previously discussed may also contribute to the discrepancy.

Knupfer et al.[252] have measured a temperature dependence in their high resolution photoemission spectra that they relate to satellites arising from phonon emission and to plasmons. To account quantitatively for their data using bulk processes, the bandwidth had to be reduced by a

[251]P. J. Benning, F. Stepniak, D. M. Poirier, J. L. Martins, J. H. Weaver, L. P. F. Chibante, and R. E. Smalley, *Phys. Rev. B* **47**, 13843 (1993).

[252]M. Knupfer, M. Merkel, M. S. Golden, J. Fink, O. Gunnarsson, and V. P. Antropov, *Phys. Rev. B* **47**, 13944 (1993).

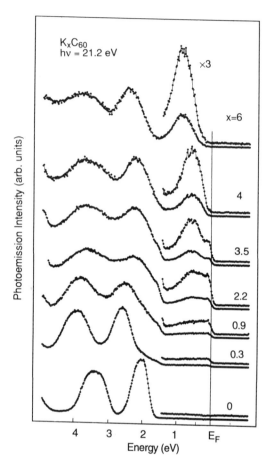

FIG. 22. Photoemission spectra from K_xC_{60} at 21 eV photon energy, for various concentrations of x determined by K $2p$ and C $1s$ core level intensities. Each curve involves contributions from more than one phase (see text). Note the sharp Fermi level cutoff in the $x = 2.2$ and 3.5 curves. From P. J. Benning, F. Stepniak, D. M. Poirier, J. L. Martins, J. H. Weaver, L. P. F. Chibante, and R. E. Smalley, *Phys. Rev. B* **47**, 13843 (1993).

factor of ~1.5 from calculated values. At present, it must be concluded that PES data of fullerides is not well understood.

17. BULK SPECTROSCOPIES

Before bulk spectra can be interpreted unambiguously, the related spectra on the C_{60} molecule itself must be understood. A primary interest is in identifying the lowest lying electronic excitations of the neutral molecules. A complication is that the lowest electron–hole creation process, involving h_u and t_{1u} states, is dipole forbidden, making this transition difficult to identify by optical absorption. Nevertheless, information is available on this and many other relatively low energy transitions.

a. Solutions

An approximation to the absorption of the isolated molecule is the absorption spectra of C_{60} molecules in solutions (benzene, n-hexane, and 3-methylpentane) that have been reported. The high resolution spectra of Leach et al.[253] span the 1.78–6.4 eV region and reveal a total of some 40 structures, comprising of electronic transitions and vibrational sidebands. Three very weak features were observed in the 1.78–1.94 eV range, likely corresponding to symmetry forbidden transitions. A strong absorption edge was observed at ~1.95 eV, with main peaks at ~2.0, 2.1, and 2.3 eV as well as at higher energies.

Earlier, Hare et al.[254] had begun at somewhat lower energies and observed a weak onset of absorption at ~1.75–1.8 eV, and their work also indicated the differences in spectra obtained in different solutions. Ajie et al.[255] earlier had reported a sharp, but weak, onset at 1.95 eV. Gasyna et al.[256] present very complex spectra for C_{60} in an argon matrix and provide tentative assignments. It is probably fair to say that absorption spectra in solutions are not well understood (at least not in their vibronic sidebands), but indicate an optically forbidden electronic absorption around 1.6 eV, with optically allowed transition beginning at ~1.95 eV.

b. Solids

The optical absorption spectrum from 0.4–6.2 eV for a thin film of C_{60} was reported by Skumanich.[257] He obtained weak absorption at ~1.6 eV,

[253] S. Leach, M. Vervloet, A. Desprès, E. Bréheret, J. P. Hare, T. J. Dennis, H. W. Kroto, R. Taylor, and D. R. M. Walton, Chem. Phys. **160**, 451 (1992).

[254] J. P. Hare, H. W. Kroto, and R. Taylor, Chem. Phys. Lett. **177**, 394 (1991).

[255] H. Ajie, M. M. Alvarez, S. J. Anz, R. D. Beck, F. Diederich, K. Fostiropoulos, D. R. Huffman, W. Krätschmer, Y. Rubin, K. E. Schriver, K. Sensharma, and R. L. Whetten, J. Phys. Chem. **94**, 8630 (1990).

[256] Z. Gasyna, P. N. Schatz, J. P. Hare, T. J. Dennis, H. W. Kroto, R. Taylor, and D. R. M. Walton, Chem. Phys. Lett. **183**, 283 (1991).

[257] A. Skumanich, Chem. Phys. Lett. **182**, 486 (1991).

interpreted as an optically forbidden transition. Some features of his films were similar to those of amorphous semiconductors, and so this data should not be considered as representative of an ordered crystal. Pichler et al.[258] reported optical absorption studies of films and tentatively assigned several peaks. A peak at 2.7 eV was assigned to the $h_u \rightarrow t_{1g}$ transition, and a strong, relatively sharp peak at 3.5 eV was assigned to the $h_g \rightarrow t_{1u}$ transition.

Matus et al.[259] have reported a study of C_{60} luminescence in films between 10 K and room temperature. Luminescence extended from 1.75–1.8 eV down to 1.4 eV, with peaks at 1.52 and 1.70 eV. The temperature dependence was interpreted in terms of a self-trapped exciton at low temperature, which becomes mobile at $T_m = 100$ K. This characteristic temperature was associated with a structural transition, with molecules rotating freely above T_m and becoming frozen in below that temperature. The luminescence spectra of Sibley et al.[260] for C_{60} in toluene solutions is peaked in the 1.3–1.55 eV range, while their bulk spectra at low temperature are concentrated in the same spectral range, but with rather different shape from the data of Matus et al.

Lof et al.[261] have presented photoemission and Auger spectra on C_{60}, and they argue that the energy difference between single-hole and single-electron excitations is ~2.3 eV, and that molecular excitons lie in the 1.5–2 eV range. This latter conclusion is consistent with the luminescence data.

Interpretation of the absorption and luminescence data has relied strongly on the calculations of excited neutral C_{60} molecules by Negri et al.[71] Their specific predictions include that the lowest energy excited state is the triplet T_{2g}. There may not yet have been any clear test of this prediction. It is optically forbidden, of course, as are the lowest nine singlet excitations. Negri et al. found that all such optically forbidden states may be activated by vibronic mixing, and they suggested that the Jahn–Teller modes should be most effective in providing such mixing. The amount of mixing by Jahn–Teller modes is strongly state dependent, and Negri et al. provide theoretical estimates of the magnitude of this coupling for several states. The excitation spectrum has also been calculated by Braga et al.[262] using the semiempirical CNDO/S-CI method, with results consistent with Negri et al.

[258]T. Pichler, M. Matus, J. Kürti, and H. Kuzmany, Solid State Commun. 81, 859 (1992).

[259]M. Matus, H. Kuzmany, and E. Sohmen, Phys. Rev. Lett. 68, 2822 (1992).

[260]S. P. Sibley, S. M. Argentine, and A. H. Francis, Chem. Phys. Lett. 188, 187 (1992).

[261]R. W. Lof, M. A. van Veenendaal, B. Koopmans, H. T. Jonkman, and G. A. Sawatzky, Phys. Rev. Lett. 68, 3924 (1992).

[262]M. Braga, A. Rosén, and S. Larsson, Z. Phys. D 19, 435 (1991).

c. Electron Energy Loss

Several electron energy loss spectroscopy (EELS) studies have been carried out for fullerite C_{60}. Gensterblum et al.[263,264,264a] presented high resolution (HREELS) studies of thin films from the far-infrared to the far vacuum ultraviolet (vuv). They identified 11 peaks below 450 meV, including 4 strong peaks. The strongest, at 66 meV (532 cm^{-1}), correlates with the strongest lines seen in reflectivity at 527 and 577 cm^{-1}. The other three strong lines (94, 156, 194 meV) correlate better with dipole inactive (Raman or silent) modes than with the other infrared active modes.

In the visible, a clear peak at 1.55 eV was interpreted as the lowest one-electron excitation (i.e., electron–hole pair). This can be compared with the LUMO–HOMO energy difference of the molecule, calculated[70,81] to be 1.55 eV, or the bandgap in the solid, calculated by Erwin[213] to be 1.2 eV within the local density approximation (for the idealized oriented structure). Several peaks in the 2–8 eV range evidently involve higher energy interband transitions. A very broad peak centered at 28 eV was interpreted as the plasmon involving all 240 $2s$–$2p$ electrons of the C_{60} molecule. The Drude formula $\omega_{p,D}^2 = 4\pi ne^2/m$ leads to a value of 21.6 eV (25% below the observed 28 eV value), reflecting that the s–p electrons do not behave like free electrons at this energy.

Similar results from the near infrared to the vuv were obtained by Sohmen et al.,[265,266] who were also able to obtain the dependence of some of the features on momentum transfer Q. The shoulder near 2 eV was found to grow in intensity up to $Q = 0.9$ Å^{-1}, while the feature at 2.7 eV decreases in intensity. Several features show dispersion of up to 0.2 eV. These authors suggested the observed broad plasmon peak (at 27 eV in their data) should be interpreted in terms of the model commonly applied to a semiconducting material with average gap E_g: $(\hbar\omega_{p,obs})^2 = E_g^2 + (\hbar\omega_{p,D})^2$. In sp^2 bonded materials, the gap would be some average transition energy between σ and σ^* bands. A value of $E_g = 16$ eV would be required to reconcile experiment with this model.

EELS measurements on a single crystal have been presented by Isaacs et al.[267] The plasmon peak at ~28 eV is isotropic within their resolution

[263]G. Gensterblum, J. J. Pireaux, P. A. Thiry, R. Caudano, J. P. Vigneron, Ph. Lambin, A. A. Lucas, and W. Krätschmer, Phys. Rev. Lett. 67, 2171 (1991).

[264]G. Gensterblum, L.-M. Yu, J. J. Pireaux, P. A. Thiry, R. Caudano, Ph. Lambin, A. A. Lucas, W. Krätschmer, and J. E. Fischer, J. Phys. Chem. Solids 53, 1427 (1992).

[264a]A. A. Lucas, J. Phys. Chem. Solids 53, 1415 (1992).

[265]E. Sohmen, J. Fink, and W. Krätschmer, Europhys. Lett. 17, 51 (1992).

[266]E. Sohmen, J. Fink, and W. Krätschmer, Z. Phys. B 86, 87 (1992).

[267]E. D. Isaacs, P. M. Platzman, P. Zschack, K. Hamalainen, and A. R. Kortan, Phys. Rev. B 46, 12910 (1992).

of 1.5 eV. They note that the plasmon energy is pushed up from its free electron value by interband transitions (as previously noted), but in diamond the plasmon peak at $\sim$31 eV is near its free electron value. The reason for this difference is not understood in detail, but the total valence bandwidth is $\sim$20% larger in diamond, and obviously fullerite has much larger inhomogeneity of the electron density than does diamond.

The complex frequency-dependent dielectric function $\varepsilon(\omega) = \varepsilon_1(\omega) + i\varepsilon_2(\omega)$ and the energy loss function $-\text{Im}[\varepsilon^{-1}(\omega)]$ have been calculated from the LDA band structures for fullerite C_{60} and for K_3C_{60} and K_6C_{60} by Ching and collaborators.[232,268,269] Standard assumptions were made, such as the neglect of electron–hole correlations and local field effects. They obtained static dielectric constants $\varepsilon_1(0) = 4.4$ and 19.2 for C_{60} and K_6C_{60}, respectively. The value for C_{60} is in excellent agreement with the value of $\varepsilon = 4.4 \pm 0.2$ reported by Hebard et al.[76] For K_3C_{60} they obtained from interband transitions only the value $\varepsilon_{1,0} = 42.2$ (intraband contributions lead to the Drude contribution that diverges as $\omega \to 0$).

The results are displayed in Figs. 23–25 and are compared with the experimental results for the loss function from EELS data of Sohmen et al.[265,266] (actually measured at $q = 0.2 \text{ Å}^{-1}$ rather than $q = 0$ as calculated). The measured structures in the loss function are reproduced remarkably well. The intensities cannot be compared without at least some phenomenological broadening of the calculated spectra; in addition, the EELS experiment (electron scattering) does not involve simply the dipole matrix elements that are included in the calculation. Note that the calculated results for K_3C_{60} are compared with a sample of nominal K concentration of $x = 2.5$, but the majority phase is likely to be K_3C_{60}.

The dipole forbidden $h_g \to t_{1u}$ in the C_{60} molecule becomes allowed (weakly) in the solid, leading to the onset of absorption, evident in the plot of ε_2, at 1.5 eV. A similar onset of the same transitions remains clear in the $n = 3$ and $n = 6$ solids, although both have transition at lower energies. In K_6C_{60}, there is a large contribution to ε_2 in the range 0.25–1.2 eV arising from transitions from the t_{1u}-derived bands to the t_{1g}-derived bands. In K_3C_{60}, the interband transitions within the t_{1u} bands give rise to strong contributions in the 0.05–0.4 eV range. These lower energy transitions steal spectral weight from higher energies; the fullerite has a much stronger weight in ε_2 in the 2.4–3.2 eV range than do the other crystals.

The peaks in the calculated loss function for the fullerite, at 3.5, 4.8,

[268]Y.-N. Xu, M.-Z. Huang, and W. Y. Ching, Phys. Rev. B **44**, 13171 (1991).
[269]W. Y. Ching, M.-Z. Huang, Y.-N. Xu, and F. Gan, Mod. Phys. Lett. B **6**, 309 (1992).

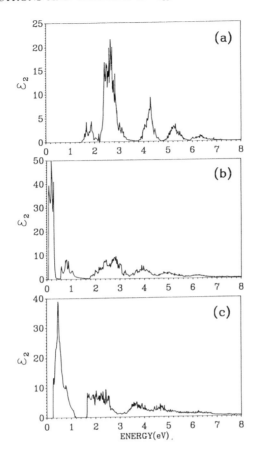

FIG. 23. The calculated imaginary part of the dielectric function of (a) fullerite C_{60}, (b) K_3C_{60}, and (c) K_6C_{60}. For the metal K_3C_{60}, only the interband part is shown; for the insulators, the predicted gaps of ~1.5 and ~0.25 eV are evident. Strong absorption below 1 eV is predicted for K_6C_{60}. From Y.-N. Xu, M.-Z. Huang, and W. Y. Ching, *Phys. Rev. B* **44**, 13171 (1991).

and 5.8 eV, correspond closely to two peaks and a shoulder in the EELS data. These three structures are broadened progressively, and shifted down somewhat, in the calculated loss function for the $n = 3$ and $n = 6$ compounds. The additional structures at lower energies are very close to those in the data. One discrepancy is that the calculated "plasmon" peak, which is at 6.8 ± 0.1 eV in all of the compounds, is at ~1 eV higher energy than seen experimentally. Xu et al.[268] suggest that some of this minor discrepancy may be removed by extending their basis set representation of the wavefunctions.

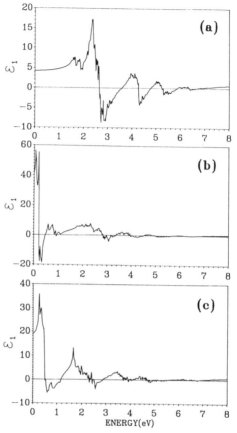

FIG. 24. The calculated real part of the dielectric function of (a) fullerite C_{60}, (b) K_3C_{60}, and (c) K_6C_{60}. For the metal K_3C_{60}, only the interband part is shown ; the intraband part would diverge at zero energy. For the insulators, the predicted zero-frequency dielectric constants are 4.4 and 19.2 for C_{60} and K_6C_{60}, respectively. From Y.-N. Xu, M.-Z. Huang, and W. Y. Ching, *Phys. Rev. B* **44**, 13171 (1991).

18. CORRELATION; MAGNETIC BEHAVIOR

The small bandwidth ~0.5–0.6 eV and the intramolecular Coulomb repulsion strength $U_{eff} \sim 1$ eV in the solid suggests effects due to correlation and invites the application of the Hubbard model to the fulleride compounds. The theoretical considerations leading to this value of U_{eff} were reviewed in Section 4.c. Whereas most theoretical work on the Hubbard model has involved a single band near half-filling, in the fullerides the interest would be in *three*, nearly half-full bands. Intersite

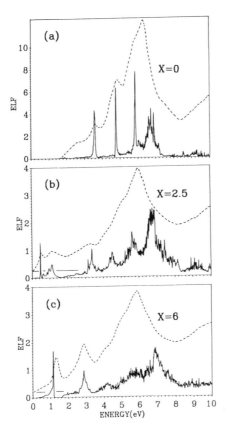

FIG. 25. The energy loss function from local density approximation calculations of Xu *et al.* for (a) fullerite C_{60}, (b) K_3C_{60}, and (c) K_6C_{60}, compared with data of E. Sohmen, J. Fink, and W. Krätschmer, *Europhys. Lett.* **17**, 51 (1992). With the exception of peak shifts of ~1 eV at the higher energies, the agreement is remarkably good. From Y.-N. Xu, M.-Z. Huang, and W. Y. Ching, *Phys. Rev. B* **44**, 13171 (1991).

interactions also might be important, assuming that these materials are not in the strongly interacting limit.

Surprisingly, little work of this kind has appeared. It is expected that the fcc lattice of molecules will frustrate charge density waves (CDW) and spin density waves (SDW) if interactions beyond nearest neighbors are negligible. Instabilities of this kind could be driven by Fermi surface nesting, and for the ordered structure of K_3C_{60} nesting was found to be substantial[218] for wavevectors $(q, 0, 0)2\pi/a$ for $0.5 < q \leq 1$. The merohedral disorder smears the bands (Section VIII) and therefore will reduce such nesting tendencies.

The ferromagnetism[29,30] of TDAE-C_{60}, which is not yet understood, suggests that other magnetic phenomena may occur in fullerides. The Hubbard ratio $U/W > 1$ suggests the possible importance of magnetic fluctuations. Bensebaa et al.[270] have used electron spin resonance (ESR) to characterize fulleride materials. They find time-dependent ESR signals for 24 hours after introduction of K, Rb, and Cs into the C_{60} crystals, but no time dependence for Na, which does not lead to superconducting crystals. For Rb_3C_{60} materials, they detect a change in the ESR intensity near T_c that they suggest indicates a transition between paramagnetic and antiferromagnetic states. Generally, the magnetic behavior of fullerides remains an open question.

VII. Electron–Phonon Coupling

19. LENGTH AND ENERGY SCALES

The transport properties of fullerides have been difficult to measure, since many of the samples are powders, and the films are extremely sensitive to air or water vapor. Some single crystals have now been prepared, however, allowing measurement of transport properties. The $n = 0$, 4, and 6 versions of A_nC_{60} have very low conductivities apparently determined by the level and type of impurities, and they will not be discussed here.

It is useful first to consider various length and energy scales associated with the fullerides. For the length scales, there are of course the interatomic distances, 1.40–1.45 Å on a molecule and ~ 3 Å between molecules, and the distance between molecule centers, ~ 10 Å. For three conduction electrons per C_{60} molecule, the average density $n = 3(4/a_B^3)$ for the fcc lattice corresponds to the homogeneous electron gas density parameter $r_s = 7.2$, where $(4\pi/3)(r_s a_B)^3 = 1/n$. This density in turn corresponds to an average electronic separation $2r_s a_B = 7.5$ Å. This value reflects an average conduction electron density one-half that of Cs ($r_s = 5.6$), which is the least dense elemental metal. The low density itself suggests the possible importance of correlation effects, quite apart from bandwidths, intramolecular repulsion energies, and the inhomogeneity of the density.

The energy scales for the corresponding homogeneous electron gas are

[270]F. Bensebaa, B. Xiang, and L. Kevan, J. Phys. Chem. 96, 10258 (1992).

$E_F^0 \equiv W_{occ} = (\hbar^2/2m)(3\pi^2 n)^{2/3} = 9$ eV, $N^0(E_F) = 0.5$ states/eV $- C_{60}$. The Fermi wavevector is $k_F^0 \sim 1.1(2\pi/a)$, and $2\pi/a$ is the distance to the Brillouin zone boundary along the cubic axis. The Fermi velocity is $v_F^0 = 6 \times 10^7$ cm/sec. The values for K_3C_{60} from LDA calculations are $W_{occ} = E_F \approx 0.25$ eV, $N(E_F) = 13$ states/eV $- C_{60}$, $\langle v_F^2 \rangle^{1/2} = 1.8 \times 10^7$ cm/sec (but v_K is very anisotropic), and the Fermi wavevectors vary with angle from $\sim \pi/a$ to $\sim 2\pi/a$.

These differences reflect the fact that fullerides definitely are not a homogeneous electron gas; rather, the electrons are concentrated in touching spherical shells rather than permeating the cell. The molecular crystal structure drastically alters the states and their dispersion from free-electron-like behavior, and it also suggests that local field effects in screening will be important. The static Thomas–Fermi screening length, from the expression for a homogeneous metal

$$\lambda_{sc} = \frac{1}{\sqrt{4\pi e^2 N(E_F)}}, \tag{19.1}$$

is $\lambda_{sc} = 0.5$ Å for K_3C_{60}, using the fcc lattice constant $a_0 = 14.2$ Å and the band structure value of the Fermi level density of states $N(E_F) = 13$ states/eV-cell. This value of screening length contrasts strongly with the very low density of the conduction electrons in the fullerides, and it remains to be determined if this value is meaningful (almost certainly it is not). [Note, however, that the corresponding screening length in the electron gas of the same average density is ~ 2.5 Å, even though the average separation is ~ 7.5 Å.]

20. FORMAL THEORY

There are at least three reasons to expect electron–phonon coupling in fullerides to harbor aspects that are unfamiliar in the context of conventional superconductors. First, the fullerides are molecular crystals, with a diversity of modes ranging from the high frequency intramolecular vibrations involving bond stretching to the medium frequency squashing modes, through the low frequency intermolecular optic modes involving relative motions of C_{60} molecules and cations or other C_{60} molecules, and finally to the very low frequency librational modes of the spherical molecules. Second, the highest vibrational frequencies (~ 0.2 eV) are of the same scale as the calculated occupied bandwidth (~ 0.25 eV), with the result that phonons do not scatter carriers *on* the Fermi surface, but rather from well below to well above the Fermi surface, in a volume that is likely to include most of the Brillouin zone. Third, the identical electronic and vibrational energy scales suggests that Migdal–Eliashberg theory[271-273] (the formal many-body theory of interacting electrons and phonons in crystals) may be insufficient because of the violation of the criteria for validity of Migdal's theorem.

This third item can be considered more quantitatively. In the conventional treatment, in which a linearization of $E_k = E_F + v_k \cdot (k - k_F)$ is invoked, the point at which the Migdal approximation[273] fails to apply[274] is when the electron velocity v_k becomes comparable with (or less than) the phonon *phase* velocity ω_q / q: $v_k q \sim \omega_q$. Taking the Fermi velocity of (orientationally ordered) $K_3 C_{60}$, $v_F = 1.8 \times 10^7$ cm/sec, the maximum value of $\hbar v_F q$ (q at the Brillouin zone boundary) is ~0.5 eV. The maximum phonon energy $\hbar \omega_{max} \sim 0.2$ eV. Thus, for the hardest vibrations the criterion $\hbar \omega \ll \hbar v_F q$ is not satisfied strictly even for the largest q; for the softer modes it will be satisfied over most of the zone. And the Fermi velocity is only the average over the Fermi surface; some fraction of the electrons will have smaller velocities, while others will have higher velocities.

Because of the similarity of $\hbar \omega_{max}$ and the bandwidth, it is not simply the bands at the Fermi surface but those within ~$\hbar \omega_{max}$ of the Fermi surface that are important, which amounts to the entire t_{1u}-derived band. Erwin[275] finds, however, that the average velocity $v(E)$ at energy E is significantly less than v_F only very near the band edges. Moreover, he finds that orientational disorder does not greatly affect the mean velocity.

A possible interpretation of the situation is that while the Migdal approximation is certainly not justified, the fact that the criterion is satisfied for some fraction of the electrons and phonons involved in the mutual scattering suggests that estimates obtained by invoking Migdal theory will be useful approximations. It is certainly not uncommon to find theoretical treatments to remain relatively accurate outside their range of strict validity.

There may, of course, be substantial necessary revisions of the theory, and formal extensions beyond the standard Migdal–Eliashberg theory

[271]G. M. Eliashberg, *Zh. Eksp. Teor. Fiz.* **38**, 966 (1960); *ibid.* **39**, 1437 (1960) [*Sov. Phys. JETP* **11**, 696 (1960); *ibid.* **12**, 1000 (1961)].
[272]D. J. Scalapino, J. R. Schrieffer, and J. W. Wilkins, *Phys. Rev.* **148**, 263 (1966).
[273]A. B. Migdal, *Zh. Eksp. Theor. Fiz.* **34**, 1438 (1958) [*Sov. Phys. JETP* **7**, 996 (1958)].
[274]W. E. Pickett, *Phys. Rev. B* **26**, 1186 (1982).
[275]S. C. Erwin, private communication.

have been suggested. Zheng and Bennemann[276] have examined the theory without linearizing the electronic structure at the Fermi level, thereby accounting for the variation of the density of states $N(E)$ much as has been done for conventional superconductors.[274,277] Not surprisingly, they find that such effects can seriously complicate the interpretation of experimental data, and specifically that the pressure dependence of T_c and the isotope effect are affected.

Pietronero and Strässler[278,279] have addressed the more serious breakdown of Migdal's theorem by considering the vertex corrections that are no longer guaranteed to be small. They provide expressions for a momentum-independent treatment of a single characteristic Einstein phonon mode ω_E. They find a strongly reduced low frequency mass renormalization $Z(\omega \to 0)$ and, partly due to this, a considerably enhanced result for T_c. Such corrections will affect the theory of the normal state as well.

21. Theoretical Estimates of Coupling Strength

The coupling of the added valence electrons to A_g symmetry modes has been discussed by Stollhoff.[280] Charge in the t_{1u} molecular orbitals counteracts the differentiation of bond lengths, and Stollhoff suggests the resulting coupling, which is linear, will dominate the electron–phonon interaction. He estimates an attractive "bond alternation" coupling of two holes on the C_{60} molecule of $-7.5\,meV$ and suggests that similar contributions to coupling hold for added electrons. Obtaining the net coupling requires consideration of electron–electron interactions as well.

A result of Allen,[281] which separates the total coupling strength λ into contributions λ_{Qv} from each phonon mode of wavevector Q and branch v, is widely used to assess the coupling strength of individual phonons. The expression is

$$\lambda = \frac{1}{3M} \sum_{v=1}^{3M} \frac{1}{N} \sum_{Q} \lambda_{Qv}, \tag{21.1}$$

[276]H. Zheng and K.-H. Bennemman, *Phys. Rev. B* **46**, 11993 (1992).
[277]S. G. Lie and J. P. Carbotte, *Solid State Commun.* **26**, 511 (1978).
[278]L. Pietronero, *Europhys. Lett.* **17**, 365 (1992).
[279]L. Pietronero and S. Strässler, *Europhys. Lett.* **18**, 627 (1992).
[280]G. Stollhoff, *Phys. Rev. B* **44**, 10998 (1991).
[281]P. B. Allen, *Phys. Rev. B* **6**, 2577 (1972).

where $3M$ is the number of branches (M is the number of atoms in the unit cell), and N is the number of states in the Brillouin zone (BZ). Expressed this way, λ is the *average* of the contributions from each phonon in the crystal. In all treatments so far, the dispersion (dependence on Q) has been neglected.

Schlüter and coworkers[161–163] and Varma et al.[160] considered the dominant coupling to arise from Jahn–Teller-related modes and suggested that the evaluation of coupling matrix elements can be carried out neglecting intermolecular hopping, except that it determines the density of itinerant conduction states $N(E_F)$. Varma et al. argued that (JT) coupling to the eight H_g modes will dominate, and they estimated the coupling for these modes. Their result was dominated by the two hardest modes (each being fivefold degenerate) at 1575 and 1428 cm^{-1}, with coupling constants (as previously defined) being $\lambda_{(1428)} = 12$ and $\lambda_{(1575)} = 4$. The net coupling strength (i.e., average over all modes) was $\lambda \sim 0.6$. With the coupling being at such high frequency ($\langle \omega \rangle \sim 1500$ cm^{-1}), this result is consistent with $T_c \sim 20$–30 K as long as μ^* is of conventional size.

Schlüter et al.[161–163] likewise emphasized the JT-active modes, using a related approximation with various phonon and electron models. Like Varma et al., they obtained strong coupling to the two hardest H_g modes, but some of the lower H_g modes (at $\sim$600–800 cm^{-1}) and the hard A_g (pentagonal pinch) mode at 1450 cm^{-1} showed strong coupling as well. Their characteristic frequency was also rather high, $\langle \omega \rangle \sim 1000$ cm^{-1}. Their characteristic frequency is smaller than that of Varma, however, and so a correspondingly larger value of λ is required to account for T_c. Since the coupling is spread over more branches, $\sim$25 rather than the 10 modes of Varma et al., the individual mode strengths are not required to be so high. This is of some importance, since a sufficiently strong coupling to a single mode will drive the mode unstable.

Several comparisons with the electron–phonon coupling in intercalated graphite, where T_c reaches a maximum of less than 2 K, gravitate to an explanation based on the curvature of the C_{60} molecule as being a key to increased electron–phonon coupling.

Mazin et al.[282] and Novikov et al.[283] have applied the rigid muffin-tin approximation to evaluate the coupling strength. This approximation is good in close-packed, high density of states metals where the change in potential due to an atom's displacement is confined to that atom. In the

[282]I. I. Mazin, S. N. Rashkeev, V. P. Antropov, O. Jepsen, A. I. Liechtenstein, and O. K. Andersen, *Phys. Rev. B* **45**, 5114 (1992).

[283]D. L. Novikov, V. A. Gubanov, and A. J. Freeman, *Physica C* **191**, 399 (1992).

fullerene solids, the openness of the structure and the poor screening makes this a less well justified approximation. On the other hand, it has the virtue of not involving ad hoc assumptions about which phonons are most important. Novikov *et al.* used their calculated Fermi surface average matrix elements and an estimated average phonon frequency of 1100 K to arrive at $\lambda = 0.5$–0.6 and $T_c \approx 16$–30 K for the superconducting fullerides.

Mazin *et al.* decomposed the Fermi surface–averaged matrix elements into radial and tangential contributions, finding that they are a factor of ~20 larger for tangential motions. They also performed frozen phonon calculations for the two A_g modes to obtain directly the change in band energies. The resulting deformation potentials were calculated to be 0.3–0.8 eV/Å along the Γ–X direction for the breathing mode, and 1–3 eV/Å for the high frequency pentagonal pinch mode. In total, they obtain a McMillan–Hopfield factor $\eta \sim 21$–36 eV/Å^2 [and $N(E_F) = 26$–48 states/eV C_{60}] for lattice constants in the range 14.1–14.6 Å. Using a four-force constant model of the dynamical lattice, they obtained coupling constants in the range $\lambda = 0.49$–0.83, which with a conventional value of $\mu^* = 0.15$ gives $T_c = 5$–44 K in just the range of experimental values.

One should also be aware that the infrared active modes also are strongly affected by the introduction of alkali atoms and resulting charge transfer, but so far they have received less theoretical attention. Rice and Choi[284] interpret this coupling as a "charged phonon" excitation such as had been discussed earlier in polymeric conductors. In fullerene-based solids, the T_{1u} phonons strongly interact with virtual $t_{1u} \rightarrow t_{1g}$ electronic excitations. The phonons thereby gain electronic oscillator strength, accounting for the giant infrared resonances reported by Fu *et al.*[155] Rice and Choi conclude that the T_{1u} modes at 537 and 1340 cm^{-1} are much more strongly coupled than the other two, which lie at 466 and 1180 cm^{-1}.

VIII. Transport, Thermodynamic, and Infrared Properties

In this section we consider electronic interactions and their manifestations, such as transport properties and superconductivity, both of which reflect the scattering of the elementary electronic excitations. The greatest attention has been given to the electron–phonon interaction. In

[284]M. J. Rice and H.-Y. Choi, *Phys. Rev. B* **45**, 10173 (1992).

addition, Coulomb scattering must be considered, and magnetic fluctuations may arise. The largest contribution to the mean free path in the fullerides may have a static origin—the intrinsic merohedral disorder.

22. D.C. TRANSPORT DATA

a. Resistivity

The report of metallic conductivity in K_xC_{60} by Haddon et al.[285] led quickly to the discovery of superconductivity and, more slowly, to the probing of the transport properties of fullerides. Early reports of the resistivity[286,287] ρ of K_3C_{60} gave a value of ~5–10 mΩ cm near room temperature. Improvements in sample quality[288,289] have improved this so far by only a factor of ~2. Xiang et al.,[289], however, succeeded in obtaining clearly metallic behavior of $\rho(T)$ in single crystals of K_3C_{60}, with $\rho(280\,\text{K}) \sim 5\,\text{m}\Omega\,\text{cm}$, a residual resistivity ratio $\rho(280\,\text{K})/\rho(20\,\text{K}) = 2$ and $d\log\rho/dT \sim 4 \times 10^{-3}/\text{K}$ at 250 K. Their data is pictured in Fig. 26. The resistivity is concave upward in the whole range, with $d\rho/dT$ approaching zero at low temperature, and there is no indication of an incipient metal–insulator transition as suggested by earlier data.[288] Maruyama et al.[286] reported metallic resistivity, with nearly linear behavior from 20–250 K, with $\rho(250\,\text{K}) \sim 14\,\text{m}\Omega\,\text{cm}$ and $d\log\rho/dT \sim 4 \times 10^{-3}/\text{K}$. Inabe et al.[290] have reported metallic resistivities of single crystals, but the temperature dependence is more nearly linear than quadratic as found by Xiang et al., and the value of ρ just above T_c is a factor of five larger.

Klein et al.[291] have inferred the resistivity from microwave surface resistance measurements at 60 GHz on pressed pellets, obtaining a factor

[285]R. C. Haddon, A. F. Hebard, M. J. Rosseinsky, D. W. Murphy, S. J. Duclos, K. B. Lyons, B. Miller, J. M. Rosamilia, R. M. Fleming, A. R. Kortan, S. H. Glarum, A. V. Makhija, A. J. Muller, R. H. Eick, S. M. Zahurak, R. Tycho, G. Dabbagh, and F. A. Thiel, *Nature* **350**, 320 (1991).

[286]Y. Maruyama, T. Inabe, H. Ogata, Y. Achiba, S. Suzuki, K. Kikuchi, and I. Ikemoto, *Chem. Lett. (Japan)* **1991**, 1849 (1991).

[287]G. P. Kochanski, A. F. Hebard, R. C. Haddon, and A. T. Fiory, *Science* **255**, 184 (1992).

[288]T. T. M. Palstra, R. C. Haddon, A. F. Hebard, and J. Zaanen, *Phys. Rev. Lett.* **68**, 1054 (1992).

[289]X.-D. Xiang, J. G. Hou, G. Briceño, W. A. Vareka, R. Mostovoy, A. Zettl, V. H. Crespi, and M. L. Cohen, *Science* **256**, 1190 (1992).

[290]T. Inabe, H. Ogata, Y. Maruyama, Y. Achiba, S. Suzuki, K. Kikuchi, and I. Ikemoto, *Phys. Rev. Lett.* **69**, 3797 (1992).

[291]O. Klein, G. Grüner, S.-M. Huang, J. B. Wiley, and R. B. Kaner, *Phys. Rev. B* **46**, 11247 (1992).

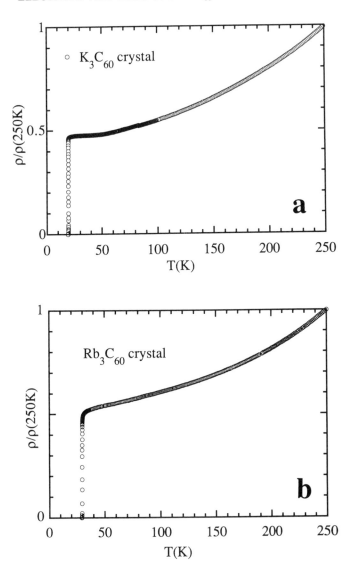

FIG. 26. Resistivity, normalized to its value at 250 K, of single crystal K$_3$C$_{60}$, showing the normal metallic behavior and the superconducting transition, which occurs at T_c = 19.8 K for this crystal. From X.-D. Xiang, J. G. Hou, G. Briceño, W. A. Vareka, R. Mostovoy, A. Zettl, V. H. Crespi, and M. L. Cohen, *Science* **256**, 1190 (1992). Copyright 1192 by the AAAS.

of 10 decrease in resistivity between 300 and 20 K, and a residual resistivity of only 0.5 mΩ cm. Both values are about a factor of five "better" than the single crystal data of Xiang *et al.*, but the method of obtaining $\rho(T)$ is much less direct. Rotter *et al.*[292] used the low frequency of infrared reflectance measurements to infer a low temperature resistivity of 0.4 mΩ cm, similar to that obtained by Klein *et al.*

Crespi *et al.*[293,294] have analyzed the data of Xiang *et al.*, noting that an $A + BT^2$ behavior fits reasonably well up to 250 K as shown in Fig. 27. The variation, however, seems to be falling below T^2 behavior beginning

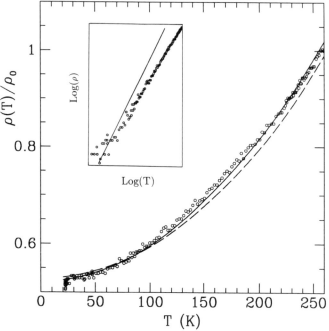

FIG. 27. The normalized resistivity of Fig. 26, showing fits to the form $A + BT^2$ (solid line) and to a modified form accounting for a temperature-dependent value of $N(E_F)$. The inset provides a comparison, in log–log fashion, of the resistivity corrected for thermal expansion to a T^2 behavior. From V. H. Crespi, J. G. Hou, X.-D. Xiang, M. L. Cohen, and A. Zettl, *Phys. Rev. B* **46**, 12064 (1992).

[292]L. D. Rotter, Z. Schlesinger, J. P. McCauley, Jr., N. Coustel, J. E. Fischer, and A. B. Smith III, *Nature* **355**, 532 (1992).
[293]V. H. Crespi, J. G. Hou, X.-D. Xiang, M. L. Cohen, and A. Zettl, *Phys. Rev. B* **46**, 12064 (1992).
[294]J. G. Hou, V. H. Crespi, X.-D. Xiang, W. A. Vareka, G. Briceño, A. Zettl, and M. L. Cohen, *Solid State Commun.* **86**, 643 (1993).

at ~200 K. Klein et al.[291] obtained a similar quality of fit to an $A + BT^2$ law with, however, quite different parameters. Crespi et al. found that phonon scattering could account for the resistivity data of Xiang et al. if the coupling is dominated by coupling to phonons at ~400 cm^{-1} as in the model of Jishi and Dresselhaus.[152] The coupling only to higher frequency modes (with average ~900–1000 cm^{-1} as calculated by Schlüter et al.[163], or with average ~1500 cm^{-1} as suggested by Varma et al.[160]) produces more curvature in the 50–250 K range than is measured. With vibrations spanning the range from 30 to 1800 cm^{-1}, it is likely that it will be necessary to take into account more aspects of the vibrational spectrum than can be treated with a single average frequency. It also has not been established that dynamic scattering and disorder scattering are distinct, i.e., that Mattheissen's rule is obeyed. In any case, until optimum sample preparation techniques are achieved and agreement between different groups is obtained, the interpretation should not be relied on too heavily.

b. Hall Coefficient: Thermopower

The Hall coefficient of K_3C_{60} films was reported by Palstra et al.[288] The Hall coefficient at low temperature is $R_H = -0.35 \times 10^{-9} \, m^3/C = -0.02 \, \Omega_c/e$, where Ω_c is the unit cell volume. Making the (incorrect) identification $R_H = 1/ne$ leads to an interpretation of a density n of 50 carriers per C_{60} molecule, clearly a nonsensical interpretation. In fact, in conventional metals R_H passes through zero at the center of a band, and that is just what seems to be the case here. Indeed, upon increasing the temperature, R_H increases linearly[288] and passes through zero around 200 K. Thus, the Hall coefficient is consistent with a half-filled metallic band, but the temperature dependence (which can arise from a number of causes) is not yet understood.

Inabe et al.[290] have reported thermopower measurements on single crystals of K_3C_{60} and Rb_3C_{60} that show clear metallic resistivity curves. For both metals the thermopower is negative and roughly linear, with a value of $-(10-15) \, \mu V/K$ at room temperature. This behavior is consistent with normal metallic behavior.

23. HEAT CAPACITY; INFRARED RESPONSE

Measurements of the heat capacity of C_{60} and its fullerides at low temperatures provides a very useful "spectroscopy" of the intermolecular vibrational and librational modes of the crystals. The data of Beyermann *et al.*[295] and Atake *et al.*[296] have been fit to models accounting for Debye-like acoustic intermolecular branches, optic intermolecular branches, and libronic modes, the latter two being treated as (one or more) Einstein modes. The fit parameters show variations: Debye temperatures of 37 and 70 K are obtained. Beyermann *et al.*'s results,[295,298] which seem to use the more physical fit, give a Debye temperature of 37 K, and librational and optic energies of 58 and 30 K, respectively. These values can be compared with the calculated values of Li *et al.* shown in Fig. 8, where the librational modes lie in the 15–30 K range and the optic modes span the 30–75 K range.

The reflectivity of the metallic alkali fullerides has been measured by several groups.[299–302] A conventional Drude behavior is seen in the far infrared, with Drude plasma energy ~1.5 eV in both K_3C_{60} (Isawa *et al.*[300]) and Rb_3C_{60} (Degiorgi *et al.*[301]). This value is comparable with the band theory value of 1.22 eV for K_3C_{60} reported by Erwin and Pickett.[218] A background dielectric constant of 5.6 was obtained by Isawa *et al.* for K_3C_{60}, and a value of 7 for Rb_3C_{60} by Degiorgi *et al.*[301] A "mid-infrared absorption" at 0.9–1.0 eV corresponds to the interband $t_{1u} \rightarrow t_{1g}$ transition energies (discussed in Section 17).

24. THEORETICAL TREATMENTS OF TRANSPORT

Electrical transport in fullerides is likely to be complicated by the orientational disorder of the molecules as well as unusual thermal lattice effects (librations) and possible electronic effects arising from the narrow bandwidth. The first theoretical prediction was by Erwin and Pickett[218] for the idealized structure with all molecules aligned (T_h point group symmetry in an fcc Bravais lattice). Applying expressions based on Fermi liquid quasi-particles as described by the local density band structures,

[295]W. P. Beyermann, M. F. Hundley, J. D. Thompson, F. N. Diederich, and G. Grüner, *Phys. Rev. Lett.* **68**, 2046 (1992).

[296]T. Atake, T. Tanaka, H. Kawaji, K. Kikuchi, K. Saito, S. Suzuki, I. Ikemoto, and Y. Achiba, *Physica C* **185–189**, 427 (1991).

[297]W. Que, *Phys. Rev. Lett.* **69**, 2736 (1992).

[298]W. P. Beyermann, J. D. Thompson, M. F. Hundley, and G. Grüner, *Phys. Rev. Lett.* **68**, 2737 (1992).

[299]S. A. FitzGerald, S. G. Kaplan, A. Rosenberg, A. J. Sievers, and R. A. S. McMordie, *Phys. Rev. B* **45**, 10165 (1992).

[300]Y. Isawa, K. Tanaka, T. Yasuda, T. Koda, and S. Koda, *Phys. Rev. Lett.* **69**, 2284 (1992).

[301]L. Degiorgi, G. Grüner, P. Wachter, S.-M. Huang, J. Wiley, R. L. Whetten, R. B. Kaner, K. Holczer, and F. Diederich, *Phys. Rev. B* **46**, 11250 (1992).

[302]L. Degiorgi, P. Wachter, G. Grüner, S.-M. Huang, J. Wiley, and R. B. Kaner, *Phys. Rev. Lett.* **69**, 2987 (1992).

they obtained a density of states (DOS) of $N(E_F) = 13.2$ states/eV-cell (0.22 states/eV-C atom), an average Fermi velocity $v_F = 1.8 \times 10^7$ cm/sec, and a Drude plasma energy $\Omega_p = 1.2$ eV. For comparison, the corresponding values for (say) Nb are {0.7 states/eV-atom, 6×10^7 cm/sec, and 4 eV}, and so these properties are different, but not wildly so (a factor of three smaller), from conventional superconducting metals. The Drude plasma energy is given by the expression

$$\Omega_p^2 = \frac{4\pi}{3} e^2 \sum_k v_k^2 \, \delta(E_k - E_F) = \frac{4\pi}{3} e^2 v_F^2 N(E_F), \qquad (24.1)$$

which depends only on Fermi surface quantities (and not on bandwidth or carrier density). This quantity describes the static (or very low frequency) response of the electronic system; for example, the magnetic field (London) penetration depth λ_L is given by $\lambda_L = c/\Omega_p$. The predicted value of λ_L for orientationally ordered K$_3$C$_{60}$ is 1600 Å. These values are listed in Table VIII and are compared with data in following sections.

Bloch–Grüneisen transport theory based on a Boltzmann treatment of scattering processes is successful in describing phonon-limited transport in conventional metals as long as successive scattering events can be treated as independent, i.e., the mean free path l is substantially larger than the interatomic separation. A Boltzmann treatment of Coulomb

TABLE VIII. CALCULATED AND EXPERIMENTALLY
DERIVED VALUES FOR VARIOUS NORMAL STATE PARAMETERS FOR
K$_3$C$_{60}$

PARAMETER	UNITS	VALUE
Bandwidth	eV	0.61
$N(E_F)$	states/eV-cell	13.2
V_F	10^7 cm/sec	1.77
$\hbar\Omega_p$	eV	1.22
Λ(clean limit)	Å	1600
Λ(dirty limit)	Å	3000–3500
l (300 K)	Å	$7\lambda_{tr}$
ρ (300 K)	$\mu\Omega$ cm	$780\lambda_{tr}$
$d \log \rho/dP$	kbar^{-1}	-0.041
R^H	10^{-8} m^3/C	0.70
$d \log R^H/dP$	kbar^{-1}	-0.003
S (300 K)	μV K^{-1}	-15.4
dS/dP (300 K)	μV K^{-1} kbar^{-1}	~ 1
$\xi(0)$	Å	45

Most results are from S. C. Erwin and W. E. Pickett, *Phys. Rev. B* **46**, 14257 (1992).

scattering is also likely to be reasonable, but accounting for orientational disorder is not so obvious, and it is discussed in what follows.

To describe phonon-limited transport, one needs to know the transport version $\alpha_{tr}^2 F(\omega)$ of the electron–phonon spectral function $\alpha^2 F(\omega)$. In the absence of specific information on the strength and spectral distribution of the coupling strength, Erwin and Pickett[303] assumed $\alpha_{tr}^2 F(\omega) \propto F(\omega)$, the phonon DOS measured with inelastic neutron scattering by Prassides et al.,[122] and evaluated the low order Bloch–Grüneisen variational expression for the resistivity $\rho(T)$. The result is shown in Fig. 28 for comparison with the single crystal data of Figs. 26 and 27. It has a somewhat superlinear behavior above ~30 K, becoming almost linear at ~200 K and above. This behavior is rather different from the measured data, but as mentioned there is not yet an experimental consensus, and merohedral disorder has been neglected.

It is necessary, of course, to take into account in the theory the effects of merohedral disorder in K_3C_{60}. Experimentally, there seems to be no

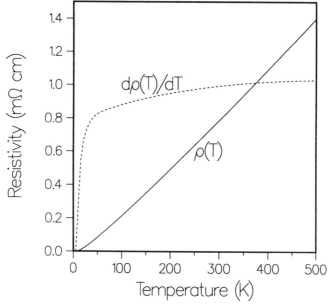

FIG. 28. Theoretical resistivity curve, using the Bloch–Grüneisen expression with electron–phonon spectral function modeled by the measured phonon density of states. The variation at low temperature is much stronger than in the data shown in Figs. 26 and 27. From S. C. Erwin and W. E. Pickett, *Phys. Rev. B* **46**, 14257 (1992).

[303]S. C. Erwin and W. E. Pickett, *Phys. Rev. B* **46**, 14257 (1992).

structural transitions below room temperature as there are in fullerite; the C_{60} molecules are statically disordered (no doubt with librational amplitudes increasing with increasing temperature). This disorder was first addressed by Gelfand and Lu,[304] who used a three-band tight-binding model to represent the band structure. The hopping amplitudes in their tight-binding model for electron transfer between molecules differed considerably depending on whether neighboring molecules were orientationally equivalent or not, allowing the possibility of large effects due to disorder.

The disorder was modeled with supercells of up to 6^3 cubic cells (864 molecules) with periodic boundary conditions, and ensemble averages were carried out. For maximal disorder, i.e., orientation of each molecule is chosen randomly from the two possibilities, the bandwidth changes very little but the peaks arising from crystalline periodicity are washed out. In fact, the resulting DOS is surprisingly close to a (somewhat rounded) rectangular form. Because of the placement of the Fermi level, the value of $N(E_F)$ decreases by only 10%. Although the effect of disorder on the DOS seems substantial, the states remain itinerant except very near the band edges.

This work was extended by Gelfand and Lu[305] to transport properties by using the electronic states in the disordered supercells in the nonzero frequency Kubo–Greenwood linear response formalism, smearing the energies of the states, and extrapolating as well as possible to zero frequency. They obtain the estimate of the residual resistivity due to orientational disorder $\rho_0 = 300\ \mu\Omega\,\text{cm}$, rather less than the lowest experimental results. They also obtained values of the disorder-limited (electronic) thermal conductivity and thermopower; the latter is found to be negative. They find that this method of calculation gives results that are not too sensitive to the amount of disorder as long as the material is not close to perfect ordering.[306]

The Hall coefficient, assuming isotropic scattering (i.e., not too low temperature) was calculated for ordered K_3C_{60} by Erwin and Pickett to be $R_H = +0.70 \times 10^{-8}\ \text{m}^3/\text{C} = 2.2\ \Omega_c/e$, in "natural" units of cell volume Ω_c per elementary charge e. The one measurement of R_H by Palstra et al.[288] gives a linearly temperature-dependent result that changes sign from negative to positive at 220 K. The calculation of the thermopower $S(T)$ is sensitive to the choice of energy dependence of the scattering time τ: assuming the scattering rate $1/\tau(E)$ is constant, a positive $S(T)$ is

[304]M. P. Gelfand and J. P. Lu, *Phys. Rev. Lett.* **68**, 1050 (1992).
[305]M. P. Gelfand and J. P. Lu, *Phys. Rev. B* **46**, 4367 (1992); erratum, *ibid.* **47**, 4149 (1993).
[306]M. P. Gelfand and J. P. Lu, *Appl. Phys. A* **56**, 215 (1993).

obtained; while the perhaps more realistic assumption $1/\tau(E) \propto N(E)$ leads to a negative thermopower much like, and similar in magnitude to, that measured by Inabe et al.[290]

IX. Superconductivity

25. Properties of the Superconducting State

To obtain information relating to the mechanism of pairing, to the symmetry and character of the order parameter, and to the supercon-ducting properties that may have technological application, it is essential to characterize the properties in the superconducting state. The response to applied electromagnetic fields is particularly important. Two items should be kept in mind: (1) in the presence of applied fields, the superconducting transition becomes broadened, and the interpretation of what the value of T_c is becomes important and affects the reported material parameters; and (2) the reported values involve theoretical interpretation and in some cases extrapolation (such as $T \rightarrow 0$).

The superconducting properties are characterized by two length scales, the coherence length ξ and the London penetration depth λ_L, and an energy scale, the superconducting gap parameter Δ. The lower and upper critical magnetic field scales H_{c1} and H_{c2} reflect these scales. The $T = 0$ superconducting coherence length ξ_0 is obtained from

$$H_{c2}(0) = \frac{\Phi_0}{2\pi\xi_0^2},\qquad (25.1)$$

where Φ_0 is the magnetic flux quantum. Besides being an important figure of merit for applications, the shape of $H_{c2}(T)$, its value at $T = 0$, and its derivative H'_{c2} at T_c are prime indicators of important material properties. The standard for conventional superconductors is given by the (weak coupling) Bardeen–Cooper–Schrieffer[307] (BCS) theory and its strong coupling generalization[272,308]

A variety of parameters relating to the superconducting state are presented in Table IX. A survey of superconducting state parameters has

[307]J. Bardeen, L. N. Cooper, and J. R. Schrieffer, *Phys. Rev.* **108**, 1175 (1957).
[308]J. Carbotte, *Rev. Mod. Phys.* **62**, 1027 (1990).

TABLE IX. REPORTED VALUES OF PARAMETERS RELATING TO THE
SUPERCONDUCTING STATE OF K_3C_{60} AND Rb_3C_{60}

PARAMETER	UNITS	K_3C_{60}	Rb_3C_{60}
T_c^{opt}	K	19.8^c	29.6
λ_L	Å	$2400,^k\ 4800,^g$	$4600,^e\ 8000^d$
		$6000,^e\ 8000^d$	
ξ_0	Å	$31,^f\ 26,^{m,k}\ 34^i$	$24,^f\ 30^i$
$\xi_{P,0}$	Å	150^m	
$\kappa = \lambda_L/\xi_0$		$82,^h\ 92^k$	~100–200
$2\Delta(T=0)$	meV	$8.8,^b\ 6.0^d$	$13.2,^a\ 7.4^d$
$2\Delta/k_BT_c$		$5.2,^b\ 3.6^d$	$5.3,^a\ 3.0^d$
$H_{c2}(4.2\ K)$	T	$30,^f\ 25,^i\ 29^j$	55^f
$-dH_{c2}/dT(T_c)$	T/K	$2.80,^f\ 2.0,^i\ 2.1^j$	$3.9,^f\ 3.9,^l\ 2.0^i$
		$3.73,^k\ 5.5^m$	

^{a}Z. Zhang, C.-C. Chen, S. P. Kelty, H. Dai, and C. M. Lieber, *Nature* **353**, 333 (1991).
^{b}Z. Zhang, C.-C. Chen, and C. M. Lieber, *Science* **254**, 1619 (1991).
^{c}X.-D. Xiang, J. G. Hou, G. Briceño, W. A. Vareka, R. Mostovoy, A. Zettl, V. H. Crespi, and M. L. Cohen, *Science* **256**, 1190 (1992).
^{d}L. Degiorgi, P. Wachter, G. Grüner, S.-M. Huang, J. Wiley, and R. B. Kaner, *Phys. Rev. Lett.* **69**, 2987 (1992).
^{e}R. Tycho, G. Dabbagh, M. J. Rosseinsky, D. W. Murphy, A. P. Ramirez, and R. M. Fleming, *Phys. Rev. Lett.* **68**, 1912 (1992).
^{f}S. Foner, E. J. McNiff, Jr., D. Heiman, S.-M. Huang, and R. B. Kaner, *Phys. Rev. B* **46**, 14936 (1992).
^{g}Y. J. Uemura, A. Keren, L. P. Le, G. M. Luke, B. J. Sternlieb, W. D. Wu, J. H. Brewer, R. L. Whetten, S.-M. Huang, S. Liu, R. B. Kaner, R. Diederich, S. Donovan, G. Grüner, and K. Holczer, *Nature* **352**, 605 (1991).
^{h}O. V. Dolgov and I. I. Mazin, *Solid State Commun.* **81**, 935 (1992).
^{i}C. E. Johnson, H. W. Jiang, K. Holczer, R. B. Kaner, and F. Diederich, *Phys. Rev. B* **46**, 5880 (1992).
^{j}G. S. Boebinger, T. T. M. Palstra, A. Passner, M. J. Rosseinsky, D. W. Murphy, and I. I. Mazin, *Phys. Rev. B* **46**, 5876 (1992).
^{k}K. Holczer, O. Klein, G. Grüner, J. D. Thompson, F. Diederich, and R. L. Whitten, *Phys. Rev. Lett.* **67**, 271 (1991).
^{l}G. Sparn, J. D. Thompson, R. L. Whetten, S.-M. Huang, R. B. Kaner, F. Diederich, G. Grüner, and K. Holczer, *Phys. Rev. Lett.* **68**, 1228 (1992).
^{m}T. T. M Palstra, R. C. Haddon, A. F. Hebard, and J. Zaanen, *Phys. Rev. Lett.* **68**, 1054 (1992).
^{n}Note that all data are not from samples with optimum values of T_c (denoted T_c^{opt}), and that some quantities can be obtained in a variety of ways.

been given by Holczer and Whetten.[309] The upper critical magnetic field $H_{c2}(T)$ measured by Foner et al.[310] for both K_3C_{60} and Rb_3C_{60} powders to high applied fields is shown in Fig. 29, where it is compared with earlier data and related to theoretical expressions. The critical fields are very high ($\sim$30 and 55 T, respectively, at low temperature according to Foner et al.), but their temperature dependences can be fit with the standard Werthamer, Helfand, Hohenberg,[311] and Maki[312] expression including Pauli paramagnetic limiting. Boebinger et al.[313] contend that H_{c2} is enhanced above the conventional theoretical value at low temperature, and suggest that strong coupling, Fermi surface anisotropy, or granularity is responsible.

These fullerene superconductors clearly are extreme type II superconductors, with Landau–Ginzburg parameter $\kappa \equiv \lambda_L/\xi \sim 100$–200. The result is a large region of "mixed state" superconductivity in which magnetic vortices penetrate the superconducting material. Due to the intrinsic merohedral disorder in the crystal lattices of these materials and to the granular nature of most samples (grain size $\sim$ 100 Å), many data may reflect the "dirty limit" in which the mean free path is comparable with or smaller than the intrinsic coherence length.

The superconducting gap $2\Delta(T)$ can be obtained from spectroscopic or thermodynamic measurements. Lieber and collaborators[314,315] have applied scanning tunneling microscopy to obtain tunneling spectra for K_3C_{60} and Rb_3C_{60}. The normalized conductance data for K_3C_{60} are shown in Fig. 30, and similar results were obtained for Rb_3C_{60}. A gap opens below T_c, and at low temperature the residual conductance inside the gap is low, as expected for a conventional (s wave) BCS superconductor. The reduced gap value of $2\Delta/k_BT_c = 5.3$ was obtained for both materials, indicating strong coupling (the weak coupling value is 3.5). The gap parameter $\Delta(T)$ was obtained from the fit to the normalized conductance, and the temperature dependence of both materials, shown in Fig. 31, is consistent with the BCS form.

The reduced gap obtained from NMR experiments[316] is $2\Delta/k_BT_c = 3$–4.

[309]K. Holczer and R. L. Whetten, Carbon 30, 1261 (1992).

[310]S. Foner, E. J. McNiff, Jr., D. Heiman, S.-M. Huang, and R. B. Kaner, Phys. Rev. B 46, 14936 (1992).

[311]N. R. Werthamer, E. Helfand, and P. C. Hohenberg, Phys. Rev. 147, 295 (1966).

[312]K. Maki, Physics 1, 127 (1964).

[313]G. S. Boebinger, T. T. M. Palstra, A. Passner, M. J. Rosseinsky, D. W. Murphy, and I. I. Mazin, Phys. Rev. B 46, 5876 (1992).

[314]Z. Zhang, C.-C. Chen, S. P. Kelly, H. Dai, and C. M. Lieber, Nature 353, 333 (1991).

[315]Z. Zhang, C.-C. Chen, and C. M. Lieber, Science 254, 1620 (1991).

[316]R. Tycho, G. Dabbagh, M. J. Rosseinsky, D. W. Murphy, A. P. Ramirez, and R. M. Fleming, Phys. Rev. Lett. 68, 1912 (1992).

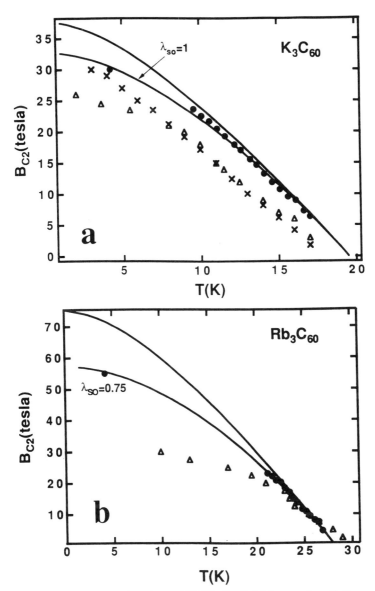

FIG. 29. Data for $H_{c2}(T)$ for (a) K_3C_{60} and (b) Rb_3C_{60}. Solid dots are from S. Foner, E. J. McNiff, Jr., D. Heiman, S.-M. Huang, and R. B. Kaner, *Phys. Rev. B* **46,** 14936 (1992); triangles are from C. E. Johnson, H. W. Jiang, K. Holczer, R. B. Kaner, and F. Diederich, *Phys. Rev. B* **46,** 5880 (1992); crosses are from G. S. Boebinger, T. T. M. Palstra, A. Passner, M. J. Rosseinsky, D. W. Murphy, and I. I. Mazin, *Phys. Rev. B* **46,** 5876 (1992). The curves are fits to the solid dots including (lower) and neglecting (upper) Pauli paramagnetic limiting. Figures from S. Foner, E. J. McNiff, Jr., D. Heiman, S.-M. Huang, and R. B. Kaner, *Phys. Rev. B* **46,** 14936 (1992).

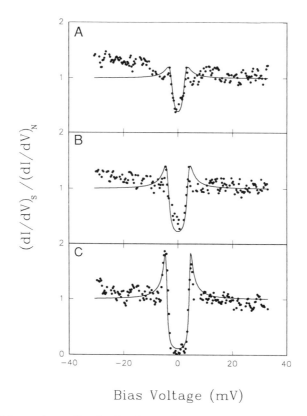

Bias Voltage (mV)

FIG. 30. Measured tunneling density of states for K_3C_{60}, showing the development of the superconducting energy gap as the temperature is lowered from $T/T_c = 0.83$ (a), 0.55 (b), 0.25 (c). The solid lines are fits to the data using a BCS form. From Z. Zhang, C.-C. Chen, and C. M. Lieber, *Science* **254**, 1619 (1991). Copyright 1991 by the AAAS.

Reflectivity measurements[317] conducted above and below T_c show a BCS-like behavior with $2\Delta/k_B T_c = 3.6$ and 3.0 for K_3C_{60} and Rb_3C_{60}, respectively. Both of these measurements therefore indicate weak coupling. The discrepancy with the tunneling results of Lieber *et al.* is not understood, but good tunneling data are considered to provide the definitive answer. It should also be noted that there is considerable controversy whether fullerene superconductors should be considered, and interpreted as, "clean limit" or "dirty limit" according to whether the mean free path is larger than or smaller than the superconducting length.

[317]L. D. Rotter, Z. Schlesinger, J. P. McCauley, Jr., N. Coustel, J. E. Fischer, and A. B. Smith III, *Nature* **355**, 532 (1992).

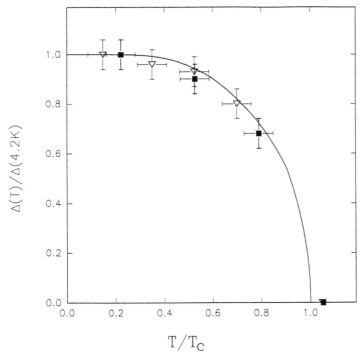

$$T/T_C$$

FIG. 31. The measured values of $\Delta(T)/\Delta(4.2\,\mathrm{K})$ for K$_3$C$_{60}$ (solid symbols) and Rb$_3$C$_{60}$ (open symbols), where the gap is determined from data such as in Fig. 30. The solid line is the BCS result. From Z. Zhang, C.-C. Chen, and C. M. Lieber, *Science* **254**, 1619 (1991). Copyright 1991 by the AAAS.

The quality of the samples has to be considered to begin to answer these questions.

The specific heat jump at T_c, which is not easy to obtain because of the large specific heat due to the intermolecular vibrations and librations, has been measured by Ramierez *et al.*[318] to be $\Delta C/T_c = 68 \pm 13\,\mathrm{mJ/mole\,K^2K_3C_{60}}$. A conventional interpretation is

$$\frac{\Delta C}{T_c} = [1.43 + g(\lambda)]\gamma_0(1 + \lambda_{\mathrm{tot}}), \qquad (25.2)$$

where $\gamma_0 = (\pi^2/3)k_{\mathrm{B}}^2 N(E_{\mathrm{F}})$ is the Sommerfeld coefficient of the electronic specific heat, $g(\lambda)$ is a strong coupling correction depending on the

[318]A. P. Ramirez, M. J. Rosseinsky, D. W. Murphy, and R. C. Haddon, *Phys. Rev. Lett.* **69**, 1687 (1992).

pairing strength λ, and λ_{tot} is the total mass enhancement due to all dynamical processes. For comparison, using the band value of $N(E_F) \sim 13$ states/eV $-$ C_{60} one obtains $1.43\gamma_0 = 31$ mJ/mole K^2, indicating either strong coupling enhancements $1 + \lambda_{tot}$ of ~ 2 or a larger value of $N(E_F)$.

From the measured magnetic susceptibility, Ramirez et al.[318] obtain (after corrections for core contributions but none for orbital contributions) from the spin susceptibility $\chi_s = \mu_B^2 N_\chi(E_F)$ values of $N_\chi(E_F) = 28$ (respectively, 38) states/eV-C_{60} for K_3C_{60} (respectively, Rb_3C_{60}). These values, which are also about a factor of ~ 2 larger than the band theory values, involve enhancements only from magnetic excitations and not from nonmagnetic excitations, including phonons. LDA calculation by Antropov et al.[44] of the magnetic enhancement gives enhancement factors of 1.53 and 1.67 for K_3C_{60} and Rb_3C_{60}, respectively.

Ramirez et al. inferred from their analysis of the data that the electron–phonon coupling strength was weak ($\lambda \sim 0.5$) and therefore that the coupling must be only to high frequency modes $\omega > 1000$ K. Other analysis has led to the opposite point of view. Kresin[319] used the values of T_c, the isotope exponent $\alpha = 0.37$, and NMR-derived densities of states to argue that only a strong coupling interpretation is viable: $\lambda = 1.5$ (2.1) for K_3C_{60} (Rb_3C_{60}), and $\langle \omega \rangle \sim 250$ cm^{-1} ~ 300 K. Mazin et al.[320] also note that several experiments indicate strong coupling, but they also recognize the direct evidence for strongly coupled high frequency modes. In particular, they use data for $H'_{c2}(T_c)$ to obtain $\Delta C/T_c$ values in the range 230–380 mJ/mole K^2, much larger than reported by Ramirez et al.[318] They provide a consistent interpretation of these observations in terms of a model with coupling at both low frequency (~ 40 cm^{-1}) and at high frequency (~ 700–1100 cm^{-1}). The low frequency coupling contributes a large strength $\lambda_{lo} \sim 2.7$ and therefore strong coupling effects, but contributes little to T_c because of the low frequencies that are involved. The high frequency coupling strength $\lambda_{hi} = 0.5$ accounts for the superconductivity. The Eliashberg equations can be solved for a model electron–phonon spectral function $\alpha^2 F$ of this sort, and the results account for much of the data that have been discussed. Dolgov and Mazin[321] have summarized several of the relationships arising from Ginzburg–Landau theory and analyzed several experimental results. They note that both $2\Delta/k_B T_c$ from tunneling and magnetic measurements imply a strong-coupling value $\lambda \sim 1.5$–2. Erwin and Pickett[218] had also noted that LDA band structure values of v_F and $N(E_F)$ combined with experimental data on the

[319]V. Kresin, Phys. Rev. B 46, 14883 (1992).
[320]I. I. Mazin, O. V. Dolgov, A. Golubov, and S. V. Shulga, Phys. Rev. B 47, 538 (1993).
[321]O. V. Dolgov and I. I. Mazin, Solid State Commun. 81, 935 (1992).

coherence length and energy gap implied strong coupling. Dolgov and Mazin suggested strongly coupled soft modes (rotational, $\omega \sim 5$–8 meV) could account for strong coupling effects without influencing T_c appreciably.

So far there has been little discussion of a gap parameter of any but fully symmetric, singlet type (s wave), and the simplest s-wave gap seems consistent with most data. An exception is the model of Mele and Erwin,[322] in which they construct an anisotropic singlet, d-wave gap function, which nevertheless has no nodes. Since there is a true gap in their model, it may not be straightforward to distinguish it from the s-wave case (which may itself have an anisotropic gap). The model leads to distinguishing characteristics, however, such as a non-BCS temperature dependence of the gap and a distinct quasi-particle density of states in the superconducting state.

26. ISOTOPE EFFECT

The isotope shift of T_c is widely regarded as the defining signature of superconducting pairing resulting from phonons (more generally, any lattice degrees of freedom). This expectation survives in spite of the fact that (1) numerous calculations demonstrate that the isotope shift is not a simple quantity to interpret, and (2) the isotope shift is not understood *quantitatively* in any nonelemental superconductor. Nevertheless, a clear isotope shift indicates atomic motions are involved, while a vanishing effect is most easily accounted for with a nonlattice pairing mechanism.

The experimental measurements to date are given in Table X, where it can be seen that the reports vary widely. The substitution of ^{12}C with ^{13}C was not particularly well controlled in the experiments of Ramirez *et al.*,[323] Ebbesen *et al.*,[324,325] or Zakhidov *et al.*,[326] who obtained (mostly for the Rb compound) widely differing results: isotope exponents α from $\sim$0.4–2. The early results of Chen and Lieber,[327] obtaining $\alpha = 0.30$,

[322]E. J. Mele and S. C. Erwin, *Phys. Rev. B* **47**, 2948 (1993).

[323]A. P. Ramirez, A. R. Kortan, M. J. Rosseinsky, S. J. Duclos, A. M. Mujsce, R. C. Haddon, D. W. Murphy, A. V. Makhija, S. M. Zahurak, and K. B. Lyons, *Phys. Rev. Lett.* **68**, 1058 (1992).

[324]T. W. Ebbesen, J. S. Tsai, K. Tanigaki, J. Tabuchi, Y. Shimakawa, Y. Kubo, I. Hirosawa, and J. Mizuki, *Nature* **355**, 630 (1992).

[325]T. W. Ebbesen, J. S. Tsai, K. Tanigaki, H. Hiura, Y. Shimakawa, Y. Kubo, I. Hirosawa, and J. Mizuki, *Physica C* **203**, 163 (1992).

[326]A. A. Zakhidov, K. Imaeda, D. M. Petty, K. Yakushi, H. Inokuchi, K. Kikuchi, I. Ikemoto, S. Suzuki, and Y. Achiba, *Phys. Lett. A* **164**, 355 (1992)).

[327]C.-C. Chen and C. M. Lieber, *J. Am. Chem. Soc.* **114**, 3141 (1992).

TABLE X. REPORTED VALUES OF THE CARBON ISOTOPE SHIFT OF
T_c FOR K_3C_{60} AND Rb_3C_{60}

K_3C_{60}	$^{13}C_{60}$ (%)	Rb_3C_{60}	$^{13}C_{60}$ (%)	Reference
		0.37 ± 0.05	75 ± 5	a
		1.4 ± 0.5	20–45	b
0.30 ± 0.06	99			c
1.3 ± 0.3	60 ± 12	2.1 ± 0.3	60 ± 12	d
		0.35; 0.7	55; 50	e

[a] A. P. Ramirez, A. R. Kortan, M. J. Rosseinsky, S. J. Duclos, A. M. Mujsce, R. C. Haddon, D. W. Murphy, A. V. Makhija, S. M. Zahurak, and K. B. Lyons, *Phys. Rev. Lett.* **68**, 1058 (1992).
[b] T. W. Ebbesen, J. S. Tasi, K. Tanigaki, J. Tabuchi, Y. Shimakawa, Y. Kubo, I. Hirosawa, and J. Mizuki, *Nature* **335**, 620 (1992).
[c] C.-C. Chen and C. M. Lieber, *J. Am. Chem. Soc.* **114**, 3141 (1992).
[d] A. A. Zakhidov, K. Imaeda, D. M. Petty, K. Yakushi, H. Inokuchi, K. Kikuchi, I. Ikemoto, S. Suzuki and Y. Achiba, *Phys. Lett. A* **164**, 355 (1992).
[e] C.-C. Chen and C. M. Lieber, *Science* **259**, 655 (1993).
[f] The percentage of ^{13} Substitution, or the range studied, is noted. See text, especially for discussion of the results in Ref. e.

involved 99% substitution of ^{13}C and seemed to leave little ambiguity for K_3C_{60}, except for the unresolved question of why the other groups found such varied results.

This question was clarified by a subsequent experiment of Chen and Lieber,[328] who repeated the experiments with two distinct substitutions of ~50% ^{13}C. In one case, each C_{60} molecule had ~50% of each isotope; while in the other case, each molecule was entirely composed of one or the other of the isotopes, and a 50% mixture of molecules of different mass were used to make the solid. Optical measurements of the solid confirmed that the IR active phonons appeared as expected, mass-averaged in the first case and mass-differentiated in the second. Different shifts of T_c were obtained. In the homogeneous sample, where all molecules have both isotopes, the isotope exponent is 0.3 ± 0.05 (consistent with their earlier report). For the inhomogeneous case, with each molecule comprising one isotope, the shift of T_c is anomalously large.

Chakravarty *et al.*,[329] whose model for superconductivity does not include phonons, suggest an isotope dependence of the t_{1u} bandwidth due to the mass dependence of the structure arising from variation in the size of zero-point motion. Chen and Lieber suggest their unexpected result gives evidence that intermolecular modes are strongly involved in the

[328] C.-C. Chen and C. M. Lieber, *Science* **259**, 655 (1993).
[329] S. Chakravarty, S. A. Kivelson, M. I. Salkola, and S. Tewari, *Science* **256**, 1306 (1992).

superconducting pairing. In retrospect, it should not be so surprising that the isotope effect is not simple. In the conventional situation, where it is taken for granted that the isotope effect is understood, all phonon frequencies are much smaller that the electronic energy scale ("E_F" or "W") and every mode contributes to the isotope shift by the same ratio, and simply scaling the phonon spectrum is justified. In the fullerides, it appears that the hardest phonons are about equal to E_F, and the isotope shift in this case is not known (no formally justified theory exists). Some of the softer phonons are enough lower than the Fermi energy that the conventional contribution might be expected from them. The point is that different phonons may give rise to contributions of different sizes or even different signs of isotope shift. The experiment of Chen and Lieber provides an important constraint on what the correct theory must predict.

27. COULOMB PSEUDOPOTENTIAL μ^*

Because the conduction band in the fullerides is so narrow and also because the phonon and electron energy scales are so similar, there has been interest in understanding whether the "Coulomb pseudopotential" that appears in Eliashberg theory is changed greatly from its value $\mu^* = 0.10$–0.15 in conventional superconducting metals. In conventional metals, the average Coulomb repulsion parameter $\mu = N(E_F)\langle V \rangle$ is reckoned to be of the order of unity. Here $\langle V \rangle$ is an appropriate average screened Coulomb interaction between states involved in the pairing. Retardation of this interaction arising from the relatively slow dynamics of the ions leads to a Coulomb pseudopotential given *roughly* by

$$\mu^* = \frac{\mu}{1 + \mu \log(\omega_{el}/\omega_{ph})} \qquad (27.1)$$

where ω_{el} and ω_{ph} are respectively electronic and phonon energy scales (electronic bandwidth and maximum phonon frequency, say).

The question of how different the fullerides may be has been addressed in greatest detail by Gunnarsson and collaborators.[330,331] These authors first estimated the renormalization of μ by all but the t_{1u} bands by evaluating the T matrix for a model of the undoped system, estimating only a 20–30% reduction. Intraband screening within the t_{1u} bands was found to be very effective in renormalizing μ, however, and upon

[330]O. Gunnarsson and G. Zwicknagl, *Phys. Rev. Lett.* **69**, 957 (1992).
[331]O. Gunnarsson, D. Rainer, and G. Zwicknagl, *Int. J. Mod. Phys. B* **6**, 3993 (1992).

considering vertex corrections as well they estimate that $\mu^* = 0.3$–0.4 is likely. This value is larger than in conventional metals but not so much as to affect most of the current models of superconductivity. The required value of electron–phonon coupling strength to attain $T_c = 30$ K might be nearer $\lambda = 1$ rather than the $\lambda = 0.7$ necessary if $\mu^* = 0.15$ (if $\langle \omega \rangle$ is large).

A different point of view was suggested by Chakravarty et al.,[332] who considered a two-band model (e.g., one narrow and one wide) with small matrix element for Coulomb scattering between the bands. Their result corresponds, in the limit that scattering between the bands is very small, to Eq. (27.1) with ω_{el} equal to the t_{1u} bandwidth, and therefore with little renormalization. A difference between their procedure and that of Gunnarsson et al. is that Chakravarty et al. base their estimates of parameter values on intra*atomic* quantities whereas Gunnarsson et al. use only intra*molecular* and solid state properties.

28. Proposed Mechanisms

Consistent with the reasonably conventional behavior observed of the superconducting state, nearly all of the models proposed so far assume that superconductivity results from the pairing of electrons (rather than what could arise in some more exotic picture). Possible pairing mechanisms include the dynamical lattice, an electronic mechanism arising from Coulomb repulsion, or a magnetic mechanism involving spin fluctuations or spin pairing. The observed isotope effect, discussed in Section 26, guarantees some strong involvement of the lattice degrees of freedom. It is convenient to separate the possibilities into those that intimately involve lattice vibrations and those that arise solely from electronic interactions and for which the lattice can be treated as static.

a. *Lattice Degrees of Freedom*

Zhang et al.[333] made estimates of the intramolecular Coulomb repulsion, the pentagonal pinch A_g mode coupling, the Jahn–Teller distortion energy, and electronic energy change due to displacement of the cations. They concluded that coupling due to the cation motions, which were modeled by a Einstein mode, could validate a negative-U Hubbard model, where the gain in energy of a second electron on a C_{60} molecule due to cation displacement could overcome the intramolecular repulsion.

[332]S. Chakravarty, S. Khlebnikov, and S. Kivelson, *Phys. Rev. Lett.* **69**, 212 (1992).
[333]F. C. Zhang, M. Ogata, and T. M. Rice, *Phys. Rev. Lett.* **67**, 3452 (1991).

If the negative-U scenario applies, the next question is whether Cooper pairing will be favored over charge density wave (CDW) formation, that is, real space ordering of paired electrons. They concluded that the fcc arrangement of molecules would frustrate CDW formation sufficiently that superconductivity would result.

This specific model (though not of necessity the negative-U Hubbard model generally) has suffered from the discovery of the considerable C isotope shift of T_c just discussed, since the fullerene molecules themselves are considered as rigid and the intramolecular vibrations were supposed to be unimportant. Nevertheless, their arguments concerning coupling of cation motions to the conduction electrons contain some truth, and reflect a kind of electron–phonon coupling that is somewhat reminiscent of the strong coupling of cation and oxygen anion modes found both experimentally and theoretically in the copper oxide high temperature superconductors. So far there have been few attempts to calculate the strength of this coupling. Zhang et al.[334] considered Madelung contributions to electron–phonon coupling and found no evidence that such contributions would be large.

A number of workers have considered a conventional model of pairing resulting from molecular phonons that are treated within standard Migdal–Eliashberg theory. Both Schlüter and coworkers[161–163] and Varma et al.[160] considered the dominant coupling to arise from Jahn–Teller-related modes, and obtained $\lambda \sim 0.6$–0.8. The coupling is confined mainly to high frequency modes ($\langle \omega \rangle \sim 1000$–$1500 \, \text{cm}^{-1}$), and this picture accounts for $T_c \sim 20$–$30 \, \text{K}$ as long as $\mu^* \sim 0.15$. A problem with this scenario is that, for the phonon frequencies and bandwidths that are envisioned, the suppositions that validate Migdal–Eliashberg theory are violated. Mazin et al.[282] and Novikov et al.[283] likewise have followed the Migdal–Eliashberg theory but do not rely so strongly on specific phonons to obtain the requisite coupling strength.

Loktev and Pashitskiĭ[335] also suggest the importance of Jahn–Teller vibrations and provide perhaps the only suggestion that consideration of the band structure of the solid may be crucial. They suggest that the multivalley nature of the band structure may play a part in enhancing electron–phonon coupling. These authors do not address the complication of merohedral disorder, which according to Gelfand and Lu[304–306] and to Erwin[275] leads to strong broadening (if not destruction) of the band structure based on the orientationally ordered idealization.

[334]W. Zhang, H. Zheng, and K. H. Bennemann, *Solid State Commun.* **82**, 679 (1992).
[335]V. M. Loktev and E. A. Pashitskiĭ, *JETP Lett.* **55**, 478 (1992) [*Pis'ma Zh. Eksp. Teor. Fiz.* **55**, 465 (1992)].

Johnson et al.[336] have suggested a dynamic Jahn–Teller mechanism based on a strongly coupled, low frequency H_g mode.

A variety of other possible ideas related to pairing have been suggested. Alexandrov[337] has suggested that the behavior of $H_{c2}(T)$ is more characteristic of a bipolaronic model of Bose condensation of small bipolarons, without specifying the character of the (bi)polaronic distortions. Wilson[338] suggested the pairing to arise from the (dynamic) breaking of the short "double" bond b_{hh}. Tachibana et al.[339] have also invoked the Jahn–Teller molecular modes, but arrive at something resembling a negative-U picture involving dynamic local distortions of the molecule. These latter two proposals seem to be opposed by the calculations of Hayden and Mele,[172] Harigaya,[169,170] and Friedman,[168] all of whom found that equatorial stretched bonds were energetically favorable to locally broken (or stretched) bonds.

The observation that the usual assumptions of Migdal–Eliashberg theory are not satisfied in the fullerides has led to suggestions that the superconductivity actually arises from, or is considerably enhanced by, nonadiabatic lattice response. Pietronero and Strässler[278,279] have suggested a theoretical treatment of the vertex corrections and obtain a strong enhancement of T_c from nonadiabatic terms. For optimal T_c, they suggest (a) a material should have the Fermi level lying within a narrow band, (b) nonadiabatic causes should be distinct from the interactions giving the dispersion of the bands, and (c) there should be a broad background of bands to provide a substantial reduction of $\mu \to \mu^*$. A molecular crystal with dominant electron–intramolecular mode coupling satisfies these criteria.

Asai and Kawaguchi[340] have explored a formalism for handling such effects on the C_{60} molecule, and obtain an enhancement of ~ 10–30% in the coupling strength from nonadiabatic contributions. Unlike most other treatments, they obtain from their model the strongest coupling (even from the adiabatic terms) from the lowest three H_g modes. Zheng and Bennemann[276] have solved the Eliashberg equations for the case where the Fermi energy is comparable with the phonon energy scale, but without considering vertex corrections. The results for the renormalization function $Z(\omega)$ and T_c are different from the usual forms and may be consistent with the behavior of the fullerides.

The unusual variation in frequency, linewidth, and oscillator strength

[336]K. H. Johnson, M. E. McHenry, and D. P. Clougherty, Physica C 183, 319 (1991).
[337]A. S. Alexandrov, Pis'ma Zh. Eksp. Teor. Fiz. 55, 195 (1992) [JETP Lett. 55, 189 (1992)].
[338]J. A. Wilson, Physica C 182, 1 (1991).
[339]A. Tachibana, S. Ishikawa, and T. Yamabe, Chem. Phys. Lett. 201, 315 (1993).
[340]Y. Asai and Y. Kawaguchi, Phys. Rev. B 46, 1265 (1992).

of the infrared active T_{1u} modes as the alkaline atoms are incorporated into fullerite indicates there is substantial coupling between these vibrations and the conduction electrons in the fullerides. The one theory for these phenomena is the "charged phonon" picture of Rice and Choi[284] (see Section 21), the principal feature of which is that the T_{1u} phonons become highly polarizable as a result of mixing with low energy interband $t_{1u} \rightarrow t_{1g}$ transitions. Rice and Choi noted that this interaction should contribute to superconducting pairing, but no theory of the effect has been developed.

Several times it has been observed that for nearly all of the alkali fulleride compounds T_c follows an approximately linear variation with lattice constant:

$$T_c = (dT_c/da)(a - a_0). \tag{28.1}$$

The relationship extrapolates[174] to $T_c = 0$ at $a_0 = 13.75$ Å, with a slope of ~50 K/Å. In addition, the electronic properties, and most notably $N(E_F)$, vary linearly as well. Since $N(E_F)$ increases with lattice constant, as does T_c, there is a natural explanation of the superconductivity as arising from phonon-mediated pairing in which the variation in coupling strength is dominated by the change in $N(E_F)$. In fact, such an observation applies to any coupling mechanism that pairs carriers at or near the Fermi surface and whose strength is proportional to $N(E_F)$. In fact, the relative variation in lattice constant in all fullerides is only ~4%, and most properties are likely to vary linearly in this range.

b. *Electronic Degrees of Freedom*

The possibility that electronic degrees of freedom might be responsible for pairing was suggested by Chakravarty et al.[105,106] (see Section 4.c). If intramolecular electron correlation results in two electrons preferring to be on a single C_{60} molecule rather than on separate molecules, then the "negative-U" model of superconductivity, such as is usually modeled by the Hubbard model, becomes a real possibility. Lal and Joshi[341] argue the inapplicability of the simple Hubbard model and outline an electronic multiband mechanism. Jansen et al.[342] propose an electronic indirect-exchange coupling via the short b_{hh} bonds.

Probably the most original suggestion is that of Friedberg et al.,[111] not only for the superconducting state but for possible correlated behavior in the normal state. Their model of "parity doublets," arising from the

[341]R. Lal and S. K. Joshi, *Solid State Commun.* **80**, 937 (1991).
[342]L. Jansen, R. Block, and E. Lombardi, *Physica C* **182**, 17 (1991).

specific characters of the t_{1u} and t_{1g} LUMOs and their near degeneracy in the C_{60} molecule, was discussed in Section II. In their picture, two electrons on a molecule can form either a "scalar" state, or "pseudoscalar" paired state with bosonic properties. In A_3C_{60}, with three particles on average on each molecule, the superconducting order parameter can be either scalar, in which case the quasi-particle spectrum has the typical BCS form, or it can be pseudoscalar, for which superconductivity results from Bose–Einstein condensation.

X. Summary

I have tried to touch on a wide variety of the properties of C_{60}-based materials that are being studied, to convey the breadth of research and give an idea of the degree of understanding of these materials at this time (early 1993).

In both the electronic and vibrational aspects, there are many important questions to be resolved. In fact, since relevant electronic and vibrational energy scales (and therefore time scales) appear to be about equal, there is even a question whether many of the properties of fullerides can be interpreted as either "electronic" or "phononic" as is done for conventional metals and superconductors, or whether a hybrid quantity must be considered from the beginning. It is probably of little use to try to summarize which properties are "well understood" and which are "not understood." A number of properties are rather well characterized from the experimental viewpoint, however, viz. the structural and many of the dynamical properties, and much good data exists in the thermodynamic and spectroscopic areas.

This chapter reflects the solid groundwork, both experimental and theoretical, that has emerged, mostly within the last two years. It should be evident that many of the important questions are far from settled. For the superconductivity, for example, a consistent picture in terms of the conventional BCS theory has arisen, except that there is no formal justification for the theory in this class of materials. These materials promise to provide a stimulating area of research for several years to come.

ACKNOWLEDGMENTS

I happily acknowledge the cooperation of the many people who allowed the use of their figures. I have benefitted from discussions with a number of colleagues, especially with S. C. Erwin on the structural, electronic, and transport properties of these materials. Discussions with M. R.

Pederson, D. A. Papaconstantopoulos, J. F. Feldman, and A. A. Quong have been important in helping to prepare this article. I especially acknowledge the support of the Office of Naval Research (Contract No. N00014-93-WX-24005), which has been crucial to my participation in research in these novel fullerene materials. Also, considerable assistance was provided by the "Complete Buckminsterfullerene Bibliography," originally kept by R. E. Smalley (Rice University) but now passed on to D. R. Huffman and F. A. Tinker (Arizona State University). I am indebted to S. C. Erwin and B. I. Dunlap for a careful reading of the manuscript.

SOLID STATE PHYSICS, VOLUME 48

Physical Properties of Metal-Doped Fullerene Superconductors

Charles M. Lieber
Zhe Zhang

Divisions of Applied Sciences and Department of Chemistry
Harvard University
Cambridge, Massachusetts

I. Introduction

The discovery of superconductivity in potassium-doped C_{60} at 18 K by Hebard and coworkers[1] was an unexpected observation that has caused an explosion of condensed matter research. In the brief two-year period since this report, the field of fullerene superconductivity has undergone a remarkable development with the identification of the superconducting phase in potassium-doped C_{60}, the discovery of new alkali metal–doped C_{60} materials having critical transition temperatures (T_c) exceeding 30 K, and the elucidation of many of the key normal and superconducting state properties of these materials.[2,3] Furthermore, the alkali metal–doped fullerenes now represent the highest-T_c molecular superconductor, and were it not for the discovery of superconductivity in copper oxide

[1] A. F. Hebard, M. J. Rosseinsky, R. C. Haddon, D. W. Murphy, S. H. Glarum, T. T. M. Plastra, A. P. Ramirez, and A R. Kortan, *Nature* **350**, 600 (1991).
[2] A. F. Hebard, *Physics Today* **45**, 26 (1992).
[3] R. C. Haddon, *Accts. Chem. Res.* **25**, 127 (1992).

materials several years earlier they would be the highest-T_c superconductors, period.

The key component of the fullerene superconductors is the molecular cluster C_{60}, or Buckminsterfullerene. C_{60} and other fullerene clusters were first detected experimentally in the mass spectroscopy studies of Smalley and coworkers.[4–10] To reconcile the unusual stability of the C_{60} cluster, Smalley and coworkers proposed that the 60 carbon atoms formed a truncated icosahedron (soccer ball) structure in which each of the 60 carbon atoms was equivalent (Fig. 1).

The name Buckminsterfullerene (C_{60}) or, more generally, fullerene to describe the family of carbon clusters (e.g., C_{60}, C_{70}, C_{84} ...) comes from the architect R. Buckminster Fuller whose geodesic domes inspired the original structure proposal for C_{60}.[11] The original techniques used by Smalley and others to generate fullerenes for their gas phase studies did not produce isolable quantities of C_{60} and thus were inadequate for materials research. This situation changed rapidly, however, with the development by Krätschmer and Huffman[12] of a simple method for preparing macroscopic quantities of C_{60} [for details see "Preparation of Fullerenes and Fullerene-Based Materials" in this volume]. This synthetic development represents a key advance, since it has been the availability of large quantities of pure, solid C_{60} that has fueled the explosion of research in this field. In addition to the fullerene-based superconductors, which represent the focus of this chapter, materials exhibiting interesting optical and magnetic properties are also emerging from this work.[13–17]

In this chapter we will review the field of fullerene superconductivity

[4]H. W. Kroto, J. R. Heath, S. C. O'Brien, R. F. Curl, and R. E. Smalley, *Nature* **318**, 162 (1985).

[5]S. C. O'Brien, J. R. Heath, R. F. Curl, and R. E. Smalley, *J. Chem. Phys.* **88**, 220 (1988).

[6]J. R. Heath, R. F. Curl, and R. E. Smalley, *J. Chem. Phys.* **87**, 4236 (1987).

[7]R. F. Curl and R. E. Smalley, *Science*, **242**, 1017 (1988).

[8]E. A. Rohlfing, D. M. Cox, and A. Kaldor, *J. Chem. Phys.* **81**, 3322 (1984).

[9]H. Kroto, *Science* **242**, 1139 (1988).

[10]R. E. Smalley, *Accts. Chem. Res.* **25**, 98 (1992).

[11]R. B. Fuller, "Inventions: The Patented Works of Buckminster Fuller" St. Martin's Press, New York, 1983.

[12]W. Kratschmer, L. D. Lamb, K. Fostiropoulos, and D. R. Huffman, *Nature* **347**, 354 (1990).

[13]H. Yonehara and C. Pac, *Appl. Phys. Lett.* **61**, 575 (1992).

[14]Y. Wang, *Nature* **356**, 585 (1992).

[15]B. Miller, J. M. Rosamilia, G. Dabbagh, R. Tycko, R. C. Haddon, A. J. Muller, W. Wilson, D. W. Murphy, and A. F. Hebard, *J. Am. Chem. Soc.* **113**, 6291 (1991).

[16]Y. Wang and L. T. Cheng, *J. Phys. Chem.* **96**, 1530 (1992).

[17]P. M. Allemand, K. C. Khemani, A. Koch, F. Wudl, K. Holczer, S. Donovan, G. Grüner, and J. D. Thompson, *Science* **253**, 301 (1991).

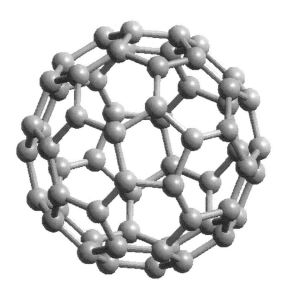

FIG. 1. Molecular structure of Buckminsterfullerene, C_{60}.

with an emphasis on the intensely studied materials K_3C_{60} and Rb_3C_{60}. Our goal is to provide condensed matter researchers with an up to date status report on the key properties of these new superconductors for which a reasonable consensus has been reached among researchers. In addition, we will comment on important open issues of fullerene superconductivity that await resolution. The structure of the chapter will be as follows: First, we will briefly review the structural and electronic properties of C_{60} and alkali metal–doped C_{60} solids. With this background information in hand, we will turn to a synopsis of normal state resistivity studies and a review of measurements of the critical fields, penetration depth, and coherence length in these materials. The role of sample granularity will also be discussed. Second, we will focus on investigations that directly address the microscopic mechanism of super-conductivity in these new solids. We will begin this section with a brief overview of key models proposed to explain fullerene superconductivity. Within the context of these models, we will then overview (1) the dependence of T_c on lattice constant, (2) the energy gap, (3) phonons, and (4) the isotope effect. Throughout this chapter we will assume that the reader has an introductory level understanding of classical superconductors.[18]

[18]M. Tikham, "Introduction to Superconductivity". McGraw-Hill, New York, 1975.

II. Normal and Superconducting State Phenomenology

1. CRYSTAL AND ELECTRONIC STRUCTURES

a. *Crystal Structures*

On the basis of extensive structural studies, it is now well established that solid C_{60} forms a face-centered cubic (fcc) lattice with a lattice constant of 14.17 Å at room temperature (Fig. 2).[19–25] In this structure the distance between nearest-neighbor C_{60} clusters is 10 Å, and thus the intercluster separation (diameter of $C_{60} = 7.1$ Å) is 2.9 Å. This intercluster separation is 0.45 Å less than the 3.35-Å interplanar separation in graphite. In addition, there are sizable empty holes, which constitute 26% of the total cell volume, within the fcc C_{60} lattice. There are two tetrahedral holes and one octahedral hole with radii of 1.12 and 2.06 Å, respectively, per C_{60} molecule. A more detailed account of the structural properties of solid C_{60} can be found in the chapter by Axe, Moss, and Neumann in this volume.

Early studies at Bell Laboratories showed that exposure of C_{60} to alkali metal vapor resulted in the uptake of alkali metal into the lattice with a concomitant increase in conductivity.[26] In the case of potassium, the potassium-doped fullerene solid was also found to exhibit superconductivity below 18 K.[1]

Although neither the stoichiometry nor structure of the superconducting phase was known initially, this group proposed that the potassium intercalated into the octahedral and/or tetrahedral holes in the lattice.

[19] J. E. Fischer, P. A. Heiney, and A. B. Smith, *Accts. Chem. Res.* **25**, 115 (1992).

[20] S. Liu, Y.-J. Lu, M. M. Kappes, and J. A. Ibers, *Science* **254**, 408 (1991).

[21] P. A. Heiney, J. E. Fisher, A. R. McGhie, W. J. Romanow, A. M. Denenstein, J. P. McCauley, and A. B. Smith, *Phys. Rev. Lett.* **66**, 2911 (1911).

[22] J. E. Fisher, P. A. Heiney, A. R. McGhie, W. J. Romanow, A. M. Denenstein, J. P. McCauley, and I. A. B. Smith, *Science* **252**, 1288 (1991).

[23] W. I. F. David, R. M. Ibberson, J. C. Matthewman, K. Prassides, T. J. S. Dennis, J. P. Hare, H. W. Kroto, R. Taylor, and D. R. M. Walton, *Nature* **353**, 147 (1991).

[24] D. A. Neumann, J. R. D. Copley, R. L. Cappelletti, W. A. Kamitakahara, R. M. Lindstrom, K. M. Creegan, D. M. Cox, W. J. Romanow, N. Coustel, J. P. McCauley, Jr., N. C. Maliszewskyj, J. E. Fischer, and A. B. Smith III, *Phys. Rev. Lett.* **67**, 3808 (1991).

[25] J. Q. Li, Z. X. Zhao, D. B. Zhu, Z. Z. Gan, and D. L. Yin, *Appl. Phys. Lett.* **59**, 3108 (1991).

[26] R. C. Haddon, A. F. Hebard, M. J. Rosseinsky, D. W. Murphy, S. J. Duclos, K. B. Lyons, B. Miller, J. M. Rosmilia, R. M. Fleming, A. R. Kortan, S. H. Glarum, A. V. Makhija, A. J. Muller, R. H. Eick, S. M. Zahurak, R. Tycko, G. Dabbagh, and F. A. Thiel, *Nature* **350**, 321 (1991).

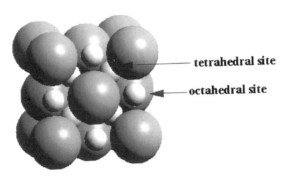

tetrahedral site

octahedral site

FIG. 2. Model illustrating the packing of individual fullerene clusters in the fcc lattice of solid C_{60} (top). Model illustrating the packing of alkali metal ions into the tetrahedral and octahedral holes in the fcc M_3C_{60} lattice (bottom). The C_{60} clusters are represented by gray shaded spheres and the alkali metal ions by smaller, light gray shaded spheres.

Subsequent studies have shown that the stoichiometries of the potassium[27,28] and rubidium-doped[29-31] C_{60} superconductors are K_3C_{60} and Rb_3C_{60}, respectively. Indeed, all of the known alkali metal–doped superconductors have the same M_3C_{60} stoichiometry. The structure of

[27]K. Holczer, O. Klein, S.-M. Huang, R. B. Kaner, K.-J. Fu, R. L. Whetten, and F. Diederich, Science 252, 1154 (1991).
[28]P. W. Stephens, L. Mihaly, P. L. Lee, R. L. Whetten, S.-M. Huang, R. Kaner, F. Deiderich, and K. Holczer, Nature 351, 632 (1991).
[29]C.-C. Chen, S. P. Kelty, and C. M. Lieber, Science 253, 886 (1991).
[30]R. M. Fleming, A. P. Ramirez, M. J. Rosseinsky, D. W. Murphy, R. C. Haddon, S. M. Zahurak, and A. V. Makhija, Nature 352, 787 (1991).
[31]P. W. Stephens, l. Mihaly, J. B. Wiley, S.-M. Huang, R. B. Kaner, F. Diederich, R. L. Whetten, and K. Holczer, Phys. Rev. B 45, 543 (1992).

this M_3C_{60} phase was first elucidated by Stephens and coworkers[28] for K_3C_{60} and shown to be a simple derivative of the undoped fcc C_{60} solid. Specifically, the three alkali metal ions per C_{60} reside in the one octahedral and two tetrahedral holes in the lattice (Fig. 2).

The M_3C_{60} superconductors in contrast to the copper oxide materials, represent a structurally limited family of materials in that only the 3:1 stoichiometry is known to be conducting and superconducting (for alkali metals). Although there are other structurally known alkali metal–doped fullerene materials, such as the body-centered tetragonal (bct) phases M_4C_{60}[19,32] and the body-centered cubic (bcc) phase M_6C_{60},[19,33] these phases are insulating. The relatively simple structure and stoichiometry of the M_3C_{60} family of superconductors, however, does eliminate many of the problems of chemical inhomogeneity that have plagued physical measurements for the copper oxide superconductors. A summary of structurally characterized alkali metal–doped M_3C_{60} superconductors is given below in Table I. This table shows that there are a number of specific alkali metal–doped materials with the M_3C_{60} stoichiometry, where M is either a single alkali metal or mixture of alkali metals.[29,30,34–36] More recently, superconductivity has also been observed in the alkaline earth–metal–doped C_{60} materials Ca_5C_{60}[37] and Ba_6C_{60}.[38] To date, few physical measurements of the superconducting state properties of these new alkaline-earth metal–doped C_{60} solids have been made, and thus we will not discuss these materials further in this chapter.

b. *Electronic Structure*

The common structure and stoichiometry of the alkali metal–doped C_{60} superconductors indicates that a single picture can be used to describe the electronic structure of these materials. Here we briefly

[32]R. M. Fleming, M. J. Rosseinsky, A. P. Ramirez, D. W. Murphy, J. C. Tully, R. C. Haddon, T. Siegrist, R. Tycko, S. H. Glarum, P. Marsh, G. Dabbagh, S. M. Zahurak, A. V. Makhija, and C. Hampton, *Nature* **352**, 701 (1991).

[33]O. Zhou, J. E. Fischer, N. Coustel, S. Kycia, Q. Zhu, A. R. McGhie, W. J. Romanow, J. P. McCauley, A. B. Smith, and D. E. Cox, *Nature* **351**, 461 (1991).

[34]O. Zhou, R. M. Fleming, D. W. Murphy, M. J. Rosseinsky, A. P. Ramirez, R. B. van Dover, and R. C. Haddon, *Nature* **362**, 433 (1993).

[35]M. J. Rosseinsky, D. W. Murphy, R. M. Fleming, R. Tycko, A. P. Ramirez, T. Siegrist, G. Dabbagh, and S. E. Barrett, *Nature* **356**, 416 (1992).

[36]K. Tanigaki, I. Hirosawa, T. W. Ebbesen, J. Mizuki, Y. Shimakawa, Y. Kubo, J. S. Tsai, and S. Kuroshima, *Nature* **356**, 419 (1992).

[37]A. R. Kortan, N. Kopylov, S. Glarum, E. M. Gyorgy, A. P. Ramirez, R. M. Fleming, F. A. Thiel and R. C. Haddon, *Nature* **355**, 529 (1992).

[38]A. R. Kortan, N. Kopylov, S. Glarum, E. M. Gyorgy, A. P. Ramirez, R. M. Fleming, O. Zhou, F. A. Thiel, P. L. Trevor, and R. C. Haddon, *Nature* **360**, 566 (1992).

TABLE I. LATTICE CONSTANTS AND TRANSITION
TEMPERATURES   OF   THE   ALKALI   METAL–DOPED C_{60}
SUPERCONDUCTORS

MATERIAL	FCC LATTICE CONSTANT (Å)	T_c (K)
Na_2RbC_{60}	14.028	2.5
Na_2CsC_{60}	14.133	11
K_3C_{60}	14.253	19.2
K_2RbC_{60}	14.299	21.8
K_2CsC_{60}	14.292	24
KRb_2C_{60}	14.364	26
Rb_3C_{60}	14.436	29.4
$(NH_3)_4Na_2CsC_{60}$	14.473	29.6
Rb_2CsC_{60}	14.493	31.3

outline a simplified picture of the electronic structure of C_{60} and its solid compounds; the chapter by Pickett in this volume provides a detailed review of the electronic structure of these materials.

Because the interactions between individual C_{60} clusters within the solid are relatively weak (i.e., the closest intercluster distance is 2.9 Å) compared with the intracluster bonding, it is possible to obtain a useful picture by first considering the isolated molecular (cluster) states and then using these molecular orbitals to derive the energy bands and density of states for the bulk solid.[2] Qualitatively, the key features of the electronic structure of C_{60} can be derived by considering the atomic $p(\pi)$ orbitals that radiate from each of 60 equivalent carbon atoms. Semiempirical Hückel calculations provide the relative energy ordering of the molecular orbitals derived from these atomic π orbitals (Fig. 3). After filling these molecular orbitals with the 60 available π electrons, it is evident that C_{60} has a closed shell ground state where the highest occupied molecular orbital (HOMO) has h_u symmetry and contains 10 electrons, and the lowest unoccupied molecular orbital (LUMO) has t_{1u} symmetry and can hold up to 6 electrons.

The band structure and density of states can be inferred from this simple molecular orbital picture. Specifically, five valence bands are derived from the fivefold degenerate h_u-HOMO, and three conduction bands arise from the threefold degenerate LUMO of C_{60}. Details of the energy dispersions for these bands can be found in the chapter by Pickett in this volume. The insulating properties of solid C_{60} are thus a consequence of the fact that the h_u-derived bands are filled and the t_{1u} bands are empty. The gap in the density of states between the h_u-derived

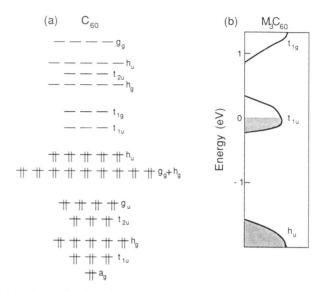

FIG. 3. (a) Hückel molecular orbital diagram that illustrates symmetry, degeneracy, and filling of the molecular orbitals on an individual C_{60} cluster. Electron pairs filling individual states are indicated by pairs of vertical parallel lines. (b) Schematic diagram of the density of states for M_3C_{60} solid in which C_{60} is orientationally disordered. The t_{1u} derived energy band is half-filled.

valence and the t_{1u}-derived conduction bands in solid C_{60} is approximately 1.7 eV.[39]

The dramatic increase in conductivity upon doping solid Buckminsterfullerene with three alkali metals per C_{60} is readily explicable in terms of the basic picture that has been presented. Specifically, the alkali metal dopants donate charge ($1e^-$/metal) to the t_{1u}-derived energy bands since an alkali metal s orbital (e.g., K $4s$) lies significantly higher in energy than the t_{1u}-derived conduction bands. The M_3C_{60} materials are thus metals with a half-filled t_{1u} band.[40–44] A schematic illustration of the density of states illustrating this point is shown in Fig. 3. In addition, it is readily evident using this simple model why the M_6C_{60} materials are insulators: The t_{1u}-derived band becomes filled with the donation of six

[39]S. Saito and A. Oshiyama, *Phys. Rev. Lett.* **66**, 2637 (1991).
[40]S. C. Erwin and W. E. Pickett, *Science* **254**, 842 (1991).
[41]D. L. Novikov, V. A. Gubanov, and A. J. Freeman, *Physica C* **191**, 399 (1992).
[42]M. P. Gelfand and J. P. Lu, *Phys. Rev. Lett.* **68**, 1050 (1992).
[43]J. L. Martins and N. Troullier, *Phys. Rev. B* **46**, 1766 (1992).
[44]S. Satpathy, V. P. Antropov, O. K. Andersen, O. Jepsen, O. Gunnarsson, and A. I. Liechtenstein, *Phys. Rev. B* **46**, 1773 (1992).

electrons per C_{60}.[45] The insulating properties of M_4C_{60}, however, cannot be explained by this simple picture (i.e., it should be metallic).

To summarize, alkali metal doping is believed to result in complete charge transfer and a half-filled t_{1u}-derived band for the M_3C_{60} superconducting stoichiometry. Importantly, the width of this band is expected to be narrow since there is only weak overlap between adjacent C_{60} clusters within the lattice. Theoretical calculations have suggested that the bandwidth is approximately 0.6 eV.[40] Experimental determinations of this important parameter have yielded considerably greater uncertainty, with estimates ranging from 0.1 to >1 eV.[46–51] Because the bandwidth has important implications for the mechanism of superconductivity in these materials (to be discussed), it remains an important objective of experimental research to determine unambiguously the width of the t_{1u}-derived energy band.

2. CONDUCTIVITY AND DIMENSIONALITY

Resistivity and magnetoresistance measurements have played an important, but often controversial, role in determining details of the normal and superconducting states of the M_3C_{60} materials.[26,52–55] To date, much of this work has been carried out using doped thin film materials.[26,52,55] The first resistivity measurements on K-doped C_{60} films found a minimum of 2–5 mΩ cm at room temperature.[26] Upon cooling these films, the resistivity increases ca. $2\times$ before undergoing a transition to a zero resistance state at 5 K. The transition temperature determined by

[45]S. C. Erwin and M. R. Pederson, *Phys. Rev. Lett.* **67**, 1610 (1991).

[46]C. T. Chen, L. H. Tjeng, P. Rudolf, G. Meigs, J. E. Rowe, J. Chen, J. P. McCauley, Jr., A. B. Smith III, A. R. McGhie, W. J. Romanow, and E. W. Plummer, *Nature* **352**, 603 (1991).

[47]P. J. Benning, F. Stepniak, D. M. Poirier, J. L. Martins, J. H. Weaver, L. P. F. Chibante, and R. E. Smalley, *Phys. Rev. B* **47**, 13843 (1993).

[48]G. K. Wertheim and D. N. E. Buchanan, *Phys. Rev. B* **47**, 12912 (1993).

[49]P. J. Benning, J. L. Martins, J. H. Weaver, L. P. F. Chibante, and R. E. Smalley, *Science* **252**, 1417 (1991).

[50]G. Sparn, J. D. Thompson, S.-M. Huang, R. B. Kaner, F. Diederich, R. L. Whetten, G. Grüner, and K. Holczer, *Science* **252**, 1829 (1991).

[51]L. D. Rotter, Z. Schlesinger, J. P. McCauley, Jr., N. Coustel, J. E. Fischer, and A. B. Smith III, *Nature* **355**, 532 (1992).

[52]G. P. Kochanski, A. F. Hebard, R. C. Haddon, and A. T. Fiory, *Science* **255**, 184 (1992).

[53]X.-D. Xiang, J. G. Hou, G. Briceño, W. A. Vareka, R. Mostovoy, A. Zettl, V. H. Crespi, and M. L. Cohen, *Science* **256**, 1190 (1992).

[54]X.-D. Xiang, J. G. Hou, V. H. Crespi, A. Zettl, and M. L. Cohen, *Nature* **361**, 54 (1993).

[55]T. T. M. Palstra, R. C. Haddon, A. F. Hebard, and J. Zaanen, *Phys. Rev. Lett.* **68**, 1054 (1992).

magnetization measurements, however, is significantly higher, 19.2 K. There are two important implications of these studies. First, even the most conducting samples are close to the Mott metal/insulator limit. Second, the T_c's determined by resistivity measurements may be reduced by film granularity. Further analysis of these initial data is not warranted, however, due to the uncertainty in sample quality (e.g., stoichiometry and crystallinity).

More recently, the Bell Laboratories group has carried out two detailed transport studies of potassium-doped C_{60} films prepared under carefully controlled ultrahigh vacuum conditions.[52,55] First, Kochanski and coworkers have prepared a series of $K_x C_{60}$ ($x = 1$–6) films in which the value of x was determined to ±2% and the crystalline grain size was 60 Å.[52] In agreement with the earlier work, they found that the minimum resistivity in the films was 2.2 mΩ cm and showed that this corresponded to the doping level appropriate for the superconducting phase, $x = 3$. Assuming that there is complete charge transfer and thus three carriers per C_{60}, the effective scattering time, τ, can be written as

$$\tau = m/\rho n e^2 \tag{2.1}$$

and is of the order 10^{-15} sec. This yields an apparent mean free path, $l = v_f \tau$, of only 2 Å using the theoretically derived[40] Fermi velocity (v_f) of 2×10^{-7} cm/sec. Because this mean free path is significantly less than the intercluster separation (10 Å), this analysis is probably an inadequate explanation of the relatively high resistivity in these materials.

An alternative analysis of these transport data assumes that conductivity is dominated by the granularity of the system. This idea is reasonable since (1) the grain size in the polycrystalline thin films is small (≈ 60 Å) and (2) alkali metal doping naturally leads to insulating phases surrounding conducting $K_3 C_{60}$ grains. Within this framework, conduction is activated due to the charging energy, E_c. Because potassium doping ($x = 0$–6) changes the conducting grain size, E_c will also depend on x:

$$E_c = \frac{e^2}{2\pi\varepsilon D} \frac{2D}{d + 2\delta}, \tag{2.2}$$

where D is the grain size and δ is the gap between grains. The resistivity depends exponentially on E_c:[52,56]

$$\rho \propto \exp(E_c/2kT). \tag{2.3}$$

[56]B. Abeles, *Appl. Solid State Sci.* **6**, 1 (1976).

This model provides a good fit to the experimental data of Kochanski and coworkers, and it yields a grain size at $x = 3$ of 75 Å. The close agreement of this value with grain size determined by x-ray diffraction strongly supports the validity of this interpretation by the Bell Laboratories group. These results suggest that it will be important to account for sample granularity in the analysis of critical parameters of the superconducting state.

The thin film resistivity data for K_3C_{60} has also been used to investigate the dimensionality of this system.[55] Above, T_c, superconducting fluctuations may be observed as an excess in the normal state conductivity (paraconductivity, σ'). The scaling of the paraconductivity near T_c is strongly dependent on the dimensionality and can thus be used to assess this important idea:

$$\sigma' \propto t^{(d-4)/2}, \tag{2.4}$$

where $t = (T - T_c)/T_c$ and d is the dimension of the system.[57,58] Plots of $\log \sigma'$ vs. $\log t$ yield a straight line with slope $= -2$ for $t < 0.4$, and thus the critical dimension of the system is zero. For $t > 0.4$, which corresponds to temperatures greater than the T_c derived from magnetic measurements, there is, however, a crossover to 3D behavior. Notably, zero-dimensional fluctuations are the signature of weakly coupled superconducting grains, and thus this analysis is completely consistent with the granular picture derived from doping studies of K_xC_{60}.[52]

An important issue regarding studies of thin film (and polycrystalline bulk) samples that should be addressed in the future is whether granularity is intrinsic to these materials. For example, can single-crystal-like films be grown at elevated temperatures, and if so, can such films be doped homogeneously? It may be that the process of doping, which initially proceeds through the formation of C_{60}, M_3C_{60}, and M_6C_{60} phases, intrinsically leads to granular materials. The development of better quality thin film materials will undoubtedly lead to a deeper understanding of the intrinsic properties of these materials.

A somewhat different picture of the conductivity has been inferred from transport measurements made on K- and Rb-doped C_{60} single crystals by Zettl and coworkers.[53,54] First, they have found that the resistivity decreases with decreasing temperature from 300 K to T_c, and second, they have determined that the T_c's determined from these data are the same (not lower) than those determined magnetically for K_3C_{60} and Rb_3C_{60}. Initial analyses of σ' for a K-doped sample showed no

[57]W. J. Skocpol and M. Tinkham, *Rep. Prog. Phys.* **38**, 1049 (1975).
[58]L. G. Aslamasov and A. I. Larkin, *Phys. Lett. A* **26**, 238 (1968).

evidence for fluctuation effects near T_c, although in a later study both K- and Rb-doped C_{60} materials were reported to show fluctuation conductivity. Analysis of the excess conductivity using the Aslamazo and Larkin[57,58] scaling relation (Eq. 2.3) and the Maki–Thompson[59] expression (which also considers pair breaking effects) leads to the conclusion that both materials are 3D conductors/superconductors; there is no evidence for granularity in their data.

These single crystal data seem to indicate that granularity in the film measurements is not intrinsic to the M_3C_{60} materials. There are, however, several caveats to the crystal results. First, the absolute room-temperature resistivity of the single crystals is about the same, $2\,m\Omega\,cm$, as that determined in the granular films. In addition, the stoichiometry and doping homogeneity of the single crystals have not been addressed. In the future it will be important to determine the stoichiometry and the fraction of bulk superconducting material. This latter point might be addressed via magnetization measurements. Last, although transport in the thin film samples appear to differ significantly (0D vs. 3D) from the single crystals, it is interesting to note that the crossover to 3D fluctuation effects in the thin films occurs in the same temperature range in which 3D fluctuations are observed in the single crystals. To summarize, these thin film and single crystal data indicate that in the limit of ideal samples M_3C_{60} behaves as 3D metal; however, for most practical thin film and bulk samples studied to date, sample granularity must be accounted for in the analysis of the normal state and superconducting state data.

3. Critical Fields, Penetration Depth, and Coherence Length

The upper and lower critical fields (H_{c2} and H_{c1}, respectively), the magnetic penetration depth (λ), and the superconducting pair coherence length (ξ) are parameters essential to characterizing any superconductor. Here we will review several studies that have addressed these critical parameters in K_3C_{60} and Rb_3C_{60} materials. The general picture to emerge from this work is that the metal-doped fullerene materials are extreme type II superconductors, $\kappa = \lambda/\xi \gg 1$, and in this regard are very much like the high temperature copper oxide superconductors.

The first measurements of H_{c2} and H_{c1} were made on K_3C_{60} materials by Holczer and coworkers.[60] They carried out temperature-dependent

[59]K. Maki and R. S. Thompson, *Phys. Rev. B* **39**, 2767 (1989).
[60]K. Holczer, O. Klein, G. Grüner, J. D. Thompson, F. Diederich, and R. L. Whetten, *Phys. Rev. Lett.* **67**, 271 (1991).

dc-magnetization studies of bulk polycrystalline K_3C_{60} samples at fields up to 5 T. The upper critical field at $T = 0$, $H_{c2}(0)$, was then determined from an extrapolation using the Werthamer–Helfand–Hohenberg (WHH) expression[61]

$$H_{c2}(0) = 0.69 \frac{\partial H_{c2}}{\partial T} \Big|_{T_c} T_c, \tag{3.1}$$

This extrapolation yields $H_{c2}(0) = 50$ T.

The upper critical field corresponds to a field at which there is one flux quantum, Φ_0, per region of a Copper pair. Hence, H_{c2} provides a direct measure of critical length scale for the superconducting state:

$$H_{c2} = \frac{\Phi_0}{2\pi\xi_{GL}^2}, \tag{3.2}$$

where ξ_{GL} is the Ginzburg–Landau coherence length. Using the extrapolated upper critical field, one finds that $\xi_{GL} = 25$ Å. Hence, the coherence length in the alkali metal–doped fullerene superconductors appear to be short and comparable with the values obtained for 2D copper oxide materials and not other 3D metals such as Nb ($\xi(Nb) = 400$ Å).

In addition, Holczer and coworkers carried out an analysis of their low-field data on the K_3C_{60} samples to determine H_{c1} and λ. The lower critical field at $T = 0$, $H_{c1}(0)$, obtained by extrapolating H_{c1} vs. T using the empirical relation

$$h_{c1}(T) = H_{c1}(0)[1 - (T/T_c)^2], \tag{3.3}$$

was ca. 130 Oe. The penetration depth can then be determined from $H_{c1}(0)$ and ξ using[18]

$$H_{c1}(0) = \frac{\Phi_0}{4\pi\lambda^2} \ln\left(\frac{\lambda}{\xi}\right) \tag{3.4}$$

This expression yields a value for $\lambda(0) = 2400$ Å. Hence, $\kappa = 96$, justifying the assertion that the M_3C_{60} materials are extreme type II superconductors.

There are, however, several important comments to make about the foregoing values of H_{c1}, H_{c2}, λ, and ξ_{GL}. First, the data used to obtain $H_{c2}(0)$ cover a limited range of fields relative to the extrapolated value. Second, in the case of $H_{c1}(0)$ it is difficult to assign unambiguously the values of $H_{c1}(T)$ since the M vs. H data defines a nearly parabolic curve at

[61]N. R. Werthamer, E. Helfand, and P. C. Hohenberg, *Phys. Rev.* **147**, 295 (1966).

low fields rather than the expected linear one.[62] Third, sample granularity, which can affect significantly the measured values of H_{c2} and H_{c1}, was not accounted for in these studies.

The first serious attempt to account for sample granularity was made by Palstra and coworkers at Bell Laboratories.[55] Analysis of magnetoresistance measurements made in fields up to 12.5 T (2.5 × larger than used in the magnetization studies) using (3.1) results in an extrapolated upper critical field $H_{c2}(0) \approx 47$ T and a Ginzburg–Landau coherence length $\xi_{GL} = 26$ Å. While these results agree quite well with the magnetization studies, it was also recognized that the sample granularity enhances the observed upper critical field. Specifically, in the clean limit (mean free path, $l \gg \xi$), $\xi_{GL} = \xi_0$, where ξ_0 is the Pippard coherence length. In the dirty limit, however, the Ginzburg–Landau coherence length is reduced by the short mean free path of the conduction electrons:

$$\xi_{GL} = 0.85(\xi_0 \cdot l)^{1/2}. \qquad (3.5)$$

Assuming that the critical field is a property of single grains, Palstra and coworkers determined that $\xi_0 \approx 150$ Å.

Finally, the penetration depth has also been estimated from other measurements. Uemura and coworkers[63] have used muon spin relaxation (μSR) measurements to determine λ as $T \to 0$.[63] In contrast to previous magnetization measurements, μSR is a relatively direct probe of the penetration depth since the μSR relaxation rate, σ, which is determined directly from experiment, is proportional to the inverse square of the penetration depth, λ^{-2}. Studies of polycrystalline K_3C_{60} samples yield $\lambda(0) \approx 4800$ Å.[63] This value is nearly two times larger than that determined from dc-magnetization measurements. Infrared reflectivity measurements made in the normal and superconducting states have also been used to estimate λ.[51,64] Studies from two different groups yield values of 5000 ± 1000 Å[51] and 8000 ± 500 Å.[64] These values of λ are consistent with extreme type II behavior, although clearly a consensus has not yet been reached for the value of the penetration depth.

Summarizing this section, the M_3C_{60} materials are extreme type II superconductors. These materials have upper critical fields as large as 50 T and correspondingly short coherence lengths with $\xi_{GL}(0) \approx 25$ Å. In

[62]V. Buntar, U. Eckern, and C. Politis, *Mod. Phys. Lett. B* **136**, 1037 (1992).
[63]Y. J. Uemura, A. Keren, L. P. Le, G. M. Luke, B. J. Sternlieb, W. D. Wu, J. H. Brewer, R. L. Whetten, S. M. Huang, S. Lin, R. B. Kaner, F. Diederich, S. Donovan, G. Grüner, and K. Holczer, *Nature* **352**, 605 (1991).
[64]L. Degiorgi, P. Wachter, G. Grüner, S.-M. Huang, J. Wiley, and R. B. Kaner, *Phys. Rev. Lett.* **69**, 1992.

addition, the magnetic penetration depth is large, ranging from 2400 to 8000 Å. Uncertainty in these length scales is likely due to variable sample quality (i.e., doping inhomogeneity) and granular effects. In the future, it will be important to improve these measurements using higher quality samples so that the intrinsic material properties can be unambiguously defined.

III. Mechanisms and Other Physical Properties

In Section II we reviewed the normal and superconducting state phenomenology of the alkali metal–doped fullerene superconductors. For the remainder of this chapter we will focus on experimental and theoretical studies of the microscopic mechanism of superconductivity in the M_3C_{60} materials. First, theoretical models based on electron–phonon- and electron–electron-mediated pairing will be introduced. Our emphasis will be to highlight features of these models that can be tested experimentally. The remaining sections of the chapter will then review key experimental studies that help to distinguish between the proposed theoretical models

4. MODELS FOR FULLERENE SUPERCONDUCTIVITY

Theoretical models put forth to explain superconductivity in the fullerenes range from the conventional electron–phonon-mediated pairing model of Bardeen, Copper, and Schreiffer (BCS)[65–72] to models in which pairing is mediated by electron correlation effects.[73–77]

[65]J. Bardeen, L. N. Cooper, and J. R. Schrieffer, *Phys. Rev.* **108**, 1175 (1957).

[66]C. M. Varma, J. Zaanen, and K. Raghavachari, *Science* **254**, 989 (1991).

[67]M. Schluter, M. Lannoo, M. Needels, G. A. Baraff, and D. Tománek, *Phys. Rev. Lett.* **68**, 526 (1992).

[68]I. I. Mazin, O. V. Dolgov, A. Golubov, and S. V. Shulga, *Phys. Rev. B* **47**, 538 (1993).

[69]I. I. Mazin, S. N. Rashkeev, V. P. Antropov, O. Jepsen, A. I. Liechtenstein, and O. K. Andersen, *Phys. Rev. B* **45**, 5114 (1992).

[70]V. Z. Kresin, *Phys. Rev. B* **46**, 14883 (1992).

[71]R. A. Jishi and M. S. Dresselhaus, *Phys. Rev. B* **45**, 2597 (1992).

[72]F. C. Zhang, M. Ogata, and T. M. Rice, *Phys. Rev. Lett.* **67**, 3452 (1991).

[73]S. Chakravarty, M. P. Gelfand, and S. Kivelson, *Science* **254**, 970 (1991).

[74]S. Chakravarty, S. A. Kivelson, M. I. Salkola, and S. Tewari, *Science* **256**, 1306 (1992).

[75]S. Chakravarty and S. Kivelson, *Europhys. Lett.* **16**, 751 (1991).

[76]S. R. White, S. Chakravarty, M. P. Gelfand, and S. A. Kivelson, *Phys. Rev. B* **45**, 5062 (1992).

[77]G. Baskaran and E. Tosatti, *Curr. Sci.* **61**, 33 (1991).

a. *Phonon-Mediated Pairing*

The transition temperature for the BCS model of superconductivity is represented by[65]

$$T_c = 1.6\omega_{ph} \exp(-1/(\lambda - \mu^*)) \qquad (4.1)$$

in the weak-coupling limit where $\lambda \ll 1$. The electron-phonon coupling parameter $\lambda = N(E_F)V$, where $N(E_F)$ is the density of states at the Fermi level and V is the coupling matrix to phonons with characteristic energy ω_{ph}. T_c is reduced in real materials by electron–electron repulsion that is accounted for by the renormalized Coulomb pseudopotential, μ^*:[78,79]

$$\mu^* = \frac{\mu}{1 + \mu \, \ln(E/\omega_{ph})}, \qquad (4.2)$$

where E is the smaller of the Fermi energy and the plasma frequency.

The weak-coupling BCS expression for T_c was also extended a number of years ago to strong coupling by Eliashberg[80] and McMillan.[81] A particularly useful expression developed by McMillan, which is valid for $\lambda \leq 1.5$, is

$$T_c = \frac{\langle \omega \rangle}{1.2} \exp\left(\frac{-1.04(1 + \lambda)}{\lambda - \mu^* - 0.62\lambda\mu^*}\right), \qquad (4.3)$$

where $\langle \omega \rangle$ is a logarithmic average of the phonon frequencies.

Electron–phonon coupling has been used to describe successfully superconductivity in a number of metallic compounds including Al (weak coupling) and Pb (strong coupling). The alkali metal–doped fullerene materials possess many differences from conventional metals (e.g., narrow conduction bandwidths and high frequency phonons); however, electron–phonon models still represent one logical starting point to describe superconductivity. Electron–phonon models specific to the fullerene superconductors have focused on coupling mediated by (1) high frequency intramolecular modes,[66,67] (2) a combination of high frequency intramolecular and low frequency intercluster modes,[68] and (3) low

[78]P. Morel and P. W. Anderson, *Phys. Rev.* **125**, 1263 (1962).
[79]O. Gunnarsson and G. Zwicknagl, *Phys. Rev. Lett.* **69**, 957 (1992).
[80]G. M. Eliashberg, *Sov. Phys. JETP* **11**, 696 (1960).
[81]W. L. McMillan, *Phys. Rev.* **167**, 331 (1968).

frequency C_{60}/alkali-metal optical modes.[72] Here we briefly summarize key features of several of these models.

Varma and coworkers[66] have developed a model to determine electron–phonon coupling parameters from the properties of single C_{60} clusters, and they have used these results in conjunction with the McMillan equation (4.3) to calculate T_c.[66] They justify the treatment of only intramolecular vibrations by asserting that because the electron–phonon coupling is proportional to the bandwidth and the covalent splitting of C_{60} states (i.e., the intramolecular band) is of order 10 eV versus the intermolecular band width of ≈ 0.5 ev, the intramolecular electron–phonon coupling will dominate in these materials. Within the context of this assumption, they showed that the coupling constant could be written as

$$\lambda = \frac{5}{6} N(E_F) \sum_m \frac{g_m^2}{M \omega_m^2},\qquad (4.4)$$

where the sum is over m intramolecular vibrations and g_m is the mth intramolecular deformation potential. Using semiempirical quantum chemical calculations, they found that only the two highest frequency tangential H_g modes at 1428 and 1575 cm^{-1} couple strongly to electrons. From the relation (4.4) they estimated that $\lambda = 0.3$–0.9, where the spread arises from uncertainty in $N(E_F)$; this range of λ corresponds to intermediate strength coupling. Finally, Varma and coworkers also investigated the dependence of T_c on $N(E_F)$ using their calculated value of λ and estimates for μ^*. An essential prediction of this analysis is that T_c should depend linearly on $N(E_F)$ for T_c above ≈ 20 K and $\mu^* \le 0.2$.

A similar theoretical approach was also taken by Schluter and coworkers.[67] In this study, they considered only coupling to intramolecular phonons. They justified the neglect of intermolecular phonons in the pairing interaction with an argument similar to that by Varma *et al.* A combination of local density approximation–density functional and semiempirical tight binding calculations were used to evaluate the coupling matrix, $V_m = g_m^2/M \omega_m^2$ for eight H_g and two A_g intramolecular phonons. In contrast to Varma *et al.*, they found that there is significant coupling both to the lower frequency radial modes and the high frequency tangential modes; the overall λ ($= 5/6 N(E_F \sum_m V_m)$ of 0.6 is, however, similar to the other work. Finally, it was pointed out that for the fullerene superconductors there is the unique situation that the electron–phonon coupling constant ($\lambda = N(E_F)V$) factors into an intercluster term $N(E_F)$, which is determined by the bandwidth, and the intramolecular term V, which is determined by single cluster phonon modes.

More recently, Mazin et al.[68] have proposed a somewhat different strong-coupling electron–phonon model.[68] In contrast to Varma and coworkers[66] and Schluter et al.,[67] they propose that coupling is mediated by both the high frequency intramolecular phonons ($\omega_{hf} \approx 1000 \, cm^{-1}$) and very low frequency intermolecular modes ($\omega_{lf} \approx 40 \, cm^{-1}$). In this "two-peak model," the electron–phonon coupling constant is factored into two components, where there is very strong coupling to the low frequency modes, $\lambda_{lf} \approx 2.7$, and moderate coupling to the high frequency modes, $\lambda_{hf} \approx 0.5$. Furthermore, the high frequency coupling constant should factor into intramolecular (V) and intermolecular ($N(E_F)$) components, but the low frequency constant cannot be factored. The important point about this two-peak model is that the inclusion of strong coupling to a low frequency mode provides a means for consistently explaining many experimental results (as will be discussed).[63,82-86]

Last, it is important to comment on several questions raised regarding the applicability of phonon models in general.[87] First, for the narrow band ($W \approx 0.5 \, eV$) fullerene superconductors, the Fermi energy ($E_F \approx W/2 \approx 2000 \, cm^{-1}$) is comparable with the intramolecular phonon energy ($\omega_{ph} \approx 1500 \, cm^{-1}$) proposed to mediate pairing. This leads to a breakdown of Migdal's approximation and thus the calculation of T_c.[82,88,89] Second, there is considerable controversy concerning appropriate values for the renormalized Coulomb pseudopotential, μ^*.[79] This parameter is essential for all calculations of T_c, and thus an unambiguous estimate of its value is clearly needed in the future.

b. *Electronic Models for Superconductivity*

A very different nonphonon approach for explaining superconductivity in the alkali metal–doped fullerenes centers on consideration of electron correlation effects in these materials.[73-75] This work is motivated in part by the fact that C_{60} is a narrow band material in which the important physics may be dominated by electron–electron interactions. Several explicit models have been developed to explain electron pairing and

[82]G. S. Boebinger, T. T. M. Palstra, A. Passner, M. J. Rosseinsky, D. W. Murphy, and I. I. Mazin, *Phys. Rev. B* **46**, 5876 (1992).

[83]Z. Zhang, C.-C. Chen, S. P. Kelty, H. Dai, and C. M. Lieber, *Nature* **353**, 333 (1991).

[84]Z. Zhang, C.-C. Chen, and C. M. Lieber, *Science* **254**, 1619 (1991).

[85]Z. Zhang and C. M. Lieber, *Mod. Phys. Lett. B* **5**, 1905 (1991).

[86]R. Tycko, G. Dabbagh, M. J. Rosseinsky, D. W. Murphy, A. P. Ramirez, and R. M. Fleming, *Phys. Rev. Lett.* **68**, 1912 (1992).

[87]P. W. Anderson, unpublished results.

[88]A. B. Migdal, *Sov. Phys. JETP* **7**, 996 (1958).

[89]M. Grabowski and L. J. Sham, *Phys. Rev. B* **29**, 6132 (1984).

superconductivity via electron–electron interactions;[73–77] however, we will illustrate the general features of these models by reviewing the work of Chakravarty and coworkers.[73]

The basic problem in any electron–electron-mediated pairing model is getting an effective attractive interaction between electrons despite their inherently repulsive microscopic interaction. Qualitatively, the underlying idea behind pair binding arises from the resonating valence bond (RVB) picture.[90] For a strongly correlated system, the RVB state implies a separation of spin and change such that the electron is composite object made up of two quasi-particles: an *eon* of charge e and no spin, and a spinon of spin $1/2$ and no change.

Within the context of the RVB picture, Chakravarty and coworkers defined Φ_0 and $\Phi_n = \Phi_0 + E_n$ as the ground state energy of the neutral molecule (C_{60}) and the energy of the molecule with n electrons, respectively. The pair binding energy, E_{pair}, is then

$$E_{pair} = 2E_1 - E_2 = 2\Phi_1 - \Phi_2 - \Phi_0, \qquad (4.5)$$

where $E_{pair} > 0$ implies an attractive pairing. In terms of charge–spin separation, the energy to add one electron to the molecule, $E_1 = \Phi_1 - \Phi_0$, is the sum of the eon creation energy, E_e, the spinon creation energy, E_s, and the eon–spinon interaction energy, V_{es}:

$$E_1 = E_e + E_s + V_{es}. \qquad (4.6)$$

Similarly, the energy to add two electrons to the molecule in the spin-singlet state required for superconductivity is $2E_e$ under the assumption that the eon–eon and spinon–spinon interaction energies are negligible. Hence, the pair binding energy can be written

$$E_{pair} = 2(E_s + V_{es}), \qquad (4.7)$$

and when $E_s + V_{es} > 0$ singlet pairing should be attractive. That is, the energy to add two electrons to one C_{60} is less than the energy to add a single electron to each of two clusters.

---

[90]P. W. Anderson, *Science* **235**, 1196 (1987).

Chakravarty and cowkers[73] have quantified this picture using the microscopic Hubbard model. In terms of the on-site Coulomb repulsion U, and the hopping matrix element t, the pair binding energy can be obtained using second-order perturbation theory:

$$E_p = -0.05U + 0.015\frac{U^2}{t} + \ldots \qquad (4.8)$$

Notably, for reasonable values of U and t they have found that $E_p > 0$; that is, pairing is attractive.

The attractive pairing energy suggests that electron correlation on C_{60} may give rise to superconductivity. Assuming that the conduction bandwidth W is large compared with E_p, the superconducting transition temperature can be expressed using a BCS-like expression:

$$T_c \propto \exp(-W/E_p), \qquad (4.9)$$

where E_p plays a role similar to V in the electron–phonon models. An important prediction that can be inferred from this model is that T_c should respond to intermolecular perturbations (through W) and intramolecular changes (through E_p) independently of each other. Other predictions as well as criticisms of this model have been reviewed.[73,74]

5. TRANSITION TEMPERATURE VERSUS LATTICE CONSTANT

Soon after the discovery of superconductivity in K_3C_{60} at 19 K,[1] two groups reported that Rb-doped C_{60}, Rb_xC_{60}, exhibited superconductivity at the remarkably high temperature of 29 K.[27,91] Subsequent studies reported simultaneously by our group[29] and Bell Laboratories[30] showed that K- and Rb-doped materials had the same stoichiometry (M_3C_{60}) and structure (Fig. 2). Studies of a series of $(K_{1-x}Rb_x)_3C_{60}$ have provided significant insight into the mechanism of superconductivity in these materials.

[91]M. J. Rosseinsky, A. P. Ramirez, S. H. Glarum, D. W. Murphy, R. C. Haddon, A. F. Hebard, T. T. M. Palstra, A. R. Kortan, S. M. Zahurak, and A. V. Makhija, *Phys. Rev. Lett.* **66**, 2830 (1991).

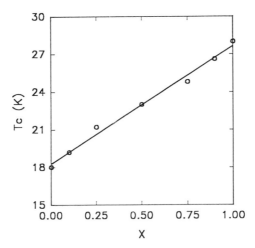

FIG. 4. Plot of T_c versus the Rb:K ratio x in $(K_{1-x}Rb_x)_3C_{60}$ solids. The fcc lattice constant increases from 14.253 to 14.436 as x increases from 0 to 1.

Chen and coworkers[29] found that T_c increased linearly with increasing x in an extensive series of $K_{1-x}Rb_x$-doped fullerene superconductors (Fig. 4). On the basis of these data Chen *et al.* suggested that substitution of Rb for K expands the fcc lattice (a "negative pressure" effect) and thereby systematically reduces the conduction bandwidth W. Because the total number of states in the t_{1u}-derived conduction band is constant, a reduction in W will increase $N(E_F)$. Hence, the linear relationship shown in Fig. 4 corresponds in effect to a systematic increase in T_c with increasing $N(E_F)$.

This picture was first quantified by the work of Fleming *et al.*[30] They found, in agreement with the work of Chen, that T_c increased linearly with the x, and they showed explicitly that T_c increased linearly with the fcc lattice constant, a_0. Fleming and coworkers further used semiempirical band structure calculations to estimate the density of states $N(E_F)$ as a function of a_0. They found an approximately 11% increase in $N(E_F)$ on going from K_3C_{60} to Rb_3C_{60}.

A number of additional experiments have been reported that are consistent with these results. First, Sparn *et al.*[50] and Schirber *et al.*[92] obtained related results from studies of the pressure dependence of T_c in polycrystalline K_3C_{60} materials. Both groups found that T_c decreased linearly with increasing pressures up to 20 kbar. The pressure coefficient,

[92]J. E. Schirber, D. L. Overmyer, H. H. Wang, J. M. Williams, K. D. Carlson, A. M. Kini, U. Welp, W.-K. Kwok, *Physica C* **178**, 137 (1991).

dT_c/dp, was similar in both studies: -0.78 and -0.63 K/kbar. In these studies, it was suggested that increasing pressure reduced the lattice constant monotonically, and that this reduction in lattice constant broadened W and reduced $N(E_F)$. An explicit calculation of the reduction in lattice constant with pressure, and consequently, the relationship between a_0 and T_c, was not made in these initial studies. Subsequent high pressure diffraction investigations of both K_3C_{60} and Rb_3C_{60} have determined the reduction in lattice constant with increasing pressure.[93] An analysis of these results seems to overlap reasonably well with the studies of the $(K_{1-x}Rb_x)_3C_{60}$ materials previously described. Hence, increasing pressure, which reduces a_0, will broaden the t_{1u}-derived conduction band and reduce $N(E_F)$.

In addition, there have been several measurements of $N(E_F)$ for K_3C_{60} and Rb_3C_{60} materials; these studies provide a direct check of the lattice constant effect proposed in the foregoing.[94,95] Tycko and coworkers[94] have used nuclear magnetic resonance (NMR) relaxation measurements to estimate the ratio of $N(E_F)$ for Rb_3C_{60} and K_3C_{60}.[94] Their analysis suggests that the Rb-doped material has 1.3–1.4 greater density of states at the Fermi level than K_3C_{60}. This observation is consistent with the calculations of Fleming et $al.$;[30] however, the magnitude of increase is $\approx 20\%$ larger. Thermoelectric power measurements made on K- and Rb-doped crystals indicate that ratio of $N(E_F)$ for Rb:K is 1.5–1.8.[95] The increase in $N(E_F)$ is consistent with the trend suggested by the lattice constant studies; however, the absolute change is much larger. This large change suggests that the simple picture may not provide an adequate explanation of the observed results. Finally, we note the recent work of Zhou et $al.$,[34] who investigated the increase in T_c for Na_2CsC_{60} ($T_c = 10.5$ K) upon formation of an ammonia intercalation compound $(NH_3)_4Na_2CsC_{60}$ ($T_c = 29.6$ K).[34] The lattice constant of the $(NH_3)_4Na_2CsC_{60}$ compound, 14.47, is nearly the same as that of Rb_3C_{60} ($T_c = 29.2$ K). Despite the large chemical difference between these two compounds, their similar T_c values appear to be defined solely by the lattice constant or interaction between neighboring C_{60} clusters in the lattice.

The results that have been presented appear to provide strong support for the intramolecular electron–phonon mediated pairing models de-

[93]O. Zhou, G. B. M. Vaughan, Q. Zhu, J. E. Fischer, P. A. Heiney, N. Coustel, J. P. McCauley, Jr., and A. B. Smith III, $Science$ **255**, 833 (1992).

[94]R. Tycko, G. Dabbagh, M. J. Rosseinsky, D. W. Murphy, R. M. Fleming, A. P. Ramirez, and J. C. Tully, $Science$ **253**, 884 (1991).

[95]T. Inabe, H. Ogata, Y. Maruyama, Y. Achiba, S. Suzuki, K. Kikuchi, and I. Ikemoto, $Phys.$ $Rev.$ $Lett.$ **69**, 3797 (1992).

scribed in Section 4. In these models, the expression for T_c factors the electron–phonon coupling constant λ into an intramolecular component V and an intermolecular component $N(E_F)$. In these models, the coupling matrix V is set by the intrinsic properties of single C_{60} clusters, and thus variations in T_c only arise from changes in $N(E_F)$. Hence, the changes in T_c due to variations in lattice constant[29,30] and pressure[50,92] support strongly these intramolecular phonon models. The results discussed here, however, also are completely consistent with pairing mediated by electron–electron correlation.[73–75] Specifically, the mean-field expression for T_c (relation (4.9)) in the electron correlation model is also factored into an intramolecular component (E_{pair}) and an intermolecular component (W). Since E_{pair} is determined by electron correlation on a single C_{60} cluster, variations in T_c for M_3C_{60} solids can be ascribed to variations in the bandwidth W. Because W and $N(E_F)$ depend similarly on lattice constant and pressure, the foregoing experimental studies cannot distinguish these fundamentally different models.

In addition, it is important to recognize that there are deviations from the linear relationship between T_c and the lattice constant. For example, Na_2CsC_{60} and Na_2RbC_{60} have been reported to have values of T_c of about 10 and 2.5 K, respectively.[35,36] Extrapolation of the T_c versus lattice constant results, previously described, to the measured lattice constants of these materials predicts higher values of T_c than determined experimentally. At present, these deviations cannot be explained by a simple factoring of the dependence of T_c into independent intramolecular and intermolecular contributions. The NMR[94] and thermoelectric power[95] measurements also deviate from the ideal intramolecular phonon picture since the large changes in the ratio of $N(E_F)$ suggest that low frequency phonons are also important in the pairing interaction. Clearly, further work will be needed to address these issues.

6. ENERGY GAP

The superconducting energy gap, 2Δ, is a measure of the energy scale binding Cooper pairs and thus is a parameter fundamental to models describing superconductivity in the alkali metal–doped fullerenes. Within the context weak coupling the BCS theory of superconductivity, it is possible to define a universal value for the reduced energy gap,

$$\frac{2\Delta}{kT_c} = 3.53, \tag{6.1}$$

that is independent of the details of the superconducting material.[65]

Correspondingly, values of 2Δ greater than $3.5kT_c$ indicate that the coupling is strong. Hence, measurements of 2Δ can be used to define the coupling regime for a material and thereby the appropriate energy scale of the excitations that mediate superconductivity. To date, the energy gap has been determined using a variety of techniques including tunneling,[83-85] infrared,[51,64,96] and NMR spectroscopies.[86] Here we will critically review the results from these different experiments.

a. *Tunneling*

Tunneling spectroscopy is a particularly attractive technique for probing the energy gap since the conductance, dI/dV, determined from current voltage, $I-V$, curves provides a direct measure of the density of states:

$$\frac{dI}{dV} = \int N_s(E)N_n |M|^2 \frac{df(E + eV)}{dV} dE, \qquad (6.2)$$

where N_s is the density of states in the superconducting state, N_n is the normal state density of states, M is the tunneling matrix element, and $f(E)$ is the Fermi function.[97] In the limit of small bias and low temperature,

$$\frac{dI}{dV} \alpha N_s(E). \qquad (6.3)$$

Zhang and coworkers[83-85] have used tunneling spectroscopy to investigate systematically the low energy electronic states in K_3C_{60} and Rb_3C_{60} polycrystalline superconductors.[83-85] Current–voltage curves recorded from 4.2 to 20 K on K_3C_{60} and from 4.2 to 30 K on Rb_3C_{60} samples were found to exhibit features characteristic of a superconducting energy gap (Fig. 5). At 4.2 K, where thermal broadening is minimized, the $I-V$ curves exhibit low current near E_F and conductance onsets at $V \approx \pm 4$ and ± 6 meV for the K_3C_{60} and Rb_3C_{60} superconductors, respectively. Zhang and coworkers suggested that the conductance onsets corresponded to gap edges ($\pm\Delta$) in these materials.

Zhang *et al.* quantitatively assessed the magnitude of 2Δ for K- and

[96]S. A. FitzGerald, S. G. Kaplan, A. Rosenberg, A. J. Sievers, and R. A. S. McMordie, *Phys. Rev. B* **45,** 10165 (1992).

[97]E. L. Wolf, "Principles of Tunneling Spectroscopy." Oxford University Press, New York, 1989.

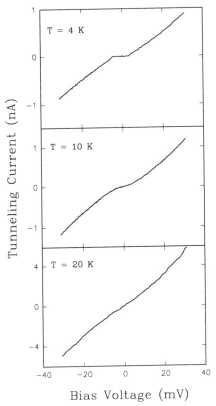

FIG. 5. Current versus voltage (I–V) curves recorded on a polycrystalline K_3C_{60} sample at 4.2, 10, and 20 K. The tunneling junction is a point contact and was formed using a low temperature scanning tunneling microscope.

Rb-doped materials by computing dI/dV and fitting the resulting curves to different models for N_s. They found that a BCS model for the density of states,[65]

$$N_s = \frac{eV}{[(eV)^2 - \Delta^2]^{1/2}}, \qquad (6.4)$$

provided a poor fit to the data. Nevertheless, the value of 2Δ determined from this fit, $5.4kT_c$, was found to exceed significantly the BCS weak-coupling limit of $3.5kT_c$.

The poor agreement between the BCS model for N_s (6.4) and the experimental data was due to significant broadening in the latter. To account for this broadening and to obtain a more reliable estimate of 2Δ,

Zhang and coworkers also fit their data to the broadened density of states model proposed by Dynes:[98]

$$N_s = \mathrm{Re}\left\{\frac{|eV - i\Gamma|}{[(eV - i\Gamma)^2 - \Delta^2]^{1/2}}\right\}. \tag{6.5}$$

The Dynes model includes a parameter Γ to account for lifetime broadening near T_c. Formally, this model is not applicable to the low temperature data of Zhang et al. ($T/T_c < 0.25$), although they used it to account phenomenologically for broadening of an unknown origin. Examples of fits to the data of both the K_3C_{60} and Rb_3C_{60} superconductors at low temperature are shown in Fig. 6. From fits using this model they found that 2Δ (K_3C_{60}) $= 5.3kT_c$ and 2Δ (Rb_3C_{60}) $= 5.2kT_c$. Importantly, these values of 2Δ exceed significantly the weak coupling limit of $3.5kT_c$ and thus suggest that electrons may couple strongly to the relevant pairing excitations in the M_3C_{60} materials.

b. Infrared

The energy gap has also been probed in several infrared reflectance (IR) spectroscopy studies. Rotter et al.[51] first used IR spectroscopy to determine 2Δ in polycrystalline Rb_3C_{60} samples. The reflectivity spectra recorded above and below T_c showed changes expected for a superconductor, although the behavior was far from ideal. They defined the frequency at which the reflectivity ratio in the superconducting and normal states, R_s/R_n, dropped rapidly to be 2Δ; using this definition, $2\Delta = 3kT_c - 5kT_c$. The considerable uncertainty in these results overlaps with both the weak-coupling BCS value for 2Δ and the larger value determined experimentally by tunneling spectroscopy. It is thus difficult to draw a strong conclusion from these results. More recently, a large range of 2Δ values, $2kT_c - 5kT_c$, was reported on the basis of IR studies of K-doped C_{60} films.[96] Although no specific conclusion about the coupling strength can be made from these studies, the authors do make the important point that sample homogeneity must be improved to obtain conclusive results.

A somewhat different result has been reported by Degiorgi et al.,[64] who used IR reflectivity to determine the energy gap in K_3C_{60} and Rb_3C_{60} polycrystalline materials. They report that the optical reflectivity exhibits a well-defined gap feature that progressively sharpens as the temperature is reduced below T_c. These clean results contrast the ill-defined gap structure in the IR studies of Rotter et al.[51] and FitzGerald et al.,[96]

---

[98]R. C. Dynes, V. Narayanamurti, and J. P. Garno, Phys. Rev. Lett. **41**, 1509 (1978).

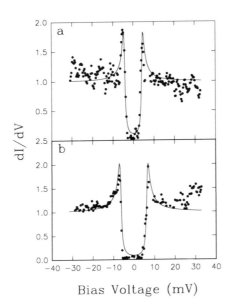

Bias Voltage (mV)

FIG. 6. Conductance (dI/dV) versus voltage curves for (a) K_3C_{60} and (b) Rb_3C_{60} at 4.2 K (filled circles). The solid lines correspond to best fits to relation (6.5) in the text. The values of the energy gap Δ and broadening function Γ in meV are $\Delta = 4.4$, $\Gamma = 0.5$ and $\Delta = 6.6$, $\Gamma = 0.6$ for K_3C_{60} and Rb_3C_{60}, respectively.

discussed already, and are believed to result from significantly improved sample quality. Notably, from the analysis of their reflectivity data Degiorgi and coworkers find that $2\Delta = 3.6kT_c$ and $3.0kT_c$ for the K_3C_{60} and Rb_3C_{60} materials, respectively. In contrast to the tunneling studies, these results appear to indiate that BCS-like weak-coupling models are appropriate for the fullerene superconductors. It is important to note, however, that the temperature dependence of the energy gap, $\Delta(T)$, determined in these IR studies deviates significantly from that predicted by BCS theory (to be discussed).

c. NMR

Finally, Tycko and coworkers[86] at Bell Laboratories have used NMR relaxation measurements to determine the magnitude of 2Δ in K_3C_{60} and Rb_3C_{60} superconductors. Fits of the ^{13}C-nuclear spin relaxation time, T_1, to an Arrhenius law,[99]

$$\frac{1}{T_1} = \omega \exp\left(-\frac{\Delta}{T}\right),\qquad (6.6)$$

[99]L. C. Hebel and C. P. Slichter, *Phys. Rev.* **113**, 1504 (1959).

yield values of 2Δ for K_3C_{60} and Rb_3C_{60} of $3.0kT_c$ and $4.1kT_c$, respectively. These reported values of 2Δ appear to be consistent with the weak-coupling prediction of BCS theory [A recent calculation[70] based on these results yields values of λ for 2.1 and 1.5 for the Rb- and K-doped materials, respectively, and thus suggests strong coupling.]. There are important concerns regarding a weak-coupling interpretation of the NMR data. First, no Hebel–Slichter peak[99] was observed in the relaxation data. Because strongly coupled superconductors are not expected to exhibit a clear Hebel–Slichter peak, it is possible that NMR results are not clear evidence for weak coupling. Along these lines, it is interesting to note that NMR probes the minimum quasi-particle excitation energy. Therefore, when the gap is not clean, the value of 2Δ determined by NMR will be smaller than the true order parameter for the system. This same criticism also holds for IR spectroscopy and will be discussed further below.

d. *Implications*

The measurements of 2Δ reported to date range from $3kT_c-5kT_c$. In comparison with work at a similar stage on the copper oxide superconductors, these fullerene gap studies have a much better agreement. Since the results range from the weak-coupling to strong-coupling regimes, however, it may not be possible to use them to distinguish critically between proposed models for fullerene superconductivity at the present time.

We thus consider whether it is possible to develop a consistent picture from these results by considering the strengths and weaknesses of the individual measurement techniques. First, it is important to consider what the techniques measure and how the polycrystalline M_3C_{60} samples might influence these measurements. As discussed in Section 6.a, tunneling spectroscopy measures the conductance, which is proportional to N_s. Uncertainty in the value of 2Δ can arise from several sources, including (1) the model used to relate N_s and Δ and (2) inhomogeneity in the surface properties of the M_3C_{60} polycrystalline samples used to form the metal/insulator/superconductor junctions. The tunneling studies of Zhang *et al.* have shown that the magnitude of 2Δ is not very sensitive to the particular model chosen to relate N_s and Δ, although N_s is clearly broadened significantly relative to an ideal BCS model. The role of sample surface inhomogeneity, which could arise from degradation of the highly reactive M_3C_{60} surfaces, was not addressed in these studies. Degradation of the superconducting properties at the sample surface should not give rise to large values of 2Δ, however, but rather will increase the conductance at zero bias voltage ($V = 0$). The relatively low conductance observed by Zhang and coworkers at $V = 0$ suggests that this (degradation) may not be a problem, although additional tunneling studies are needed to verify this work.

In contrast to tunneling spectroscopy, IR spectroscopy provides an onset frequency for the excitation of quasi-particles. In the studies we have discussed,[51,64,96] this onset was assigned to 2Δ. Uncertainty in the magnitude of 2Δ can arise from several sources, including (1) uncertainty in the onset of optical absorption and (2) the formal meaning of this optical onset. The optical onset is difficult to define since it corresponds to the deviation from 100% reflectivity. In the studies of Rotter et al.[51] and FitzGerald et al.,[96] this onset is hard to assign and thus leads to a large uncertainty in 2Δ. The report by Degiorgi and coworkers improves upon these earlier studies, since it appears to show a sharp deviation from 100% reflectivity. A more important point, which has not been addressed, is that the optical onset corresponds formally to an optical gap, ω_g. Only in the case of an ideal BCS superconductor does $\omega_g = 2\Delta$, where Δ is the order parameter that defines the energy scale for pair binding. More generally, ω_g will be less than 2Δ due to inelastic (pair breaking) scattering and/or other processes that allow for excitations at frequencies less than 2Δ.[100–102]

It is therefore interesting to consider whether the tunneling and optical data are in fact consistent but simply measure different properties (i.e., 2Δ vs. ω_g) of the M_3C_{60} superconducting state. An important indication from the IR data of Degiorgi and coworkers[64] that supports this idea is that the temperature dependence of their measured energy gap deviates strongly from the behavior predicted by BCS theory. They report that the gap opens sharply below T_c and exhibits only a weak temperature dependence. Notably, these results agree with predictions of pair breaking theory.[100–102] We have thus analyzed the previous tunneling data of Zhang and coworkers within the framework of pair breaking to explicitly determine 2Δ and ω_g, although the origin of scattering in the fullerenes is unknown. An example of a fit to N_s for Rb_3C_{60} is shown in Fig. 7. The fit to the tunneling data using this model is significantly better than observed in the previous reports[83,84] and yields $2\Delta = 5.2kT_c$ and $\omega_g = 4.0kT_c$. Notably, the value of ω_g determined from this analysis is similar to the onset reported by Degiorgi and coworkers.[64]

This analysis suggests that 2Δ may be significantly larger than $3.5kT_c$

[100] A. A. Abrikosov and L. P. Gorkov, Sov. Phys. KETP **12**, 1243 (1991).

[101] Y. Wada, Rev. Mod. Phys. **253** (1964).

[102] S. Skalski, O. Betheder-Matibet, and P. R. Weiss, Phys. Rev. **136**, 1500 (1964).

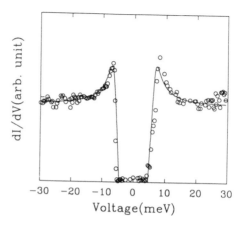

FIG. 7. Conductance (dI/dV) versus voltage data for Rb_3C_{60} at 4.2 K (open circles). The solid line corresponds to a best fit of the pair breaking model discussed in the text. The values of the energy gap Δ, optical gap ω_g, and pair breaking parameter α, are 6.6, 5.0, and 1.1 meV, respectively.

and is thus suggestive of strong coupling. Strong coupling constrains the electron–phonon models of superconductivity discussed in Section 4 in several ways. A large value of $2\Delta/kT_c$ would be inconsistent with models that consider only high frequency intramolecular phonons. Strong electron–phonon coupling with these high frequency modes would predict transition temperatures significantly higher than observed experimentally. The electron–phonon model that includes contributions from both high frequency intramolecular modes and low frequency intermolecular phonons,[68] however, would be consistent with the large values of $2\Delta/kT_c$. Finally, while these energy gap results may help to distinguish different electron–phonon models, they cannot, unfortunately, at present distinguish phonon-mediated pairing from the fundamentally different model based on pairing mediated by electron correlation.[73] Explicit predictions for the gap equation are needed for this latter model (e.g., how close the superconducting state is to a BCS-like one) in order to assess quantitatively its agreement with the experimental energy gap results.

7. PHONONS

One way in which to assess whether electron–phonon coupling or electron correlation is responsible for superconductivity is to investigate the coupling of different phonons to the conduction electrons. In solid

C_{60}, there are a number of phonon modes that should in principle be considered; these include (1) librational modes of C_{60} clusters in the lattice ($\omega = 15\,\text{cm}^{-1}$); (2) intermolecular C_{60}–C_{60} vibrational modes ($\omega \approx 40\,\text{cm}^{-1}$); (3) C_{60}–M^+ optical modes ($\omega \approx 60$–$100\,\text{cm}^{-1}$); and (4) radial and tangential intramolecular modes. The intramolecular radial modes can be characterized as low frequency (200–$800\,\text{cm}^{-1}$), while the tangential modes are higher frequency (1000–$1600\,\text{cm}^{-1}$). To date, most efforts have focused on the analysis of the radial and tangential intramolecular phonons,[103–110] since these modes are proposed to play a dominant role in intramolecular electron–phonon pairing models.

Symmetry analysis shows that only the eight H_g and two A_g intramolecular phonon modes can couple to the t_{1u} electronic states.[111] These modes, which are Raman active, have been studied by several groups using Raman and inelastic neutron scattering.[103–110] Duclos and coworkers[103] at Bell Laboratories have carried out Raman scattering experiments on C_{60}, M_3C_{60}, and M_6C_{60} (M = Na, K, Rb and Cs) to characterize the interaction of the Raman active modes with the conduction electrons. They found that K- and Rb-fullerene samples doped to the M_3C_{60} stoichiometry exhibit only three modes; these were the radial modes $H_g(1) = 265\,\text{cm}^{-1}$ and $A_g(1) = 497\,\text{cm}^{-1}$, and the tangential mode $A_g(2) = 1447\,\text{cm}^{-1}$. Interestingly, the $H_g(2)$ through $H_g(8)$ modes could not be detected in the spectra of the M_3C_{60} materials, although these phonons were clearly observed in pure C_{60} and M_6C_{60} solids. Duclos et al. suggest that the disappearance of the $H_g(2)$–$H_g(8)$ modes may be due to broadening of the spectral function as a result of electron–phonon coupling. They also note, however, that a decrease in the optical penetration depth, which occurs in the metallic M_3C_{60} state, could also explain the disappearance of these modes.

[103] S. J. Duclos, R. C. Haddon, S. Glarum, A. F. Hebard, and K. B. Lyons, *Science* **254**, 1625 (1991).
[104] M. G. Mitch, S. J. Chase and J. S. Lannin, *Phys. Rev. Lett.* **68**, 883 (1992).
[105] M. G. Mitch, S. J. Chase, and J. S. Lannin, *Phys. Rev. B* **46**, 3696 (1992).
[106] P. Zhou, K.-A. Wang, A. M. Rao, P. C. Eklund, G. Dresselhaus, and M. S. Dresselhaus, *Phys. Rev. B* **45**, 10838 (1992).
[107] P. Zhou, K.-A. Wang, Y. Wang, P. C. Eklund, M. S. Dresselhaus, G. Dresselhaus, and R. A. Jishi, *Phys. Rev. B* **46**, 2595 (1992).
[108] K.-A. Wang, Y. Wang, P. Zhou, J. M. Holden, S.-L. Ren, G. T. Hager, H. F. Ni, P. C. Eklund, G. Dresselhaus, and M. S. Dresselhaus, *Phys. Rev. B* **45** (1992).
[109] K. Prassides, J. Tomkinson, C. Christides, M. J. Rosseinsky, D. W. Murphy, and R. C. Haddon, *Nature* **354**, 462 (1991).
[110] K. Prassides, C. Christides, M. J. Rosseinsky, J. Tomkinson, D. W. Murphy, and R. C. Haddon, *Europhys. Lett.* **19**, 629 (1992).
[111] M. Lannoo, G. A. Baraff, M. Schluter, and D. Tomanek, *Phys. Rev. B* **44**, 1210 (1991).

Results similar to the study of Duclos *et al.* have also been reported by several other groups. Lannin and coworkers[104,105] have found significant broadening of the radial $H_g(2)$ and tangential $H_g(7)$ and $H_g(8)$ modes in Raman scattering studies of ultrathin Rb_xC_{60} thin films. These results suggests significant electron–phonon coupling only to three modes versus coupling to seven $H_g(2)-H_g(8)$ modes as reported by Duclos and coworkers. The Raman investigations of Zhou *et al.*[106] suggest a still different picture in that they find significant broadening of five radial and tangential H_g symmetry phonons in K_3C_{60} thin films. Finally, inelastic neutron scattering studies of C_{60} and K_3C_{60} have also been reported to show evidence for strong electron–phonon interactions.[109,110] Specifically, Prassides *et al.*[109,110] found that the $H_g(2)$ and $H_g(8)$ phonon modes disappeared in neutron scattering spectra of K_3C_{60}, and that the $H_g(1)$, $H_g(3)$, and $H_g(4)$ modes exhibited only small changes.

To summarize, Raman and inelastic neutron scattering experiments show significant spectral line broadening effects for several intramolecular H_g phonon modes. While studies reported to date do not agree on all of the specific modes that are broadened, there appears to be a consensus for at least the $H_g(2)$ and $H_g(8)$ phonon modes. Although it is reasonable to interpret these results as evidence supporting intramolecular electron–phonon-mediated pairing models for fullerene superconductivity, this interpretation should be made with caution. Electron–phonon coupling, as evidenced by spectral line broadening at fixed temperature, does not necessarily imply that these broadened phonons are important in superconductivity. Indeed, there will be electron–phonon interactions regardless of the mechanism of superconductivity. Only when the electron–phonon interaction is shown to affect the superconducting state is it reasonable to assert that phonons are responsible for pairing. In this regard, it would be interesting to investigate the Raman spectra of the M_3C_{60} superconductors as a function of temperature near T_c.[112] If the H_g modes really are responsible for pairing, they should exhibit changes in linewidth and intensity as the sample is cooled through the superconducting transition temperature.

8. Isotope Effect

A more direct method than those already discussed for investigating the role of phonons in superconductivity is to determine the magnitude of the shift in T_c upon isotopic substitution—the *isotope effect*. Historically,

---

[112]M. E. Flatté, *Phys. Rev. Lett.* **70**, 658 (1993); and M. E. Flatté, personal communication.

the observation of a suppression in T_c upon substitution of heavy isotopes of mercury in metallic mercury was a key result supporting phonon-mediated pairing as a mechanism for superconductivity.[113,114] For phonon-mediated pairing, the BCS model of superconductivity predicts that

$$T_c \propto M^{-\alpha}, \qquad (8.1)$$

where M is the ionic mass and α is the isotope shift exponent. The ideal value of α predicted by BCS theory, 0.5, has been observed in a number of simple metals.[115] Corrections due, for example, to Coulomb interactions, however, will reduce the value of α below 0.5. Hence, measurements of the $^{13}C/^{12}C$ isotope effect in M_3C_{60} superconductors should provide a key test of the mechanism of superconductivity in the fullerenes.

Not surprisingly, there have been a number of experimental invesigations of the isotope effect in the M_3C_{60} materials.[116-120] In all cases, a significant isotope effect has been observed, although the range of reported values for α is large. Careful analysis of these results suggest that the intrinsic value of α for ^{13}C-substituted C_{60} is approximately 0.3.[116-118] Larger values of α have been observed in samples containing poorly controlled mixtures of $(^{13}C_{1-x}^{12}C_x)_{60}$ and $^{12}C_{60}$. In general, one expects that a simple mass average weighting should provide an accurate extrapolation of α to 100% substitution; however, the sensitivity of the M_3C_{60} T_c to impurities and the unique cluster properties of the C_{60} superconductor (to be discussed) makes this assumption questionable. In short, only homogeneous samples possessing sharp superconducting transitions can provide a good measure of the shift in T_c. The transition broadening that is often observed in small isotopically substituted samples[119,120] can easily lead to an overestimate of the shift in T_c, and thus to a large isotope effect.

Isotope shift measurements made in our laboratory at Harvard have

[113]C. A. Reynolds, B. Serin, W. H. Wright and L. B. Nesbitt, *Phys. Rev.* **78**, 487 (1950).

[114]E. Maxwell, *Phys. Rev.* **78**, 477 (1950).

[115]E. A. Lynton, "Superconductivity". Methuen, London, 1961.

[116]C.-C. Chen and C. M. Lieber, *J. Am. Chem. Soc.* **114**, 3141 (1992).

[117]C.-C. Chen and C. M. Lieber, *Science* **259**, 655 (1993).

[118]A. P. Ramirez, A. R. Kortan, M. J. Rosseinsky, S. J. Duclos, A. M. Mujsce, R. C. Haddon, D. W. Murphy, A. V. Makhija, S. M. Zahurak, and K. B. Lyons, *Phys. Rev. Lett.* **68**, 1058 (1992).

[119]T. W. Ebbesen, J. S. Tasi, K. Tanigaki, J. Tabuchi, Y. Shimakawa, Y. Kubo, I. Hirosawa, and J. Mizuki, *Nature* **355**, 620 (1992).

[120]A. A. Zakhidov, K. Imaeda, D. M. Petty, K. Yakushi, H. Inokuchi, K. Kikuchi, I. Ikemoto, S. Suzuki, and Y. Achiba, *Phys. Lett. A* **164**, 355 (1992).

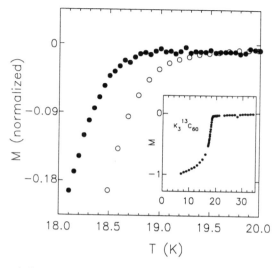

FIG. 8. High resolution temperature-dependent magnetization measurements on $K_3{}^{13}C_{60}$ (●) and $K_3{}^{12}C_{60}$ (○) samples highlighting the depression in T_c for the isotopically substituted material. The inset shows a full magnetization curve for a $K_3{}^{13}C_{60}$ sample.

avoided these experimental problems by focusing on homogeneous, 99.5% ^{13}C-substituted C_{60} solids.[116,117] An example of data obtained on a K_3C_{60} sample is shown in Fig. 8. Measurements made on both the $K_3{}^{13}C_{60}$ and $Rb_3{}^{13}C_{60}$ samples show unambiguously that the value of α for the fully substituted M_3C_{60} superconductors is 0.3.

With this reproducible result in hand, it is possible to evaluate the applicability of different models proposed to explain fullerene superconductivity. First, the observation of a significant isotope effect strongly supports the role of phonons in pairing. Using an expression derived from the McMillan equation (4.3), it is possible to evaluate whether intramolecular or intermolecular phonons are more likely to explain superconductivity in the fullerenes. Specifically, we can rewrite the McMillan equation expressing the isotope exponent α in terms of ω, λ, and μ^*:

$$\alpha = \frac{1}{2}\left\{1 - \left[\mu^* \ln\frac{\langle\omega\rangle}{1.2T_c}\right]^2 \frac{1 + 0.62\lambda}{1 + \lambda}\right\} \qquad (8.2)$$

Assuming high frequency intramolecular phonons are important and $\alpha = 0.3$, we find that $\lambda = 0.68$ and $\mu^* = 0.17$. If $\langle\omega\rangle$ is only made over the very low frequency intermolecular phonon modes, then $\lambda = 5$ and $\mu^* = 0.35$. In this latter case, the value of λ is outside the range for which the McMillan relationship (8.2) is valid. This analysis thus indicates that high frequency intramolecular phonons may provide the most conventional parameter set for the experimentally observed isotope exponent of 0.3. Kresin[70] has also argued, however, that using a theory that properly accounts for strong coupling ($\lambda > 1.5$) can provide reasonable values of λ and μ^* (2 and 0.2, respectively) assuming that only low frequency modes are important.

There are two additional complications in the interpretation of these seemingly straightforward isotope effect experiments. First, Chakravarty and coworkers[74] have shown that it is also reasonable to expect a sizable isotope effect within the context of their electron correlation model of pairing.[74] The isotope effect in their model arises from the fact that the hopping matrix element t in relation (4.8) will increase upon ^{13}C-substitution due to a reduction in zero point fluctuation of the carbon–carbon bonds; that is, the mean C–C separation decreases upon isotopic substitution. Because the pair-binding energy is inversely related to t, T_c will decrease upon ^{13}C-substitution (see (4.8) and (4.9)). The estimated magnitude of the suppression in T_c for Rb_3C_{60} is 0.2 to 0.6 K, which yields α ranging from 0.08 to 0.25. Since the largest value of α approaches the experimental isotope exponent, it may be premature to rule out this electronic model on the basis of the experimental isotope effect.

Finally, we also mention the interesting experimental result of Chen et al.[117] To investigate the role of intramolecular and intermolecular

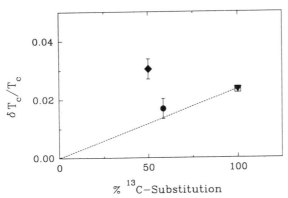

FIG. 9. Plot of the normalized isotope shift ($\delta T_c/T_c$) versus the percentage ^{13}C enrichment in the fullerene samples. The dashed line corresponds to the behavior expected for $\alpha = 0.3$. The experimental points are $Rb_3{}^{13}C_{60}$ (▼), $K_3{}^{13}C_{60}$ (□), $Rb_3({}^{13}C_{.55}{}^{12}C_{.45})_{60}$ (●), and $Rb_3[({}^{13}C_{60})_{.5}({}^{12}C_{60})_{.5}]$ (◆). Notably, the isotope shifts observed for the $Rb_3[({}^{13}C_{60})_{.5}({}^{12}C_{60})_{.5}]$ samples are significantly larger than predicted for $\alpha = 0.3$.

phonons, they have prepared and studied a series of ^{13}C-substituted materials of the general form $Rb_3(^{13}C_{1-x}{}^{12}C_x)_{60}$ and $Rb_3[(^{13}C_{60})_{1-x}$ $(^{12}C_{60})_x]$. In principle, these materials should exhibit the same isotope effect for a given value of x (assuming phonon-mediated pairing) since the materials have a similar weighted phonon density of states. Interestingly, Chen has reported that $Rb_3[(^{13}C_{60})_{1-x}(^{12}C_{60})_x]$ samples show an anomalously large isotope shift (Fig. 9). The origin of this interesting result is not known at present; however, we believe that a detailed explanation should provide critical insight into the mechanism of superconductivity in the alkali metal–doped fullerenes.

IV. Concluding Remarks

In this chapter we have attempted to provide a relatively comprehensive overview of the experimental status of the fullerene superconductors. We have emphasized (1) the normal state and superconducting state phenomenology as well as (2) experimental probes of the microscopic mechanism of superconductivity in these new molecular superconductors. As in any emerging field, there remains considerable uncertainty, and throughout this chapter we have tried to present a balanced review of accepted results and indicate where additional work is warranted. At present, many results point to a mechanism of electron–phonon-mediated pairing, but as indicated throughout Section III, electron correlation effects may still represent a viable explanation of superconductivity in these materials. Clearly, there is room (and a need) to explore more deeply the fascinating physics of these new superconductors.

Author Index

Numbers in parentheses are reference numbers and indicate that an author's work is referred to although his or her name is not cited in the text.

A

Abbate, M., 79(332), *79*
Abdul-Sada, A. K., 7(23), *7*, 17(23)
Abeles, B., 71(322), *7 1*, 358(56), *358*
Abello, L, 21(116), *21*
Abello, L., 102(116)
Abrikosov, A. A., 377(100), *377*
Abudl-sada, A. K., 122(39), *122*
Aburano, R. D., 29(184), *29*, 107(184)
Achiba, Y., 8(31-36, 38), *8*, 9(32, 44), *9*,
  10(35), 11(50-51), *11*, 26(154), *26*, 27(171),
  27, 34-35(171), 50(36), 51(32-33, 35-36),
  52(36), 84(34-35, 343, 345), *84*,
  90(376, 387), *90*, 91(376), 94(387), 124(48),
  124, 193(59-60), *193*, 228(31a-31b), *228*,
  243(75), *243*, 308(246-247), *308*,
  309(246-247), 324(286, 290), *324*, 327(290),
  328(296), *328*, 332(290), 339(326), *339-340*,
  370(95), *370*, 371(95), 381(120), *381*
Adams, G. B., 101(430), *101*, *258*, 265(145),
  265, *267*
Adelmann, P., 23(130), *23*, 27(167), *27*,
  73(329), *73*, 75(329), 79(329), 174(40),
  174, 181(47), *181*, 191(47), 194(40),
  196(40), 198-199(40), 212-213(47), 215(47),
  219(47), 280(189), *280*
Afagonov, V., 194(67), *194*, 199(67)
Aida, T., 25(146), *25*, 129(60), *129*
Ajayan, P. M., 95(392-393), *95*, 96(392),
  97(399), *97*, 101(393), 145(103), *145*,
  146(103), 147(113-114), *147*
Ajie, H., 17(82), *17*, 38(82), 119(33), *119*,
  311(255), *311*
Albouy, P. A., 194(67), *194*, 199(67)
Alers, G. B., 192(56), *192*, 194(56), 203(56)
Alexander, A. M., 73(329), *73*, 75(329),
  79(329)
Alexandrov, A. S., 344(337), *344*
Alford, J. M., 90(373, 377), *90*, 91(373, 377),
  92-93(377), 143(93), *143*, 144(93)

Alford, M., 246(100), *246*
Alford, M. J., 15(72), *15*, 90(72), 94(72)
Alivisatos, A. P., 21(113), *21*
Allaf, A. W., 2(3), *2*, 14(3), 17(87), *17*, 55(87),
  227(18), *227*
Allemand, P. M., 111(21), *111*, 350(17), *350*
Allen, B. H., 26(156), *26*, 27(156), 137(82),
  137
Allen, P. B., 321(281), *321*
Allenmand, P. M., 228(30-31), *228*,
  288(30-31), 318(30)
Almhof, J., 170(28), *170*, 171(28)
Almlöf, J., 14(67), *14*, 237(53), *237*
Altman, E. I., 29(180), *29*, 33(180), 62(180)
Alvarez, M. M., 8(29-30), *8*, 9(29), 17(82, 89),
  17, 35(201), *35*, 38(82), 39(89), 59(89),
  90(382), *90*, 92(382), 119(32-33), *119*,
  120(32), 143(94, 96), *143*, 144(94, 96),
  311(255), *311*
Amelinckx, S., 24(135), *24*, 25(135), 26(157),
  26, 27(157), 100(404), *100*, 279(187), *279*
Ameline, J.-C., 197(70), *197*
Amster, I. J., 58(276), *58*, 107(276)
Andersen, O. K., 80-81(336), *81*, 281(196),
  281, 282(199), *282*
Andersen, O., K., 289(220), *289*, 295-296(220),
  304(220)
Andersen, O. K., 322(282), *322*, 343(282),
  356, 356(44), 363(69), *363*
Anderson, E. B., 11(55), *11*
Anderson, P. W., 364(78), *364*, 366(87), *366*,
  367(90), *367*
Anderson, S. L., 16(78, 80), *16*, 90(78, 80),
  91(80)
Ando, Y., 26(153), *26*, 90(383-384), *90*,
  93(383-384), 98(400), *98*, 100(400),
  102(434), *102*, 143(97, 100), *143*,
  144(97, 100), 145(104), *145*
André, D., 194(67), *194*, 199(67)
Andreoni, W., 49(240), *49*, 90(381), *90*,
  92-93(381), 145(101), *145*, 265(143-144),

385

W

Subject Index

419

Cumulative Author Index